# RISK MANAGEMENT PROGRAM GUIDANCE FOR WAREHOUSES (40 CFR PART 68)

May 26, 2000

This document provides guidance to help owners and operators of stationary sources to determine if their processes are subject to chemical accident prevention regulation under section 112(r) of the Clean Air Act and 40 CFR part 68 and to comply with regulations. This document does not substitute for EPA's regulations, nor is it a regulation itself. Thus, it cannot impose legally binding requirements on EPA, states, or the regulated community, and may not apply to a particular situation based upon circumstances. The guidance does not represent final agency action, and EPA may change it in the future, as appropriate.

**FOR UPDATES**

To keep up-to-date on changes to this document and to the risk management program rule, and to find other information related to part 68 and accident prevention, visit EPA's Office of Emergency Management website at:

**www.epa.gov/emergencies/rmp**

# TABLE OF CONTENTS

## CHAPTER 10 IMPLEMENTATION

## CHAPTER 11 COMMUNICATION WITH THE PUBLIC

## APPENDICES

# LIST OF BOXES AND EXHIBITS

## LIST OF EXHIBITS

**LIST OF BOXES**

viii

# TABLE OF POTENTIALLY REGULATED ENTITIES

*This table is not intended to be exhaustive, but rather provides a guide for readers regarding entities likely to be regulated under 40 CFR part 68. This table lists the types of entities that EPA is now aware could potentially be regulated by this rule and covered by this document. Other types of entities not listed in this table could also be affected. To determine whether your facility is covered by the risk management program rules in part 68, you should carefully examine the applicability criteria discussed in Chapter 1 of this guidance and in 40 CFR 68.10, which is available in Appendix A of this document. If you have questions regarding the applicability of this rule to a particular entity, call the EPCRA/CAA Hotline at (800) 424-9346 (TDD: (800) 553-7672)(see Appendix C, Technical Assistance, for other sources of information).*

| Category | NAICS Codes | SIC Codes | Examples of Potentially Regulated Entities |
|---|---|---|---|
| Warehouses | 49311<br>49312<br>49319 | 4225<br>4222<br>4226 | General warehousing and storage<br>Refrigerated warehousing<br>Other warehousing and storage |

x

May 26, 2000

# INTRODUCTION

## WHY SHOULD I READ THIS GUIDANCE?

If you handle, manufacture, use, or store any of the toxic and flammable substances listed in 40 CFR §68.130 (see Appendix A of this document) above the specified threshold quantities in a process, you are required to develop and implement a risk management program rule issued by the U.S. Environmental Protection Agency (EPA). This rule, "Chemical Accident Prevention Provisions" (part 68 of Title 40 of the Code of Federal Regulations (CFR)), applies to a wide variety of facilities that handle, manufacture, store, or use toxic substances, including chlorine and ammonia and highly flammable substances such as propane. This document provides guidance on how to determine if you are subject to part 68 and how to comply with part 68. If you are subject to part 68, you must be in compliance no later than June 21, 1999, or the date on which you first have more than a threshold quantity of a regulated substance in a process, whichever is later.

This guidance is intended for warehouses that handle or store chemicals; some of these warehouses may repackage chemicals, but most limit their activities to storing substances in containers designed to meet DOT transportation regulations. Information that is not applicable to warehouses has been omitted. If your warehouse is part of a larger facility that processes or uses chemicals or stores large quantities of chemicals for its own use, there will be information that is applicable to those other operations that is not presented in this document. For those operations, you should consult the *General Guidance of Risk Management Programs* or EPA's other industry-specific guidance documents, as appropriate.

The goal of part 68 — the risk management program — is to prevent accidental releases of substances that can cause serious harm to the public and the environment from short-term exposures and to mitigate the severity of releases that do occur. The 1990 Amendments to the Clean Air Act (CAA) require EPA to issue a rule specifying the type of actions to be taken by facilities (referred to in the statute as stationary sources) to prevent accidental releases of such hazardous chemicals into the atmosphere and reduce their potential impact on the public and the environment. Part 68 is that rule.

In general, part 68 requires that:

- • Covered facilities must develop and implement a risk management program and maintain documentation of the program at the site. The risk management program will include an analysis of the potential offsite consequences of an accidental release, a five-year accident history, a release prevention program, and an emergency response program.

- • Covered facilities also must develop and submit a risk management plan (RMP), which includes registration information, to EPA no later than June 21, 1999, or the date on which the facility first has more than a threshold quantity in a process, whichever is later. The RMP provides a summary of the risk management program. The RMP will be available to federal, state, and local government agencies and the public.

- Covered facilities also must continue to implement the risk management program and update their RMPs periodically or when processes change, as required by the rule.

The phrase "risk management program" refers to all of the requirements of part 68, which must be implemented on an on-going basis. The phrase "risk management plan (RMP)" refers to the document summarizing the risk management program that you must submit to EPA.

## HOW DO I USE THIS DOCUMENT?

This is a technical guidance document designed for owners and operators of sources covered by part 68. It will help you to:

- Determine if you are covered by the rule;

- Determine what level of requirements is applicable to your covered process(es);

- Understand which specific risk management program activities must be conducted;

- Select a strategy for implementing a risk management program, based on your current state of compliance with other government rules and industry standards and the potential offsite impact of releases from your process(es); and

- Understand the reporting, documentation, and risk communication components of the rule.

This document provides guidance and reference materials to help you comply with EPA's risk management program regulations. You should view and retain this guidance as a reference document for use when you are unsure about what a requirement means. This document does not provide guidance on any other rule or part of the CAA.

---

### STATE PROGRAMS

This guidance applies to 40 CFR part 68. You should check with your state government to determine if the state has its own accidental release prevention rules or has obtained delegation from EPA to implement and enforce part 68 in your state. State rules may be more stringent than EPA's rules. They may cover more substances or cover the same substances at lower thresholds. They may also impose additional requirements. For example, California's state program requires a seismic study. See Chapter 10 for information on state implementation of part 68. Unless your state has been granted delegation, you must comply with part 68 as described in this document even if your state has different rules under state law.

---

## WHAT DO I DO FIRST?

Before developing a risk management program, you should do five things:

**(1) Determine which, if any, of your processes are covered by this program**

Only sources with a threshold quantity of a regulated substance (see 40 CFR 68.130 in Appendix A) in a "process" need to comply with part 68. "Process" is defined by the rule in § 68.3 and does not necessarily correspond with an engineering concept of process. The requirements apply only to covered processes. See Chapter 1 for more information on how to define your processes and determine if they are subject to the rule.

**(2) Determine the appropriate program level for each covered process**

Depending on the specific characteristics of a covered process and the results of the offsite consequence analysis for that process, it may be subject to one of three different sets of requirements (called program levels). See Chapter 2 for more information.

**(3) Determine EPA's requirements for the facility and each covered process**

Certain requirements apply to the facility as a whole, while others are process-specific. See Chapter 2 for more information.

**(4) Assess your operations to identify current risk management activities**

Because you probably conduct some risk management activities already (e.g., employee training, equipment maintenance, and emergency planning), you should review your current operations to determine the extent to which they meet the provisions of this rule. EPA does not expect you to redo these activities if they already meet the rule's requirements. See Chapters 5 to 8 individually for guidance on how to tell if your existing practices meet those required by EPA.

**(5) Review the regulations and this guidance to develop a strategy for conducting the additional actions you need to take for each covered process. Discuss the requirements with management and staff.**

The risk management program takes an integrated approach to assessing and managing risks and will involve most of the operations of covered processes. Early involvement of both management and staff will help develop an effective program.

## REQUIREMENTS ARE PERFORMANCE BASED

Finally, keep in mind that many of these requirements are performance-based; that is, EPA is not specifying how often you must inspect storage tanks, only that you do so in a manner that minimizes the risk of a release. This allows you to tailor you program to fit the particular conditions at your facility. The degree of complexity required in a risk management program will depend on the complexity of the facility. For example, the operating procedures for a chemical distributor are likely to be relatively brief, while those for a chemical manufacturer will be extensive. Similarly, the length of training necessary to educate employees on such procedures would be proportional to the complexity of your operating procedures. And while a facility with complex processes may benefit from a computerized maintenance tracking system, a small facility with a simpler process may be able to track maintenance activities using a logbook.

There is no one "right" way to develop and implement a risk management program. Even for the same rule elements, your program will be different from everyone else's program (even those in the same industry) because it will be designed for your specific situation and hazards — it will reflect whether your facility is near the public and sensitive environmental areas, the specific equipment you have installed, the managerial decisions that you have made previously, and other relevant factors.

## WHERE DO I GO FOR MORE INFORMATION?

EPA's risk management program requirements may be found in Part 68 of Volume 40 of the Code of Federal Regulations. The relevant sections were published in the Federal Register on January 31, 1994 (59 FR 4478) and June 20, 1996 (61 FR 31667). EPA has amended the rules several times. A consolidated copy of these regulations, including amendments through June 30, 1999, is available in Appendix A. On March 13, 2000, EPA finalized regulations to implement new rules excluding flammable substances when used as fuels or held for sale as fuel by retailers (65 FR 13,243); a copy of this final rule is also available in Appendix A.

EPA has worked with industry and local, state, and federal government agencies to assist sources in complying with these requirements. For more information, refer to Appendix C (Technical Assistance). Your local emergency planning committee (LEPC) also can be a valuable resource and can help you discuss issues with the public.

Finally, if you have access to the Internet, EPA has made copies of the rules, fact sheets, and other related materials available at www.epa.gov/emergencies/rmp. Please check the site regularly as additional materials are posted.

## IF YOU ARE NEW TO REGULATIONS

We have tried to make this document as clear and readable as possible, but if you have rarely dealt with regulations before, some of the language may seem initially odd and confusing. All regulations have their own vocabulary. A few words and phrases have very specific meanings within the regulation. Some of these are unusual, which is to say they are not used in everyday language. Others are defined by the rule in ways that vary to some degree from their everyday meaning. The following are the major regulatory terms used in this document and a brief introduction to their meaning within the context of part 68. They are defined in § 68.3 of the rule.

*"Stationary source"* basically means facility. The CAA and, thus Part 68 use the term "stationary source" and we explain it in Chapter 1. Generally, we use "facility" in its place in this document.

*"Process"* is given a broad meaning in this rule and document. Most people think of a process as the mixing or reacting of chemicals. Its meaning under this rule is much broader. It basically means any equipment, including storage vessels, and activities, such as loading, that involve a regulated substance and could lead to an accidental release. Chapter 1 discusses the definition of process under this rule in detail.

*"Regulated substance"* means one of the 140 chemicals listed in part 68.

*"Threshold quantity"* means the quantity, in pounds, of a regulated substance which, if exceeded, triggers coverage by this rule. Each regulated substance has its own threshold quantity. If you have more than a threshold quantity of a regulated substance in a process, you must comply with the rule. Chapter 1 explains how to determine whether you have a threshold quantity.

*"Vessel"* means any container, from a single drum or pipe to a large storage tank or sphere.

*"Public receptor"* generally means any place where people live, work, or gather, with the exception of roads. Buildings, such as houses, shops, office buildings, industrial facilities, the areas surrounding buildings where people are likely to be present, such as yards and parking lots, and recreational areas, such as parks, sports arenas, rivers, lakes, beaches, are considered public receptors. Chapter 2 discusses public receptors.

*"Environmental receptor"* means a limited number of natural areas that are officially designated by the state or federal government. Chapter 2 discusses this definition.

## WHAT IS A LOCAL EMERGENCY PLANNING COMMITTEE?

Local emergency planning committees (LEPCs) were formed under the Federal Emergency Planning and Community Right-to-Know Act (EPCRA) in 1986. The committees are designed to serve as a community forum for issues relating to preparedness for emergencies involving hazardous substances. They consist of representatives from local government, local industry, transportation groups, health and medical organizations, community groups, and the media. LEPCs:

- • Collect information from facilities on hazardous substances that pose a risk to the community;
- • Develop a contingency plan for the community based on this information; and
- • Make information on hazardous substances available to the general public.

Contact the mayor's office or the county emergency management office for more information on your LEPC.

# CHAPTER 1: GENERAL APPLICABILITY

## 1.1    INTRODUCTION

The purpose of this chapter is to help you determine if you are subject to Part 68, the risk management program rule.  Part 68 covers you if you are:

- • •    The owner or operator of a stationary source

- • •    That has more than a threshold quantity

- • •    Of a regulated substance

- • •    In a process.

The goal of this chapter is to make it easy for you to identify processes that are covered by this rule so you can focus on them.

This chapter walks you through the key decision points (rather than the definition items above), starting with those provisions that may tell you that you are not subject to the rule.  We first outline the general applicability provisions and the few exemptions and exclusions, then discuss which chemicals are "regulated substances." If you do not have a "regulated substance" at your site, you are not covered by this rule.  The exemptions may exclude you from the rule or simply exclude certain activities from consideration.  (Throughout this document, when we say "rule" we mean the regulations in part 68.)

We then describe what is considered a "process," which is critical because you are subject to the rule *only* if you have more than a threshold quantity in a process.  The chapter next describes how to determine whether you have more than a threshold quantity.

Finally, we discuss how you define your overall stationary source and when you must comply.  These questions are important once you have decided that you are covered.  For most facilities covered by this rule, the stationary source is basically all covered processes at your site.  If your facility is part of a site with other divisions of your company or other companies, the discussion of stationary source will help you understand what you are responsible for in your compliance and reporting.  Exhibit 1-1 presents the decision process for determining applicability.

---

### STATE PROGRAMS

This guidance applies to only 40 CFR part 68.  You should check with your state government to determine if the state has its own accidental release prevention rules or has obtained delegation from EPA to implement and enforce part 68 in your state.  State rules may be more stringent than EPA's rules.  Unless your state has been granted delegation, you must comply with part 68 as described in this document even if your state has different rules under state law.  See Chapter 10 for a discussion of state implementation of part 68.

---

# EXHIBIT 1-1
## EVALUATE FACILITY TO IDENTIFY COVERED PROCESSES

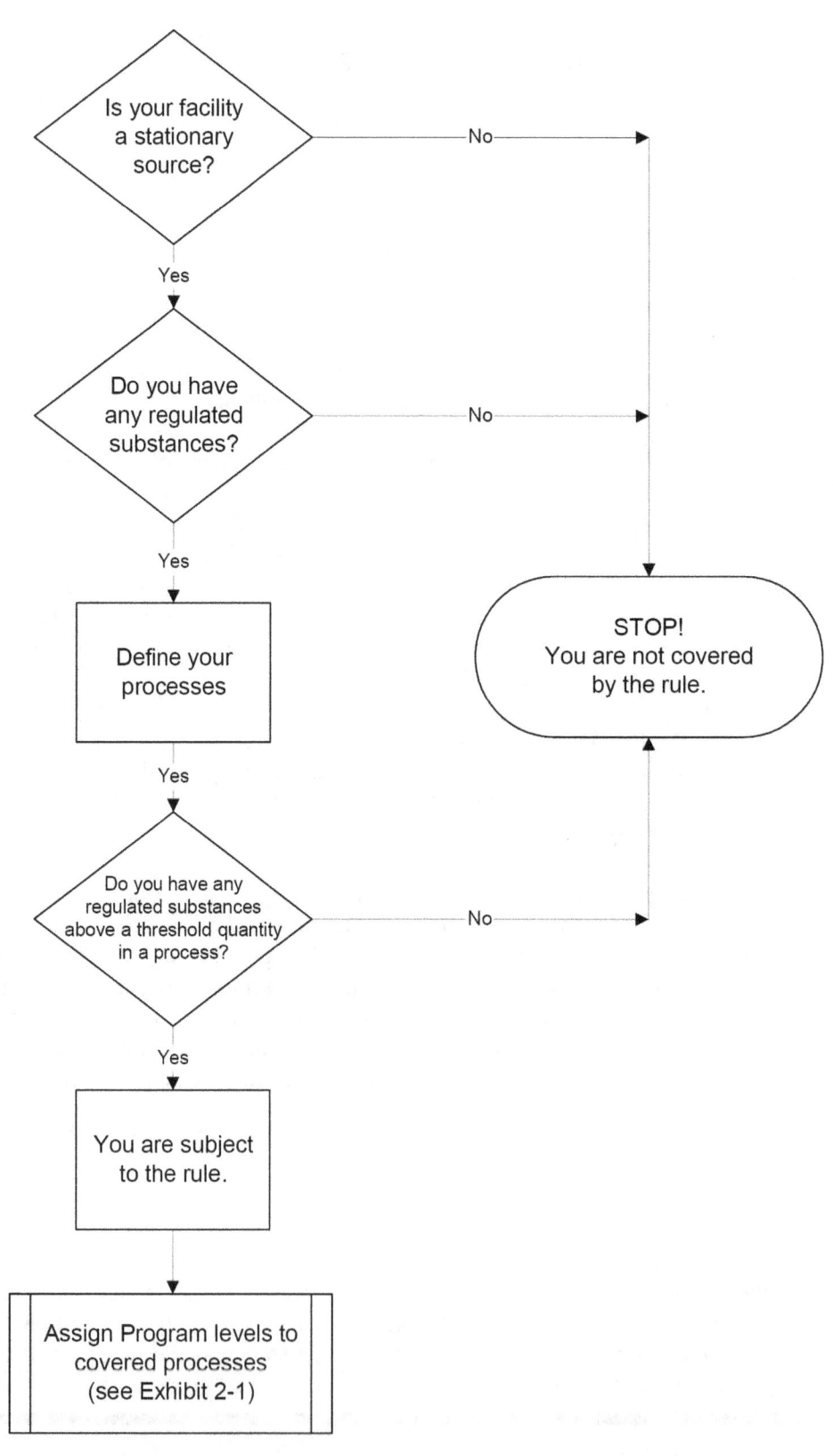

## 1.2 GENERAL PROVISIONS

The CAA applies this rule to any person who owns or operates a stationary source. "Person" is defined to include

"An individual, corporation, partnership, association, State, municipality, political subdivision of a state, and any agency, department, or instrumentality of the United States and any officer, agency, or employee thereof."

The rule, therefore, applies to all levels of government as well as private businesses.

CAA section 112(r)(2)(c) defines "stationary sources" as:

"Any buildings, structures, equipment, installations, or substance emitting stationary activities

- Which belong to the same industrial group,

- Which are located on one or more contiguous properties,

- Which are under the control of the same person (or persons under common control), and

- From which an accidental release may occur."

EPA has added some language in the rule to clarify issues related to transportation (see below).

### FARMS

The rule has only one exemption: for ammonia when held by a farmer for use on a farm. This exemption applies to ammonia only when used as a fertilizer by a farmer. It does not apply to agricultural suppliers or the fertilizer manufacturer.

### FLAMMABLE FUELS (§ 68.126)

The flammable substances listed in § 68.130 are excluded from coverage under part 68 when they are used as a fuel or held for sale as a fuel at a retail facility. A retail facility is defined as a stationary source at which more than half of the income is obtained from direct sales to end users or at which more than one-half of the fuel sold, by volume, is sold through a cylinder exchange program.

Unless your facility meets the definition of a "retail facility," if you store a listed flammable substance for purposes other than on-site use as fuel, you are potentially covered by part 68. If you store a listed flammable substance for purposes other than on-site use as fuel and also use it as a fuel, the quantity used as fuel is not covered; the quantity stored for purposes other than on-site use as fuel is potentially subject to the rule.

### TRANSPORTATION ACTIVITIES

Transportation containers used for storage not incident to transportation and transportation containers connected to equipment at a stationary source are considered part of the stationary source. Transportation containers that have been unhooked from the motive power that delivered them to the site (e.g., truck or locomotive) and left on your site for short-term or long-term storage are part of your stationary source. For example, if you have railcars on a private siding that you use as storage tanks, these railcars should be considered to be part of your source. If a tank truck is being unloaded **and** the motive power is still attached, the truck and its contents are considered to be in transportation and not covered by the rule. You should count only the substances in the piping or hosing as well as quantity unloaded. Some issues related to transportation are still under discussion with DOT.

---

### QS & AS
### STATIONARY SOURCE

**Q.** What does "same industrial group" mean?

**A.** Operations at a site that belong to the same three-digit North American Industry Classification System (NAICS) code (which has replaced the old two-digit SIC codes) belong to the "same industrial group. In addition, where one or more operations at the site serve primarily as support facilities for the main operation at the site, the supporting operations are part of the "same industrial group" as the main operation.

**Q.** What does "contiguous property" mean?

**A.** Property that is adjoining. Public rights-of-way (e.g., railroads, highways) do not prevent property from being considered contiguous. Property connected only by rights-of-way are not considered contiguous (e.g., two plants with a connecting pipeline).

**Q.** What does "control of the same person" mean?

**A.** Control of the same person refers to corporate control, not site management. If two divisions of a corporation operate at the same site, even if each operation is managed separately, they will count as one source provided the other criteria are met because they are under control of the same company.

---

### RELATIONSHIP TO OSHA PROCESS SAFETY MANAGEMENT STANDARD EXEMPTIONS

The OSHA Process Safety Management (PSM) standard exempts flammable substances stored in atmospheric storage tanks. (Other OSHA exemptions are not relevant to warehouses.) The OSHA exemptions do not apply or extend to EPA's Risk Management Program Rule. Your processes are not exempt from the Risk

Management Program simply because they qualify for one of the OSHA exemptions. Consequently, EPA covers substances stored in atmospheric storage tanks if there is more than a threshold quantity in a process.

## 1.3    REGULATED SUBSTANCES AND THRESHOLDS

The list of substances regulated under § 68.130 is in Appendix A.  Check the list carefully.  If you do not have any of these substances (either as pure substances or in mixtures above 1 percent concentration) or do not have them above their listed threshold quantities, you do not need to read any further.

The list includes 77 chemicals that were listed because they are acutely toxic; they can cause serious health effects or death from short-term exposures.  The list also covers 63 flammable gases and highly volatile flammable liquids.  The flammable chemicals have the potential to form vapor clouds and explode or burn if released.  The rule also covers flammable mixtures that include any of the listed flammables if the mixture meets the criteria for the National Fire Protection Association's (NFPA) 4 rating.

## 1.4    WHAT IS A PROCESS

The concept of "process" is key to whether you are subject to this rule.  Process is defined as:

"Any activity involving a regulated substance, including any use, storage, manufacturing, handling or on-site movement of such substances or any combination of these activities.  For the purposes of this definition, any group of vessels that are interconnected, or separate vessels that are located such that a regulated substance could be involved in a potential release shall be considered a single process."

"Vessel" means a reactor, tank, drum, barrel, cylinder, vat, kettle, boiler, pipe, hose, or other container.

The definition of process is identical to the definition of process under the OSHA PSM standard.  It is important in determining whether you have a threshold quantity of a regulated substance and what the level of requirements you must meet if the process is covered.

What does this mean to you?

- If you store a regulated substance in a single vessel in quantities above the threshold quantity, you are covered.

- If you have interconnected vessels that altogether hold more than a threshold quantity, you are covered.  The connections need not be permanent.  If two or more vessels are connected occasionally, they are considered a single process for the purposes of determining whether a threshold quantity is present.

- If you have multiple unconnected vessels, containing the same substance, you will have to determine whether they need to be considered together.

### WAREHOUSES AS A SINGLE PROCESS

Because warehouses usually consist of one large storage area, even if subdivided, and because you are likely to have the same prevention practices for the entire warehouse, you may want to consider the warehouse building a single process. You are not required to treat the warehouse as a single process; if the storage areas for regulated substances are widely separated and do not meet the criteria for co-location discussed below, you may treat each area separately. If you store chemicals outside the warehouse, they may be considered a separate process. The issue you will have to decide is whether you have more than a threshold quantity of a regulated substance to determine whether your warehouse building is a covered process. Co-location, discussed below, will probably be the key issue in determining whether your warehouse is a covered process and, if so, which chemicals must be included in your risk management program.

### SINGLE VESSELS

As a warehouse, you are unlikely to have a single storage tank of any regulated substance holding more than a threshold quantity unless you repackage chemicals. If you have a tank and it is the only place you have a regulated substance, you need not worry about the other possibilities for defining a process and can skip to the next section.

### CO-LOCATION

You must consider whether you have separate vessels that contain the same regulated substance that are located such that they could be involved in a single release. If so, you must add together the total quantity in all such vessels to determine if you have more than a threshold quantity. This possibility will be particularly important if you store a regulated substance in cylinders or barrels or other containers in a warehouse or outside in a rack. In some cases, you may have two vessels or systems that are in the same building or room. For each of these cases, you should ask yourself:

- • Would a release from one of the containers lead to a release from the other? For example, if a cylinder of a flammable substance were to rupture and burn, would the fire spread to other cylinders?

- • Would an event external to the containers, such as a fire or explosion, have the potential to release the regulated substance from multiple containers?

You must determine whether there is a credible scenario that could lead to a release of a threshold quantity.

For flammables, you should consider the distance between vessels. If a fire could spread from one vessel to others or an explosion could rupture multiple vessels, you must count all of them. For toxics, a release from a single vessel will not normally lead to a release from others unless the vessel fails catastrophically and explodes, sending metal fragments into other vessels. Co-located vessels containing toxic substances, however, may well be involved in a release caused by a fire or explosion that occurs from another source. The definition of process is predicated on the assumption that explosion will take place. In addition, a collapse of storage racks could lead to multiple vessels breaking open.

If the vessels are separated by fire walls or barricades or by group occupancy rooms that will contain the blast waves from explosions of the substances, you will not need to count the separated vessels, but you may have to count any that are in the same room if a fire could spread to involve all of the containers.

You may not dismiss the possibility of a fire spreading based on an assumption that your fire department will be able to prevent any spread. You should ask yourself how far the fire would spread if the worst happens — the fire department is slow to arrive, the water supply fails, or the fire department decides it is safer to let the fire burn itself out. If you have vessels that, when taken together, could release more than a threshold quantity in such worst-case circumstances, you should count them as a single process. At a warehouse, you will probably want to consider the nature of the other material stored. If the other materials are flammable or combustible and likely to feed a fire, you may need to be conservative in estimating how far a fire could spread. If the other materials are not combustible, a fire may be confined to part of a room.

## INTERCONNECTED VESSELS

Interconnection is unlikely to be applicable to warehouses. In general, if you have two or more vessels that contain a regulated substance and are connected through piping or hoses for the transfer of the regulated substance, you must consider the total quantity in all the connected vessels and piping when determining if you have a threshold quantity in a process. If the vessels are connected for transfer of the substance using hoses that are then removed, you still have to consider the contents of the vessels as one process, because if one vessel were to rupture while the hose was attached or the hose were to break during the transfer, you could lose the total quantity in both tanks. Therefore, you must count the quantities in both tanks and in any connecting piping or hoses. You cannot consider the presence of automatic shutoff valves or other devices that can limit flow, because these are assumed to fail for the purpose of determining the total quantity in a process.

## PROCESSES WITH MULTIPLE CHEMICALS

When you are determining whether you have a covered process, you should not limit your consideration to units that have the same regulated substance. A covered process includes any units that hold more than a threshold quantity of regulated substances and that are interconnected or co-located. Therefore, if you have four storage vessels holding four different regulated substances above their individual thresholds and they are located close enough to be involved in a single event, they are considered a single process. One implication of this approach is that if you have two vessels, each containing slightly less than a threshold quantity of the same regulated substance and located a considerable distance apart, and you have other storage or process vessels in between with other regulated substances above their thresholds, the vessels with the first substance may be part of the process involving the other vessels and other regulated substances, based on co-location.

Exhibit 1-2 provides illustrations of what may be defined as a process.

### DIFFERENCES WITH OSHA

OSHA aggregates different flammable liquids across vessels in making threshold determinations; OSHA also aggregates different flammable gases (but does not aggregate flammable liquids with flammable gases); EPA aggregates neither. Therefore, if you have three co-located or connected reactor vessels each containing 5,000 pounds of a different flammable liquid, OSHA considers that you have 15,000 pounds of flammable liquids and are covered by the PSM standard. Under EPA's rule, you would not have a covered process because you do not meet the threshold quantity for any one of the three substances. OSHA, like EPA, does not aggregate quantities for toxics as a class (i.e., each toxic substance must meet its own threshold quantity).

## 1.5  THRESHOLD QUANTITY IN A PROCESS

The threshold quantity for each regulated substance is listed in Appendix A. You should determine whether the maximum quantity of each substance in a process is greater than the threshold quantity listed. If it is, you must comply with this rule for that process. Even if you are not covered by this rule, you may still be subject to reporting requirements under the Emergency Planning and Community Right to Know Act (EPCRA).

### QUANTITY IN A VESSEL

To determine if you have the threshold quantity of a regulated substance in a vessel involved in a single process, you need to consider the maximum quantity in that vessel at any one time. You do not need to consider the vessel's maximum capacity if you never fill it to that level. Base your decision on the actual maximum quantity that you may have in the vessel. Your maximum quantity may be more than your normal operating maximum quantity; for example, if you may use a vessel for emergency storage, the maximum quantity should be based on the quantity that might be stored.

"At any one time" means you need to consider the largest quantity that you ever have in the vessel. If you fill a tank with 50,000 pounds and immediately begin using the substance and depleting the contents, your maximum is 50,000 pounds.

# EXHIBIT 1-2:  PROCESS

| Schematic Representation | Description | Interpretation |
|---|---|---|
| | 1 vessel<br>1 regulated substance above TQ | 1 process |
| | 2 or more connected vessels<br>*same* regulated substance<br>above TQ | 1 process |
| | 2 or more connected vessels<br>*different* regulated substances<br>each above TQ | 1 process |
| | pipeline feeding multiple vessels<br>total above TQ | 1 process |
| | 2 or more vessels co-located<br>*same* substance<br>total above TQ | 1 process |
| | 2 or more vessels co-located<br>*different* substances<br>each above TQ | 1 process |
| | 2 vessels, located so they won't be<br>involved in a single release<br>*same* or *different* substances<br>each above TQ | 2 processes |
| | 2 locations with regulated substances<br>each above TQ | 1 or 2 processes<br>depending on distance |
| | 1 series of interconnected vessels<br>*same* or *different* substances above TQs<br>*plus* a co-located storage vessel<br>containing flammables | 1 process |

---

## AGGREGATION OF SUBSTANCES

A toxic substance is never aggregated with a different toxic substance to determine whether a threshold quantity is present. If your process consists of co-located vessels with different toxic substances, you must determine whether each substance exceeds its threshold quantity.

A flammable substance in one vessel is never aggregated with a different flammable substance in another vessel to determine whether a threshold quantity is present. However, if a flammable mixture meets the criteria for NFPA-4 and contains different regulated flammables, it is the mixture, not the individual substances, that is considered in determining if a threshold quantity is present.

---

If you fill the vessel four times a year, your maximum is still 50,000 pounds. Throughput is not considered because the rule is concerned about the maximum quantity you could release in a single event.

### QUANTITY IN A PIPELINE

The maximum quantity in a pipeline will generally be the capacity of the pipeline (volume). In most cases, pipeline quantity will be calculated and added to the interconnected vessels.

### INTERCONNECTED/CO-LOCATED VESSELS

If your process consists of two or more interconnected vessels, you must determine the maximum quantity for each vessel and the connecting pipes or hoses. The maximum for each individual vessel and pipe is added together to determine the maximum for the process.

If you have determined that you must consider co-located containers as one process, you must determine the maximum quantity for each container and sum the quantities of all such containers.

### QUANTITY OF A SUBSTANCE IN A MIXTURE

#### *TOXICS WITH LISTED CONCENTRATION*

Four toxic substances have listed concentrations in the rule: hydrochloric acid — 37 percent or greater; hydrofluoric acid — 50 percent or greater; nitric acid — 80 percent or greater; and ammonia — 20 percent or greater.

- If you have these substances in solution and their concentration is less that the listed concentration, you do not need to consider them at all.

---

**Qs and As**
**THRESHOLD DETERMINATIONS**

**Q.** I store several different flammable liquids and products that contain flammable liquids. Do I combine them or consider each separately?

**A.** You must estimate the quantity of each separate listed flammable liquid.

**Q.** How far apart do containers have to be to be considered different processes?

**A.** There is no hard and fast rule for how great this distance should be before you do not need to consider the vessels as part of one process. Two containers at opposite ends of a large warehouse room might have to be considered as one process if the entire warehouse or room could be engulfed in a fire. Two containers separated by the same distance out of doors might be far enough apart that a fire affecting one would be unlikely to spread to the other. At a warehouse, the nature of the other materials stored should be considered; if containers are widely separated by other materials that could slow a fire's spread, the distance required to consider the containers separate processes will be much shorter than if most of the warehouse's contents are combustible. You may want to consult with your local fire department. You should then use your best professional judgment. Ask yourself how much of the regulated substance could be released if the worst happens (you have a major fire, an explosion, a natural disaster).

---

- If you have one of these four above their listed concentration, you must determine the weight of the substance in the solution and use that to calculate the quantity present. If that quantity is greater than the threshold, the process is covered. For example, aqueous ammonia is covered at concentrations above 20 percent, with a threshold quantity of 20,000 pounds. If the solution is 25 percent ammonia, you would need 80,000 pounds of the solution to meet the threshold quantity; if the solution is 44 percent ammonia, you would need 45,455 pounds to meet the threshold quantity (quantity of mixture x percentage of regulated substance = quantity of regulated substance).

Note that in a revision to part 68, EPA changed the concentration for hydrochloric acid to 37 percent or greater (see Appendix A).

*TOXICS WITHOUT A LISTED CONCENTRATION*

For toxics without a listed concentration, if the concentration is less than one percent you need not consider the quantity in your threshold determination. If the concentration in a mixture is above one percent, you must calculate the weight of the regulated substance in the mixture and use that weight to determine whether a threshold quantity is present. However, if you can measure or estimate (and document) that the partial pressure of the regulated substance in the mixture is less than 10 mm Hg, you do not need to consider the mixture. Note that the partial pressure rule does not apply to toluene diisocyanate (2,4-, 2,6-, or mixed isomers) or oleum.

EPA treats toxic mixtures differently from OSHA. Under the OSHA PSM standard, the entire weight of the mixture is counted toward the threshold quantity; under part 68, only the weight of the toxic substance is counted.

## FLAMMABLES

Flammable mixtures are subject to the rule only if there is a regulated substance in the mixture above one percent and the entire mixture meets the NFPA-4 criteria. If the mixture meets both of these criteria, you must use the weight of the entire mixture (not just the listed substance) to determine if you exceed the threshold quantity. The NFPA-4 definition is as follows:

"Materials that will rapidly or completely vaporize at atmospheric pressure and normal ambient temperature or that are readily dispersed in air, and that will burn readily. This degree usually includes:

### FLAMMABLE GASES

Flammable cryogenic materials

Any liquid or gaseous material that is liquid while under pressure and has a flash point below 73 F (22.8 C) and a boiling point below 100 F (37.8 C) (i.e., Class 1A flammable liquids)

Materials that will spontaneously ignite when exposed to air."

You do not need to consider gasoline, when in distribution or related storage for use as fuel for internal combustion engines when you determine the applicability of the rule.

## EXCLUSIONS (§ 68.115)

The rule has a number of exclusions that allow you to ignore certain sources that contain a regulated substance when you determine whether a threshold quantity is present. Note that these same exclusions apply to EPCRA section 313; you may be familiar with them if you comply with that provision.

## ARTICLES (§ 68.115(B)(4))

You do not need to include in your threshold calculations any manufactured item (as defined under 29 CFR 1910.1200(b)) that:

- • Is formed to a specific shape or design during manufacture,

- • Has end use functions dependent in whole or in part upon the shape or design during end use, and

## Qs and As
### CONSIDERATION OF PRODUCTS

**Q.** We frequently store large numbers of bottles of household ammonia and bleach. Do I have to figure out the percentage of ammonia or chlorine in each bottle?

**A.** No. Household ammonia (as a consumer product) does not meet the concentration threshold of 20 percent. Unless the concentration of ammonia in solution is 20 percent or greater, you do not need to consider the solution in your threshold determinations. Household bleach is usually a solution of water and sodium hypochlorite. Because the latter is not a listed substance, you do not need to consider it.

**Q.** We store consumer products that use butane as a propellant. Each product only has a few ounces of butane. Do we need to estimate the total amount of butane in all the products?

**A.** Listed flammable substances are excluded from coverage only if they are used as a fuel. In this case, butane is not being used as a fuel (i.e., it is not being burned to produce heat or power). As long as the butane is released from the product in normal use, you must estimate the amount of the regulated substance present. If the butane is mixed with the product, you should determine whether the product itself meets the criteria for NFPA 4. If the mixture does not meet the NFPA 4 criteria, the butane in the mixture is not counted toward the threshold.

- Does not release or otherwise result in exposure to a regulated substance under normal conditions of processing and use.

### USES (§ 68.115(b)(5))

You also do not need to include regulated substances in your calculation when in use for the following purposes:

- Use as a structural component of the stationary source;

- Use of products for routine janitorial maintenance;

- Use by employees of foods, drugs, cosmetics, or other personal items containing the regulated substances; and

- Use of regulated substances present in process water or non-contact cooling water as drawn from the environment or municipal sources, or use of regulated substances present in air used either as compressed air or as part of combustion.

## 1.6   STATIONARY SOURCE

The rule applies to "stationary sources" and each stationary source with one or more covered processes must file an RMP that includes all covered processes.

### SIMPLE SOURCES

For most facilities covered by this rule, determining what constitutes a "stationary source" is simple. If you own or lease a property, your processes are contained within the property boundary, and no other companies operate on the property, then your stationary source is defined by the property boundary and covers any process within the boundaries that has more than a threshold quantity of a regulated substance. You must comply with the rule and file a single RMP for all covered processes.

### MULTIPLE OPERATIONS OWNED BY A SINGLE COMPANY

If the property is owned or leased by your company, but several separate operating divisions of the company have processes at the site, the divisions' processes may be considered a single stationary source because they are controlled by a single company. Two factors will determine if the processes are to be considered a single source: Are the processes located on one or more contiguous properties? Are all of the operations in the same industrial group?

If your company does have multiple operations that are on the same property and are in the same industrial group, each operating division may develop its prevention program separately for its covered processes, but you must file a single RMP for all covered processes at the site. You should note that this is different from the requirements for filing under CAA Title V and EPCRA section 313 (the annual toxic release inventory), where each division could file separately if your company chose to do so.

### OTHER SOURCES

There are situations where two or more separate companies occupy the same site. The simplest of these cases is if multiple companies lease land at a site (e.g., an industrial park). Each company that has covered processes must file an RMP that includes information on its own covered processes at the site. You are responsible for filing an RMP for any operations that you own or operate.

Another possibility is that one company owns the land and operates there while leasing part of the site to a second company. If both companies have covered processes, each is considered a separate stationary source and must file separate RMPs even if they have contractual relationships, such as supplying product to each other or sharing emergency response functions.

If you and another company jointly own a site, but have separate operations at the site, you each must file separate RMPs for your covered processes. Ownership of the land is not relevant; a stationary source consists of covered processes located on the same property and controlled by a single owner.

### JOINT VENTURES

You and another company may jointly own covered processes. In this case, the legal entity you have established to operate these processes should file the RMP. If you consider this entity a subsidiary, you should be listed as the parent company in the RMP.

### MULTIPLE LOCATIONS

If you have multiple operations in the same area, but they are not on physically connected land, you must consider them separate stationary sources and file separate RMPs for each, even if the sites are connected by pipelines that move chemicals among the sites. Remember, the rule applies to covered processes at a single location.

Exhibit 1-3 provides examples of stationary source decisions.

## 1.7   WHEN MUST YOU COMPLY

If you had a covered process prior to June 21, 1999, you must comply with the requirements of part 68 no later than June 21, 1999. This means that whenever a process starts prior to June 21, 1999, you must be in compliance with the rule on June 21, 1999. You must have developed and implemented all of the elements of the rule that apply to each of your covered processes, and you must have submitted an RMP to EPA.

If the first time you have a covered process is after June 21, 1999, or you bring a new process on line after that date, you must comply with part 68 no later than the date on which you first have a more than a threshold quantity of a regulated substance in a process.

---

**Q and A**
**STATIONARY SOURCE**

**Q.** If I lease space in another building and store regulated substances above their thresholds there, must I file a separate RMP for them?

**A.** Yes, if the other building is a separate stationary source (i.e., it is not contiguous to the property where your warehouse is) you must file a separate RMP.

---

# EXHIBIT 1-3: STATIONARY SOURCE

| Schematic Representation | Description | Interpretation |
|---|---|---|
| ABC Chemicals General Chemicals Division / ABC Chemicals Plastics Division / ABC Chemicals Agricultural Chemicals Division | *same* owner<br>*same* industrial group | 1 stationary source<br>1 RMP |
| ABC Chemicals / ABC Chemicals / XYZ Gases | two owners | 2 stationary sources<br>2 RMPs<br>1 ABC<br>1 XYZ |
| ABC Chemicals / ABC Refinery / XYZ Gases | two owners<br>three industrial groups | 3 stationary sources<br>1 ABC Chemicals<br>1 ABC Refinery<br>1 XYZ Gases |
| ABC Chemicals / ABC-MNO Joint-Venture | two owners | 2 stationary sources<br>2 RMPs |
| ABC Products / ABC Products | *same* owner<br>*same* industrial group<br>contiguous property | 1 stationary source<br>1 RMP |
| Building owned by Brown Properties<br>Farm Chemicals Inc.<br>ABC Chemicals<br>Brown Property offices<br>Pet Supply Storage (no regulated substances) | two owners | 2 stationary sources<br>2 RMPs<br>1 ABC Chemicals<br>1 Farm Chemicals |

## 1.8   VARYING INVENTORIES AND PREDICTIVE FILING

As a warehouse owner, the main problem you are likely to face as you determine whether you are covered by this rule is that your inventory changes frequently. There may be periods when you have no regulated substances and other periods when you have several.   Determining your applicability under this rule on a day-to-day basis may be difficult, and in some cases, impossible.  One way to deal with this difficulty is to use predictive filing.

Predictive filing is an option that allows you to submit an RMP that includes regulated substances that may not be held at the facility at the time of submission. This option is intended to assist facilities such as chemical warehouses, chemical distributors, and batch processors whose operations involve highly variable types and quantities of regulated substances, but who are able to forecast their inventory with some degree of accuracy. Under § 68.190, you are required to update and re-submit your RMP no later than the date on which a new regulated substance is first present in a covered process above a threshold quantity.  By using predictive filing, you will not be required to update and re-submit your RMP when you receive a new regulated substance if that substance was included in your latest RMP submission (as long as you receive it in a quantity that does not trigger a revised offsite consequence analysis as provided in § 68.36).

To use predictive filing, review your inventories over the past several years and talk with your main customers to determine, to the extent possible, the kinds of materials they are planning to store at your facility.   If at some point during a year you normally receive enough vessels (drums, barrels, cylinders) to exceed a threshold quantity of a particular substance, list it on your registration in June 1999 even if you do not have it on the day you submit. If it appears, over time, that your customers will not be using your warehouse to store the substance again, you can deregister it later.  In the short run, you will be safer listing too many substances, than too few, because this approach will limit the need to resubmit your RMP every time your inventory changes.

If you have flammable mixtures at your warehouse, you may want to register them as a class rather than listing each covered flammable substance.  This approach will assure that you are in compliance with the registration requirements while limiting the effort you need to make to identify the specific substances.

If you use predictive filing, you must implement your Risk Management Program and prepare your RMP exactly as you would if you actually held all of the substances included in the RMP.  This means that you must meet all rule requirements for each regulated substance for which you file, whether or not that substance is actually held on site at the time you submit your RMP.  Depending on the substances for which you file, this may require you to perform additional worst-case and alternative-case scenarios and to implement additional prevention program elements. If you use this option, you must still update and resubmit your RMP if you receive a regulated substance that was not included in your latest RMP.  This approach will not completely eliminate the need to update your RMP, but should limit the frequency of updates. If you use this option, you must still comply with the other update requirements stated in § 68.190. RMPs must be updated when you:

- • Add a new regulated substance above its threshold (i.e., one not already reported in your latest predictive RMP submission);

- • Add a new covered process;

- • Have the program level of the process change (see Chapter 2);

- • Make a major change that requires a revised PHA or hazard review (see Chapters 6 and 7); or

- • Make a change that changes the distance to endpoint for a worst-case release by a factor of two or more.

Listing all the regulated substances you think you are likely to handle will mean more work initially (primarily more alternative release scenarios), but will limit the need for updates. As a rule of thumb, you will need to increase or decrease the quantity of a chemical in the single largest vessel by a factor of five or more to change the distance to an endpoint by a factor of two.

Predictive filing will work best when you simply store chemicals. If you repackage chemicals, you will need to complete prevention program information for each repackaging process. If you can predict which regulated substances you will repackage and can establish your prevention program, you can file predictively for that process. If, however, you have listed a regulated substance in your RMP based on expected storage, but you subsequently begin to repackage as well as store the chemical, you will need to update the RMP to reflect the new process.

### Qs & As
### COMPLIANCE DATES

**Q.** What happens if I bring a new covered process on line (e.g., install a second storage tank) after June 21, 1999?

**A.** For new covered process after the initial compliance date, you must be in compliance on the date you first have a regulated substance above the threshold quantity in that process. There is no grace period. You must develop and implement all the applicable rule elements before you start operating the new process.

**Q.** What if EPA lists a new substance?

**A.** You will have three years from the date on which the new listing is effective to come into compliance for any process that is covered because EPA has listed a new substance.

**Q.** I store 1-ton cylinders of chlorine. If I normally have 20 cylinders located together on site and register that quantity, do I need to update my RMP if I increase the number of cylinders to 200? How does this affect my worst-case scenario?

**A.** You do not necessarily need to update the RMP simply to reflect the higher quantity of chlorine. In this case, because you have not changed the size of your single largest vessel, your worst-case release scenario will not change. You will update the quantity information on your next scheduled update.

**Q.** I have stored 1-ton cylinders of chlorine together. Because of customer demand, I have started repackaging and have a tank with 40,000 pounds of chlorine. Do I need to update the RMP?

**A.** Yes, for two reasons. First, if the tank is a new process, you must update your RMP immediately; if it is part of an existing process, you must update within 6 months. Second, the 40,000-pound tank may result in the distance to endpoint for your worst-case release increasing by more than a factor of 2. If this is the case, you will need to update that change as well.

# CHAPTER 2: APPLICABILITY OF PROGRAM LEVELS

## 2.1    WHAT ARE PROGRAM LEVELS?

Once you have decided that you have one or more processes subject to this rule (see Chapter 1), you need to identify what actions you must take to comply.  The rule imposes different requirements on processes based on the potential for public impacts and the level of effort needed to prevent accidents.  EPA has set three levels of requirements that apply to covered processes:

> **Program 1:** Processes with no public receptors within the distance to the endpoint from a worst-case release and with no accidents with specific offsite consequences within the past five years are eligible for Program 1, which imposes minimal requirements on the process.

> **Program 2:** Processes not eligible for Program 1 or subject to Program 3 are placed in Program 2, which imposes a streamlined prevention program.

> **Program 3:** Processes not eligible for Program 1 and either subject to OSHA's PSM standard under federal or state OSHA programs or in ten specified North American Industry Classification System (NAICS) codes are placed in Program 3, which imposes the OSHA PSM program as the prevention program.

If you can qualify a process for Program 1, it is in your best interests to do so, even if the process is already subject to OSHA PSM.  For Program 1 processes, the implementing agency will inspect and enforce only on compliance with the minimal Program 1 requirements.  If you assign a process to Program 2 or 3 when it might qualify for Program 1, the implementing agency will inspect or enforce for compliance with all the requirements of the higher program levels.  If, however, you are already in compliance with the prevention elements of Program 2 or Program 3, you may want to use the RMP to inform the community of your prevention efforts.

See Exhibit 2-1 for a diagram of the decision rules on Program level.

### KEY POINTS TO REMEMBER

In determining program levels for your process(es), keep in mind the following:

(1)    **The program levels apply to individual processes** and generally indicate the risk management measures necessary to comply with this regulation for the process, not the facility as a whole.  The eligibility of one process for a program level does not influence the eligibility of other covered processes for other program levels.

(2)    **Any process can be eligible for Program 1**, even if it is subject to OSHA PSM or is in one of the NAICS codes.

# EXHIBIT 2-1
## EVALUATE PROGRAM LEVELS FOR COVERED PROCESSES

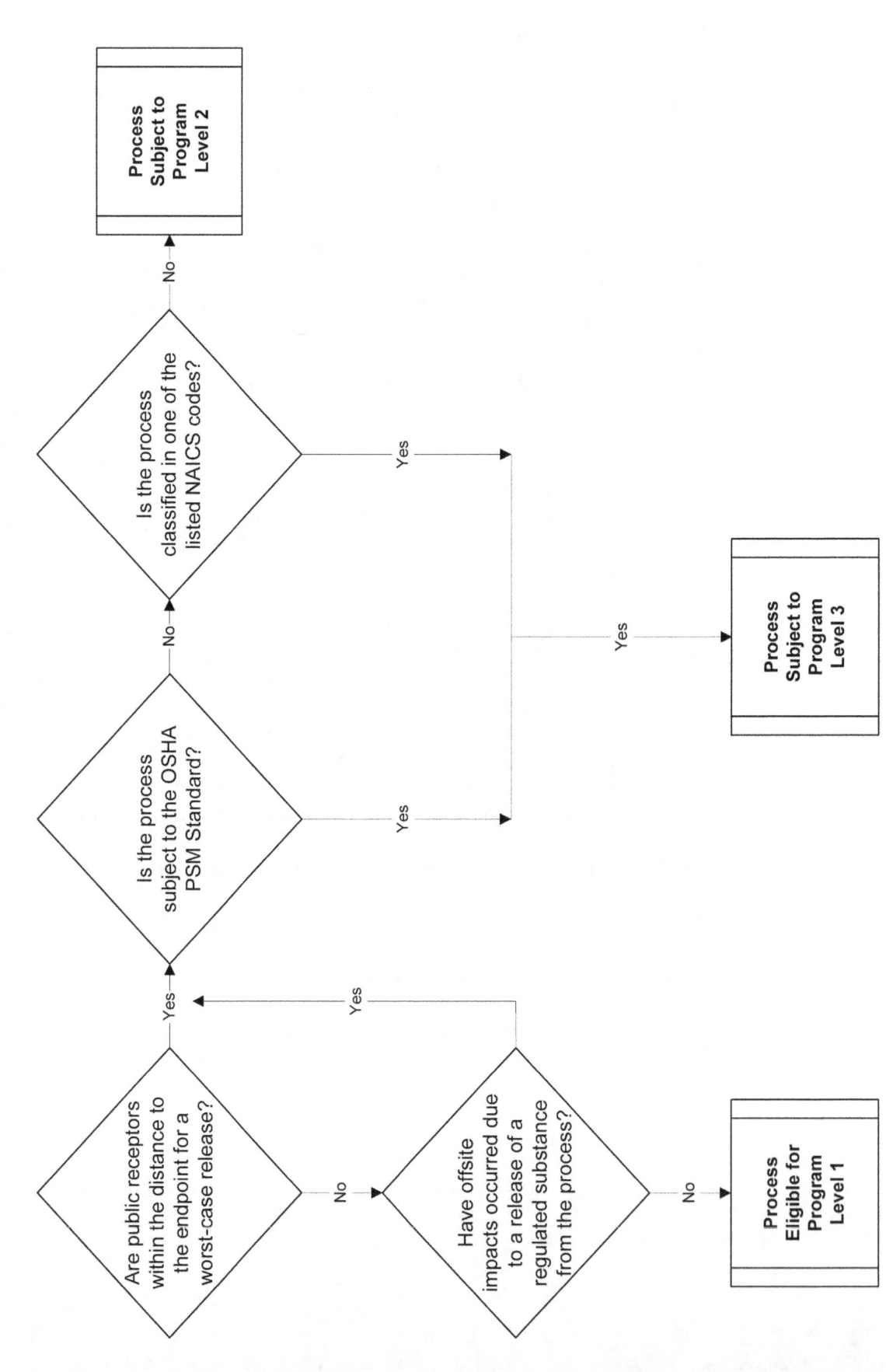

**(3)** **Program 2 is the default program level.** There are no "standard criteria" for Program 2. Any process that does not meet the eligibility criteria for either Programs 1 or 3 is subject to the requirements for Program 2.

**(4)** **Only one Program level can apply to a process.** If a process consists of multiple production or operating units or storage vessels, the highest Program level that applies to any segment of the process applies to all parts.

---

### Q & A
### PROCESS AND PROGRAM LEVEL

**Q.** My process includes a series of interconnected units, as well as several storage vessels that are co-located. Several sections of the process could qualify for Program 1. Can I divide my process into sections for the purpose of assigning Program levels?

**A.** No, you cannot subdivide a process for this purpose. The highest Program level that applies to any section of the process is the Program level for the whole process. If the entire process is not eligible for Program 1, then the entire process must be assigned to Program 2 or Program 3.

---

## 2.2   PROGRAM 1

### WHAT ARE THE ELIGIBILITY REQUIREMENTS?

Your process is eligible for Program 1 if:

**(1)** There are no public receptors within a distance to an endpoint from a worst-case release;

**(2)** The process has had no release of a regulated substance in the past five years where exposure to the substance, its reaction products, overpressures generated by explosion involving the substance, or radiant heat from a fire involving the substance resulted in offsite deaths, injuries, or response or restoration activities for exposure of an environmental receptor; and

**(3)** You have coordinated your emergency response activities with the local responders. (This requirement applies to any covered process, regardless of program level.)

### WHAT IS A PUBLIC RECEPTOR?

The rule defines **public** as "any person who is not an employee or contractor of the stationary source." Consequently, employees of other facilities that may share your site are considered members of the public even if they share the same physical location. Being "the public," however, is not the same as being a public receptor.

**Public receptors** include offsite residences, institutions (e.g., schools and hospitals), industrial, commercial, and office buildings, parks, or recreational areas inhabited or occupied by the public at any time without restriction by the stationary source. **Offsite** means areas beyond your property boundary and "areas within the property boundary to which the public has routine and unrestricted access during or outside of business hours."

For most facilities, the meaning of the definition of public receptor is straightforward. If you restrict access to your property at all times, public receptors are any occupied buildings or public gathering areas beyond your boundaries. Access restrictions include precautions such as a fully fenced site, security guards on duty at a reception area, or ID badges necessary to gain entry.

If you have unrestricted sections of your site that are predictably used by the public (e.g., ball fields or picnic areas), then these sections would also be considered public receptors. Neighboring businesses, whether commercial or industrial, are considered public receptors, as are residences, institutions such as hospitals, schools, prisons, marinas and airport terminals, public and private parking lots, golf courses, transit stations, and toll booth plazas for roads and bridges. The ability of others to restrict access to an area does not change its status as a public receptor.

Not all areas offsite are public receptors. Public roads and bridges are not considered public receptors. For other areas, you need to make a reasonable determination as to whether the public is likely to inhabit or occupy an offsite area. For example, a facility located in a remote mountainous area surrounded by unimproved forest might reasonably determine that the surrounding land is not a public receptor, even if it is infrequently traversed by hunters or fishermen. If a remote facility borders a park or wilderness area, the parts of the park, such as the campground, picnic area, or hiking trails that are likely to be occupied by the public, even if only seasonally, would be considered public receptors. Farm land may or may not be a public receptor. If farm workers are usually present, the farm land is a public receptor. If, however, the farm or ranch land is rarely occupied by workers, it may not be a public receptor. If you are in doubt about whether to consider certain areas around your facility as public receptors, you should consult with local emergency planning officials, local or state authorities, the land owners, and your implementing agency for guidance on whether such areas should be considered as public receptors.

## WHAT IS A DISTANCE TO AN ENDPOINT FROM A WORST-CASE RELEASE?

The rule establishes "endpoints" for each substance and defines a worst-case release scenario (see Chapter 4 or the *RMP Offsite Consequence Analysis Guidance* for more information). You will have to define a worst-case release (usually the loss of the total contents of your largest vessel) and either use EPA's guidance or conduct modeling on your own to determine the distance to the endpoint. Beyond that point, the effects on people are not considered to be severe enough to merit the need for additional action under this rule.

**Qs & As**
**Public Receptors**

**Q.** My processes are fenced, but my offices and parking lot for customers are not restricted. What is considered offsite?

**A.** The unrestricted areas would be considered potential public receptors.

**Q.** What is considered a recreational area?

**A.** Recreational areas would include most bodies of waters (oceans, lakes, rivers, and streams) because they are used for fishing, swimming, or boating. Areas that are predictably used by hunters, fishermen, bird watchers, children, bike riders, or hikers would be considered recreational areas. Areas where there are places for public to gather (e.g., ball fields, picnic tables, jungle gyms, hiking paths, campsites) would be considered recreational areas. Even if an area is only used during certain parts of the year for recreation, it would still be considered a recreational area. EPA recognizes that some judgment is involved in determining whether an area should be considered a public receptor. You are responsible for making a reasonable judgment. If you have doubts about whether an area can be legitimately excluded from consideration as a public receptor, EPA encourages you to consult with local officials and the community to reach an agreement on an area's status; your local emergency planning committee (LEPC) can help you with these consultations. If your facility is surrounded by undeveloped land, you may also want to consult with the land owner.

**Q.** Does public receptor cover only buildings on a property or the entire property? If the owner of the land next to my site restricts access to the land, is it still a public receptor?

**A.** Public receptors are not limited to buildings. For example, if there are houses near your property, both the houses and their yards are considered public receptors because it is likely the people will be present in both at times and would be in more danger if they were outside when a release occurred. If the owner of a neighboring property restricts access to the land, the question you will need to consider is whether that land is generally unoccupied. If your site abuts farm land where farm workers are generally present, it is considered a public receptor. If the land is undeveloped or rarely has anyone on it, but you are uncertain about whether to consider it a public receptor, you should talk with the landowner and the community to reach an agreement on its status. Because it is the landowner and members of the local community who are likely to be affected by your decision, you should involve them in the decision is you have doubts.

To define the area of potential impact from the worst-case release, draw a circle on a map, using the process as the center and the distance to the endpoint as the radius. If there are any public receptors within that area, your process is not eligible for Program 1.

---

**Q and A**
**Determining Distances**

**Q.** Our distance to the endpoint for the worst-case release is 0.3 miles. The nearest public receptor is 0.32 miles away. What tools are available to document that the public receptor is beyond the distance to the endpoint so we can qualify for Program 1?

**A.** The results of any air dispersion model (from EPA's guidance documents or other models) are not precise predictions. They represent an estimate, but the actual distances to the endpoint could be closer to or farther from the point of release. If your distance to the endpoint and distance to a public receptor are so close that you cannot document, using a USGS map, that the two points are different, it would be advisable to comply with the higher Program level. (The most detailed maps available from the US Geological Survey (scale of 1:24,000) are not accurate enough to map the distances you cite and document that the two points (which are about 100 feet apart) differ. GPS systems now have a margin of error of 22 meters (about 0.014 miles or 72 feet); if you are using a GPS system, you may be able to document that these points are different.)

## ACCIDENT HISTORY

To be eligible for Program 1, no release of the regulated substance from the process can have resulted in offsite deaths, injuries, or response or restoration activities at an environmental receptor during the five years prior to submission of your RMP. A release of the regulated substance from another process has no bearing on whether the first process is eligible for Program 1.

### WHAT IS AN INJURY?

An injury is defined as "any effect on a human that results from direct exposure to toxic concentrations, radiant heat, or overpressures from accidental releases or from the direct consequences of a vapor cloud explosion (such as flying glass, debris, and other projectiles) from an accidental release." The effect must "require medical treatment or hospitalization." This definition is taken from the OSHA regulations for the keeping of the employee injury and illness logs and should be familiar to most employers. Medical treatment is further defined as treatment, other that first aid, administered by a physician or registered professional personnel under standing orders from a physician. The definition of medical treatment will likely capture most instances of hospitalization. However, if someone goes to the hospital following direct exposure to a release and is kept overnight for observation (even if no specific injury or illness is found), that would qualify as hospitalization.

### WHAT IS AN ENVIRONMENTAL RECEPTOR?

The environmental receptors you need to consider are limited to natural areas such as national or state parks, forests, or monuments; officially designated wildlife sanctuaries, preserves, refuges, or areas; and Federal wilderness areas. All of these areas can be identified on local U.S. Geological Survey maps.

*WHAT ARE RESTORATION AND RESPONSE ACTIVITIES?*

The type of restoration and response activity conducted to address the impact of an accidental release will depend on the type of release (volatilized spill, vapor cloud, fire, or explosion), but may include such activities as:

- • •    Collection and disposal of dead animals and contaminated plant life;

- • •    Collection, treatment, and disposal of soil;

- • •    Shutoff of drinking water;

- • •    Replacement of damaged vegetation; or

- • •    Isolation of a natural area due to contamination associated with an accidental release.

If an impact occurs, such damaged vegetation, and no steps are taken to replace the vegetation, the process remains eligible for Program 1.

---

## Q & A
### ENVIRONMENTAL RECEPTORS

**Q.**  Do environmental receptors include areas that are not Federal Class I areas under the CAA?

**A.**  Yes.  The list of environmental receptors in Part 68 is not related to the Federal Class I areas under CAA section 162.  Under Part 68, national parks, monuments, and wilderness areas are not limited by size criteria.  In addition, other areas are covered; for example, national forests and state parks, monuments, and forests are environmental receptors.

---

### DOCUMENTING PROGRAM 1 ELIGIBILITY

As part of your risk management program, you must keep records of your compliance with this requirement.  For each Program 1 process, your records should include the following:

- •    The worst-case release scenario, which shall include a description of the vessel or pipeline and substance selected as worst case, assumptions and parameters used, and the rationale for selection.

---

## Qs & As
### ACCIDENT HISTORY

**Q.** What is the relationship between the accident history for Program 1 and the five-year accident history? If my process is eligible for Program 1, do I still need to do a five-year accident history.

**A.** Although both cover the previous five years, the accidental release criteria for Program 1 and the general accident history for the source are different.

• The five-year accident history is an information collection requirement that is designed to provide data on all serious accidents from a covered process involving a regulated substance held above the threshold quantity.

• In contrast, the Program 1 criteria focus on whether the process in question has the potential to experience a release of the regulated substance that results in harm to the public based on past events. Onsite effects, sheltering-in-place, and evacuations are not relevant. Therefore, it is possible that a process eligible for Program 1 may still have experienced a release that must be reported in the accident history for the source.

**Q.** A process with more than a threshold quantity of a regulated substance had an accident with offsite consequences three years ago. After the accident, we altered the process to reduce the quantity stored on site. Now the worst-case release scenario indicates that there are no public receptors within the distance to an endpoint. Can this process qualify for Program 1?

**A.** No, the process cannot qualify for Program 1 until five years have passed since any accident with the specified consequences.

**Q.** A process involving a regulated substance had an accidental release with offsite consequences two years ago. The process has been shutdown. Do I have to report anyway?

**A.** No. The release does not have to be included in your accident history. Your risk management plan only needs to address processes that have more than a threshold quantity of a regulated substance on the date you file your RMP.

---

•  •     Assumptions shall include use of any administrative controls and any passive mitigation that were assumed to limit the quantity that could be released;

•  •     Documentation of estimated quantity released, release rate, and duration of release;

•  •     The methodology used to determine distance to endpoints;

•  •     Data used to determine that no public receptor would be affected;

•  •     Information on your coordination with public responders.

## 2.3   QUICK RULES FOR DETERMINING PROGRAM 1 ELIGIBILITY

You generally will not be able to predict with certainty that the worst-case analysis for a particular process will be eligible for Program 1. Processes containing certain substances, however, may be more likely than others to be eligible for Program 1, and processes containing certain other substances may be very unlikely to be eligible for Program 1 because of the toxicity and physical properties of the substances. The information presented below may be useful in helping you to decide whether to carry out analyses of processes to determine Program 1 eligibility (accident history criteria must be met separately).

### TOXIC GASES

If you have a process containing more than a threshold quantity of any regulated toxic gas that is not liquefied by refrigeration alone (i.e., you hold it as a gas or liquefied under pressure), the distance to the endpoint estimated using EPA's required worst-case assumptions is unlikely to be less than the distance to public receptors, unless your site is very remote; these distances will generally be several miles. In some cases, however, toxic gases in processes in enclosed areas may be eligible for Program 1

### TOXIC LIQUIDS

The distance to the endpoint from the worst-case analysis for toxic liquids kept under ambient conditions may be smaller than the distance to public receptors in a number of cases. If public receptors are not found very close to the process (within ½ mile), such processes may be eligible for Program 1. Warehouses in highly developed areas are unlikely to meet this criterion for most toxic liquids; it will be more relevant to remotely located warehouses or warehouses found near the center of large acreage sites. Substances that are potential candidates to be in processes that are eligible for Program 1 are noted below.

For processes containing toluene diisocyanate (including toluene 2,4-diisocyanate, toluene 2,6-diisocyanate, and unspecified isomers) or ethylene diamine, the analysis of a spill of more than a threshold quantity into an undiked area under ambient conditions is likely to demonstrate eligibility for Program 1. If the area of the spill is diked, processes containing very large quantities of these substances may be eligible for Program 1. In addition, processes containing the following toxic liquids under ambient conditions are likely to be eligible for Program 1 if a spill would take place in a diked area and public receptors are not close to the process:

- • Chloroform
- • Cyclohexylamine
- • Hydrazine
- • Isobutyronitrile
- • Isopropyl chloroformate
- • Oleum
- • Propylene oxide
- • Titanium tetrachloride

• •      Vinyl acetate monomer

## WATER SOLUTIONS OF TOXIC SUBSTANCES

The list of regulated substances includes several common water solutions of toxic substances. Processes containing such solutions at ambient temperatures may be eligible for Program 1 (depending in some cases on the concentration of the solution), if spills would be contained in diked areas and public receptors are not located close to the process (within ½ mile). As noted above, warehouse in developed areas are highly unlikely to meet this criterion; it will be more relevant to remotely located facilities or processes found near the center of large acreage sites.

Processes containing the following water solutions may be eligible for Program 1, assuming diked areas that would contain the spill and ambient temperatures:

• •      Ammonia in solution
• •      Formaldehyde (commercial concentrations)
• •      Hydrofluoric acid (concentration 50 to 70 percent)
• •      Nitric acid (commercial concentrations)

## FLAMMABLE SUBSTANCES

Warehouses that handle only regulated flammable substances are likely to be eligible for Program 1, unless there are public receptors within a very short distance. If you have a process containing up to about 20,000 pounds (twice the threshold quantity) of a regulated flammable substance (other than hydrogen), your process is likely to be eligible for Program 1 if you have no public receptors within about 400 yards (1,200 feet) of the process. If you have up to 100,000 pounds in a process (ten times the threshold quantity), the process may be eligible for Program 1 if there are no public receptors within about 700 yards (2,000 feet). In general, it would be worthwhile to conduct a worst-case analysis for any processes containing flammables to determine Program 1 eligibility, unless you have public receptors very close to the process. You must be able to demonstrate, through your worst-case analysis, that every process you claim as Program 1 meets the criteria.

## 2.4    PROGRAM 3

Any covered process that is not eligible for Program 1 and meets one of the two criteria specified below is covered by Program 3 requirements. Program 3 sets risk management measures, including compliance with the OSHA PSM Standard, for an eligible covered process.

### WHAT ARE THE ELIGIBILITY CRITERIA FOR PROGRAM 3?

Your process qualifies for Program 3 if:

• •      Your process does not meet the eligibility requirements for Program 1, and

• •    Either

• •    Your process is subject to OSHA PSM (federal or state); or

• •    Your process is in one of ten NAICS codes specified by EPA.

### WHAT IS THE OSHA PSM STANDARD?

The OSHA Process Safety Management standard (codified at 29 CFR 1910.119) is a formal set of procedures in thirteen management areas designed to protect worker health and safety from accidental releases. As with EPA's rule, they apply to a range of facilities that have more than a threshold quantity of a listed substance in a process. All processes subject to this rule and the OSHA PSM standard (federal or state) and not eligible for Program 1 are assigned to Program 3 because the Program 3 prevention program is identical to the elements of the PSM standard. If you are already complying with OSHA PSM for a process, you probably will need to take few, if any, additional steps and develop little, if any, additional documentation to meet the requirements of the Program 3 prevention elements (see Chapter 7 for a discussion of differences between Program 3 prevention and OSHA PSM). EPA placed all covered OSHA PSM processes in Program 3 to eliminate the possibility of imposing overlapping, inconsistent requirements on the same process.

Processes covered by OSHA PSM may include equipment, activities, and regulated substances, particularly flammables used as fuels, that in other circumstances are exempted under the OSHA PSM standard.

### WHAT ARE THE TEN NAICS CODES?

Program 3 requirements are applicable to a covered process if the process involves an activity in one of ten manufacturing NAICS codes: 32211 (pulp mills), 32411 (petroleum refineries), 32511 (petrochemical manufacturers), 325181 (chlor-alkali manufacturers), 325188 (al other inorganic chemicals manufacturers), 325192 (other cyclic crude and intermediate manufacturers), 325199 (all other basic organic chemical manufacturers), 325211 (plastics and resins manufacturers), 325311 (nitrogen fertilizer manufacturers), and 32532 (pesticide and other agricultural chemicals manufacturers). These codes are all for manufacturing and, therefore, are not relevant to warehouses. Even if your warehouse is a support activity for a chemical manufacturer, it is not considered to be in a manufacturing NAICS code for the purposes of determining program level. Appendix B provides a list of NAICS codes for industries that may be subject to part 68.

## 2.5    PROGRAM 2

Program 2 is considered a default program level because any covered process that is not eligible for Program 1 and Program 3 requirements is, by default, covered by Program 2 requirements. Program 2 sets risk management measures, including a streamlined accident prevention program, for an eligible covered process. Your process(es) are likely to be in Program 2 if:

- You are a publicly owned facility in a state that does not have a delegated OSHA program.

- You use or store the regulated acids in solution, and your activities do not fall into one of the ten specified NAICS codes.

- You store regulated liquid flammable substances in atmospheric storage tanks and they are not being used as a fuel.

### WHAT ARE THE ELIGIBILITY CRITERIA FOR PROGRAM 2?

Your process is eligible for Program 2 if:

- Your process does not meet the eligibility requirements for Program 1;

- Your process is not subject to OSHA PSM (federal or state).

Exhibit 2-2 provides a summary of the criteria for determining Program level.

## 2.6    DEALING WITH PROGRAM LEVELS

### WHAT IF I HAVE MULTIPLE PROGRAM LEVELS?

If you have more than one covered process, you may be dealing with multiple program levels in your risk management program.

If your facility has multiple processes subject to different program requirements, you will need to treat each group of processes in the same program level (and potentially each process) separately from the other processes and program level requirements. Nevertheless, you must submit a single RMP for all covered processes. At the same time, if you prefer, you may choose to adopt the most stringent applicable program level requirements for all covered processes. For example, you have three covered processes: one eligible for Program 1 and two subject to Program 3. You may find it administratively easier to follow the Program 3 requirements for all three covered processes. Remember that this is only an option; we expect that most sources will comply with the set of program level requirements for which each process is eligible.

### CAN THE PROGRAM LEVEL FOR A PROCESS CHANGE?

If a covered process meets the requirements for a new program level, you must re-evaluate the requirements for the process. If you are switching to another program level, this change must be reflected in an updated RMP that must be submitted within six months of the change that altered the program level for the covered process. If the process no longer qualifies as a covered process (e.g., as a result of a change in the quantity of the regulated substance in the process), then you will need to "deregister" the process; see Chapter 9 for more information. Typical examples of switching program levels include:

| EXHIBIT 2-2 PROGRAM LEVEL CRITERIA | | |
|---|---|---|
| **Program 1** | **Program 2** | **Program 3** |
| No accidents in the previous five years that resulted in any offsite:<br><br>Death<br>Injury<br>Response or restoration activities at an environmental receptor | The process is not eligible for Program 1 or subject to Program 3. | Process is not eligible for Program 1. |
| AND | | AND |
| No public receptors in worst-case circle. | | Process is subject to OSHA PSM. |
| AND | | OR |
| Emergency response coordinated with local responders. | | Process is classified in NAICS code<br>32211 (pulp mills)<br>32411 (petroleum refineries)<br>32511 (petrochemical manufacturers)<br>325181 (chlor-alkali manufacturers)<br>325188 (all other inorganic chemicals manufacturers)<br>325192 (other cyclic crude and intermediate manufacturers)<br>325199 (all other basic organic chemical manufacturers)<br>325211 (plastics and resins manufacturers)<br>325311 (nitrogen fertilizer manufacturers)<br>32532 (pesticide and other agricultural chemicals manufacturers) |

---

**Qs & As**
**OSHA**

**Q.** If my state administers the OSHA program under a formal delegation from the federal OSHA, does that mean that my processes subject to OSHA PSM under state rules are in Program 3?

**A.** Yes (as long as the process does not qualify for Program 1). Any process for which a facility is complying with PSM, under federal or state rules, is considered to be in Program 3.

**Q.** I am a publicly owned facility in a state with a delegated OSHA program. Why are my processes considered to be in Program 3 when the same process in a state where federal OSHA runs the program are in Program 2.

**A.** Federal OSHA cannot impose its rules on state or local governments, but when OSHA delegates its program to a state for implementation, the state imposes the rules on itself and local governments. Because these governments are complying with the identical OSHA PSM rules imposed by federal OSHA, they are subject to Program 3. They are already substantially in compliance with the Program 3 prevention program to meet their obligations under the state OSHA rules. State and local governments in non-state-plan states are not subject to any OSHA rules and must comply with Program 2.

*MOVING UP*

**From Program 1 to Program 2 or 3.** You have a covered process subject to Program 1 requirements. A new residential development results in public receptors being located within the distance to the endpoint for a worst-case release for that process. The process is, thus, no longer eligible for Program 1 and must be evaluated to determine whether Program 2 or Program 3 applies. You must submit a revised RMP within six months of the program level change, indicating and documenting that your process is now in compliance with the new program level requirements.

**From Not Covered to Program 1, 2 or 3.** You have a process that was not originally covered by part 68, but, due to an expansion in production, the process holds an amount of regulated substance that now exceeds the threshold quantity. You must determine which Program level applies and come into compliance with the rule by June 21, 1999, or by the time you exceed the threshold quantity, whichever is later.

**From Program 2 to Program 3.** You have a process that involves a regulated substance above the threshold that is not in one of the ten NAICS codes specified for Program 3 and that had not been subject to OSHA PSM. However, due to one of the following OSHA regulatory changes, the process is now subject to the OSHA PSM standard:

• •    An OSHA PSM exemption applicable to your process has been eliminated, or

- The regulated substance has been added to OSHA's list of highly hazardous substances.

As a result, the process becomes subject to Program 3 requirements and you must submit a revised RMP to EPA within six months, indicating and documenting that your process is now in compliance with the Program 3 requirements.

*SWITCHING DOWN*

**From Program 2 or 3 to Program 1.** At the time you submit your RMP, you have a covered process subject to Program 2/3 requirements because it experienced an accidental release of a regulated substance with offsite impacts four years ago. Subsequent process changes have made such an event unlikely (as demonstrated by the worst-case release analysis). One year after you submit your RMP, the accident will no longer be included in the five-year accident report for the process, so the process is eligible for Program 1. If you elect to qualify the process for Program 1, you must submit a revised RMP within six months of the program level change, indicating and documenting that the process is now in compliance with the new program level requirements.

**From Program 1, 2 or 3 to Not Covered.** You have a covered process that has been subject to part 68 requirements, but due to a reduction in production, the amount of a regulated substance it holds no longer exceeds the threshold. Therefore, the process is no longer a covered process. You must submit a revised RMP within six months indicating that your process is no longer subject to any program level requirements.

## 2.7   SUMMARY OF PROGRAM REQUIREMENTS

Regardless of the program levels you assign to your processes, you must complete a five-year accident history for each process (see Chapter 3) and submit an RMP that covers all processes (see Chapter 9). Exhibit 2-3 diagrams the requirements in general and Exhibit 2-4 lists them in more detail.

### PROGRAM 1

For each Program 1 process, you must conduct and document a worst-case release analysis. You must coordinate your emergency response activities with local responders and sign the Program 1 certification as part of your RMP submission.

### PROGRAMS 2 AND 3

For all Program 2 and 3 processes, you must conduct and document at least one worst-case release analysis to cover all toxics and one to cover all flammables. You must also conduct one alternative release scenario analysis for each toxic and one for all flammables. See Chapter 4 or the *RMP Offsite Consequence Analysis Guidance* for specific requirements. You must coordinate your emergency response activities with local responders and, if you use your own employees to respond to releases, you must develop and implement an emergency response program.  See Chapter 8 for more details.

For each Program 2 process, you must implement all of the elements of the Program 2 prevention program: safety information, hazard review, operating procedures, training, maintenance, compliance audits, and incident investigations.  See Chapter 6 for more details.

For each Program 3 process, you must implement all of the elements of the Program 3 prevention program: process safety information, process hazard analysis, standard operating procedures, training, mechanical integrity, compliance audits, incident investigations, management of change, pre-startup reviews, contractors, employee participation, and hot work permits.  See Chapter 7 for more details.

# EXHIBIT 2-3
# DEVELOP RISK MANAGEMENT PROGRAM AND RMP

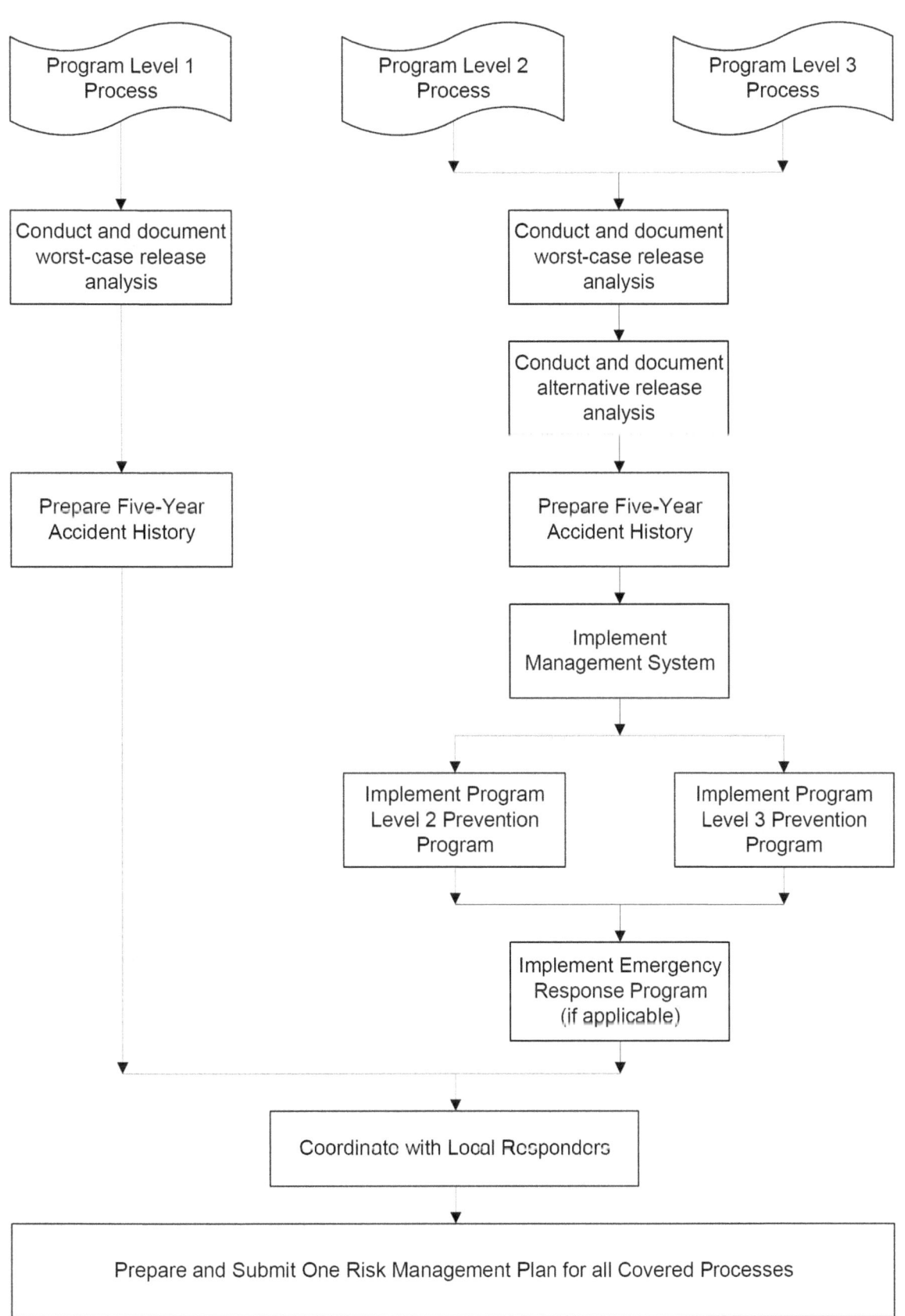

| EXHIBIT 2-4 | | |
| :---: | :---: | :---: |
| **COMPARISON OF PROGRAM REQUIREMENTS** | | |
| **Program 1** | **Program 2** | **Program 3** |
| Worst-case release analysis | Worst-case release analysis | Worst-case release analysis |
| | Alternative release analysis | Alternative release analysis |
| 5-year accident history | 5-year accident history | 5-year accident history |
| | Document management system | Document management system |
| Prevention Program | | |
| Certify no additional prevention steps needed | Safety Information | Process Safety Information |
| | Hazard Review | Process Hazard Analysis. |
| | Operating Procedures | Operating Procedures |
| | Training | Training |
| | Maintenance | Mechanical Integrity |
| | Incident Investigation | Incident Investigation |
| | Compliance Audit | Compliance Audit |
| | | Management of Change |
| | | Pre-Startup Review |
| | | Contractors |
| | | Employee Participation |
| | | Hot Work Permits |
| Emergency Response Program | | |
| Coordinate with local responders | Develop plan and program (if applicable) and coordinate with local responders | Develop plan and program (if applicable) and coordinate with local responders |
| Submit One Risk Management Plan for All Covered Processes | | |

# CHAPTER 3: FIVE-YEAR ACCIDENT HISTORY

The five-year accident history involves an examination of the effects of any accidental releases of one or more of the regulated substances from a covered process in the five years prior to the submission of a Risk Management Plan (RMP). A five-year accident history must be completed for each covered process, including the processes in Program 1, and all accidental releases meeting specified criteria must be reported in the RMP for the process.

Note that a Program 1 process may have had an accidental release that must be included in the five-year accident history, even though the release does not disqualify the process from Program 1. The accident history criteria that make a process ineligible for Program 1 (certain offsite impacts) do not include other types of effects that require inclusion of a release in the five-year accident history (on-site impacts and more inclusive offsite impacts). For example, an accidental release may have led to worker injuries, but no other effects. This release would not bar the process from Program 1 (because the injuries were not offsite), but would need to be reported in the five-year accident history. Similarly, a release may have resulted in damage to foliage offsite (environmental damage), triggering reporting, but because the foliage was not part of an environmental receptor (e.g., national park or forest) it would not make the process ineligible for Program 1.

## 3.1     WHAT ACCIDENTS MUST BE REPORTED?

The five-year accident history covers only certain releases:

- The release must be from a covered process and involve a regulated substance held above its threshold quantity in the process.

- The release must have caused at least one of the following:

    - On-site deaths, injuries, or significant property damage (§68.42(a)); or

    - Known offsite deaths, injuries, property damage, environmental damage, evacuations, or sheltering in place (§68.42(a)).

If you have had a release of a regulated substance from a process where the regulated substance is held below its threshold quantity, you do not need to report that release even if the release caused one of the listed impacts or if the process is covered for some other substance. You may choose to report the release in the five-year accident history, but you are not required to do so.

## 3.2     WHAT DATA MUST BE PROVIDED?

The following information should be included in your accident history for every reported release;

**Time.** Indicate the time at which the release began.

**Release duration.** Indicate the approximate length of time of the release in minutes.

**Chemical(s).** Indicate the regulated substance(s) released. Use the name of the substance as listed in § 68.130 rather than a synonym (e.g., ethylene oxide rather than oxirane). If the release was of a flammable mixture, list the primary regulated substances in the mixture if feasible; if the contents of the mixture are uncertain, list it as a flammable mixture.

**Quantity released.** Estimate the amount of each substance released in pounds. The amount should be estimated to two significant digits, or as close to that as possible. For example, if you estimate that the release was between 850 and 900 pounds, provide a best guess. We realize that you may not know precise quantities. For flammable mixtures, you may report the quantity of the mixture, rather than that of the individual regulated substances.

**Release event.** Indicate which of the following release events best describes your accident. Check all that apply:

◆      *Gas Release.* A gas release is a release of the substance as a gas (rather than vaporized from a liquid). If you hold a gas liquefied under refrigeration, report the release as a liquid spill.

◆      *Liquid Spill/ Evaporation.* A liquid spill/evaporation is a release of the substance in a liquid state with subsequent vaporization.

◆      *Fire.* A fire is combustion producing light, flames, and heat.

◆      *Explosion.* An explosion is a rapid chemical reaction with the production of noise, heat, and violent expansion of gases.

◆      *Uncontrolled/Runaway Reaction.* A release event caused by an uncontrolled chemical reaction that generates excessive heat, pressure, or harmful reaction products. Such events may involve highly exothermic chemical reactions, self-reactive substances (e.g., substances that undergo polymerization), unstable, explosive, or spontaneously combustible substances, substances that react strongly with water or other contaminants, oxidizers, peroxide-forming substances, or other types of chemical reactions that generate harmful products or byproducts. This category of release event may often occur in conjunction with one of the previous categories. In such cases, be sure to check this category in addition to any other applicable release event category (e.g., explosion). The burning of ordinary flammable substances is not typically included in this category.

**Release source.** Indicate all that apply.

◆ *Storage Vessel.* A storage vessel is a container for storing or holding gas or liquid. Storage vessels include transportation containers being used for on-site storage.

◆ *Piping.* Piping refers to a system of tubular structures or pipes used to carry a fluid or gas.

◆ *Process Vessel.* A process vessel is a container in which substances under certain conditions (e.g., temperature, pressure) participate in a process (e.g., substances are manufactured, blended to form a mixture, reacted to convert them into some other final product or form, or heated to purify).

◆ *Transfer Hose.* A transfer hose is a tubular structure used to connect, often temporarily, two or more vessels.

◆ *Valve.* A valve is a device used to regulate the flow in piping systems or machinery. Relief valves and rupture disks open to release pressure in vessels.

◆ *Pump.* A pump is a device that raises, transfers, or compresses fluids or that attenuates gases by suction or pressure or both.

◆ *Joint.* The surface at which two or more mechanical components are united.

◆ *Other.* Specify other source of the release.

Weather conditions at time of event (if known). This information is important to those concerned with modeling the effects of accidents. Reliable information from those involved in the incident or from an on-site weather station is ideal. However, this rule does not require your facility to have a weather station. If you do not have an onsite weather station, use information from your local weather station, airport, or other source of meteorological data. To the extent possible, complete the following:

◆ *Wind Speed and Direction.* Wind speed is an estimate of how fast the wind is traveling. Indicate the speed in miles per hour. Wind direction is the direction from which the wind comes. For example, a wind that blows from east to west would be described as having an eastern wind direction. You may describe wind direction as a standard compass reading such as "Northeast" or "South-southwest."

You may also describe wind direction in degrees—with North as zero degrees and East as 90 degrees. Thus, northeast would represent 45 degrees and south-southwest would represent 202.5 degrees. Abbreviations for the wind direction such as NE (for northeast) and SSW (for south-southwest) are also acceptable.

◆ *Temperature.* The ambient temperature at the scene of the accident in degrees Fahrenheit. If you did not keep a record, you can use the high (for daytime releases) or low (for nighttime releases) for the day of the release. Local papers publish these data.

◆ *Stability Class.* Depending on the amount of incoming solar radiation as well as other factors, the atmosphere may be more or less turbulent at any given time. Meteorologists have defined six atmospheric stability classes, each representing a different degree of turbulence in the atmosphere. When moderate to strong incoming solar radiation heats air near the ground, causing it to rise and generating large eddies, the atmosphere is considered unstable, or relatively turbulent. Unstable conditions are associated with stability classes A and B. When solar radiation is relatively weak, air near the surface has less of a tendency to rise and less turbulence develops. In this case, the atmosphere is considered stable or less turbulent with weak winds. The stability class is E or F. Stability classes D and C represent conditions of neutral stability or moderate turbulence respectively. Neutral conditions are associated with relatively strong wind speeds and moderate solar radiation. Exhibit 3-1 presents the stability classes associated with wind speeds, time of day, and cloud cover.

◆ *Precipitation Present.* Precipitation may take the form of hail, mist, rain, sleet, or snow. Indicate "yes" or "no" based on whether there was any precipitation at the time of the accident.

◆ *Unknown.* If you have no record for some or all of the weather data, indicate "unknown" for any missing item. We realize that you may not have weather data for accidents that occurred in the past. You should, however, collect these data for any future accidents.

## EXHIBIT 3-1
## ATMOSPHERIC STABILITY CLASSES

| SURFACE WIND SPEED AT 10 METERS | | DAY | | | NIGHT | |
|---|---|---|---|---|---|---|
| | | Incoming Solar Radiation | | | Thinly Overcast or ≥ 4/8 low cloud | ≤ 3/8 Cloud |
| Meters per second | Miles per hour | Strong* | Moderate | Slight** | | |
| < 2 | <4.5 | A | A-B | B | | |
| 2-3 | 4.5-7 | A-B | B | C | E | F |
| 3-5 | 7-11 | B | B-C | C | D | E |
| 5-6 | 11-13 | C | C-D | D | D | D |
| >6 | >13 | C | D | D | D | D |

\* Sun high in the sky with no clouds.
\*\* Sun low in the sky with no clouds.

**On-site impacts.** Complete the following about on-site effects.

◆  *Deaths.* Indicate the number of on-site deaths that are attributed to the accident or mitigation activities. On-site deaths means the number of employees, contract employees, offsite responders, or others (e.g., visitors) who were killed by direct exposure to toxic concentrations, radiant heat, or overpressures from accidental releases or from indirect consequences of a vapor cloud explosion from an accidental release (e.g., flying glass, debris, other projectiles). You should list employee/contractor, offsite responder, and other on-site deaths separately.

◆  *Injuries.* An injury is any effect that results either from direct exposure to toxic concentrations, radiant heat, or overpressures from accidental releases or from indirect consequences of a vapor cloud explosion (e.g., flying glass, debris, other projectiles) from an accidental release and that requires medical treatment or hospitalization. You should list injuries to employees and contractors, offsite responders, and others separately.

Medical treatment means treatment, other than first aid, administered by a physician or registered professional personnel under standing orders from a physician.

Your Log of Work-Related Injuries and Illnesses (OSHA Form 300) and Injury and Illness Incident Report (OSHA Form 301) will help complete these items for employees.

◆  *Property Damage.* Estimate the value of the equipment or business structures (for your business alone) that were damaged by the accident or mitigation activities. Record the value in American dollars. Insurance claims may provide this information. Do **not** include any losses that you may have incurred as a result of business interruption.

**Known offsite impacts.** These are impacts that you know or could reasonably be expected to know of (e.g., from media reports or from reports to your facility) that occurred as a result of the accidental release. You are not required to conduct an additional investigation to determine offsite impacts.

---

### Qs & As
### PROPERTY DAMAGE

**Q.** What level of offsite property damage triggers reporting?

**A.** Any level of known offsite property damage triggers inclusion of the accident in the five-year accident history. You are not required to conduct a survey to determine if such damage occurred, but if you know, or could reasonably be expected to know (e.g., because of reporting in the newspapers), that damage occurred, you must include the accident.

◆    *Deaths.* Indicate the number of offsite deaths that are attributable to the accident or mitigation activities. Offsite deaths means the number of community members who were killed by direct exposure to toxic concentrations, radiant heat, or overpressures from accidental releases or from indirect consequences of a vapor cloud explosion from an accidental release (e.g., flying glass, debris, other projectiles).

◆    *Injuries.* Indicate the number of injuries among community members. Injury means any effect that results either from direct exposure to toxic concentrations, radiant heat, or overpressures from accidental releases or from indirect consequences of a vapor cloud explosion from an accidental release (e.g., flying glass, debris, other projectiles) and that requires medical treatment or hospitalization.

◆    *Evacuated.* Estimate the number of members of the community who were evacuated to prevent exposure that might have resulted from the accident. A total count of the number of people evacuated is preferable to the number of houses evacuated. People who were ordered to move simply to improve access to the site for emergency vehicles are not considered to have been evacuated.

◆    *Sheltered.* Estimate the number of members of the community who were sheltered-in-place during the accident. Sheltering-in-place occurs when community members are ordered to remain inside their residence or place of work until the emergency is over to prevent exposure to the effects of the accidental release. Usually these orders are communicated by an emergency broadcast or similar method of mass notification by response agencies.

◆    *Environmental Damage.* Indicate whether any environmental damage occurred and specify the type. The damage to be reported is not limited to environmental receptors listed in the rule. Any damage to the environment (e.g., dead or injured animals, defoliation, water contamination) should be identified. You are **not**, however, required to conduct surveys to determine whether such impact occurred. Types of environmental damage include:

   ▹    Fish or animal kills.

   ▹    Lawn, shrub, or crop damage (minor defoliation).

   ▹    Lawn, shrub, or crop damage (major defoliation).

   ▹    Water contamination.

   ▹    Other (specify).

**Initiating event.** Indicate the initiating event that was the immediate cause of the accident, if known. If you conducted an investigation of the release, you should have identified the initiating event.

◆ *Equipment Failure.* A device or piece of equipment failed or did not function as designed. For example, the vessel wall corroded or cracked.

◆ *Human Error.* An operator performed a task improperly, either by failing to take the necessary steps or by taking the wrong steps.

◆ *Weather Conditions.* Weather conditions, such as lightning, hail, ice storms, tornados, hurricanes, floods, or high winds, caused the accident.

◆ *Unknown.*

**Contributing factors.** These are factors that contributed to the accident, but were not the initiating event. If you conducted an investigation of the release, you may have identified factors that led to the initiating event or contributed to the severity of the release. Indicate all that apply.

◆ *Equipment Failure.* A device or piece of equipment failed to function as designed, thereby leading to or worsening the accidental release.

◆ *Human error.* An operator performed an operation improperly or made a mistake leading to or worsening the accidental release.

◆ *Improper Procedures.* The procedure did not reflect the proper method of operation, the procedure omitted steps that affected the accident, or the procedure was written in a manner that allowed for misinterpretation of the instructions.

◆ *Overpressurization.* The process was operated at pressures exceeding the design working pressure.

◆ *Upset Condition.* Incorrect process conditions (e.g., increased temperature or pressure) contributed to the release.

◆ *By-pass Condition.* A failure occurred in a pipe, channel, or valve that diverts fluid flow from the main pathway when design process or storage conditions are exceeded (e.g., overpressure). By-pass conditions may be designed to release the substance to restore acceptable process or storage conditions and prevent more severe consequences (e.g., explosion).

◆ *Maintenance Activity/Inactivity.* A failure occurred because of maintenance activity or inactivity. For example, the storage racks remained unpainted for so long that corrosion caused the metal to fail.

◆       *Process Design.* A failure resulted from an inherent flaw in the design of the process (e.g., pressure needed to make the product exceeds the design pressure of the vessel).

◆       *Unsuitable Equipment.* The equipment used was incorrect for the process. For example, the forklift was too large for the corridors.

◆       *Unusual Weather Conditions.* Weather conditions, such as lightning, hail, ice storms, tornados, hurricanes, floods, or high winds contributed to the accident.

◆       *Management Error.* A failure occurred because management did not exercise its managerial control to prevent the accident from occurring. This is usually used to describe faulty procedures, inadequate training, inadequate oversight, or failure to follow existing administrative procedures.

**Whether offsite responders were notified.** If known, indicate whether response agencies (e.g., police, fire, medical services) were contacted.

**Changes introduced as a result of the accident.** Indicate any measures that you have taken at the facility to prevent recurrence of the accident. Indicate all that apply.

◆       *Improved/ Upgraded Equipment.* A device or piece of equipment that did not function as designed was repaired or replaced.

◆       *Revised Maintenance.* Maintenance procedures were clarified or changed to ensure appropriate and timely maintenance including inspection and testing (e.g., increasing the frequency of inspection or adding a testing method).

◆       *Revised Training.* Training programs were clarified or changed to ensure that employees and contract employees are aware of and are practicing correct safety and administrative procedures.

◆       *Revised Operating Procedures.* Operating procedures were clarified or changed to ensure that employees and contract employees are trained on appropriate operating procedures.

◆       *New Process Controls.* New process designs and controls were installed to correct problems and prevent recurrence of an accidental release.

◆       *New Mitigation Systems.* New mitigation systems were initiated to limit the severity of accidental releases.

◆       *Revised Emergency Response Plan.* The emergency response plan was revised.

◆       *Changed Process.* Process was altered to reduce the risk (e.g., process chemistry was changed).

◆       *Reduced Inventory.* Inventory was reduced at the facility to reduce the
        potential release quantities and the magnitude of the hazard.

◆       *Other.*

◆       *None.* No changes initiated at facility as a result of the accident (e.g.,
        because none were necessary or technically feasible). There may be some
        accidents that could not have been prevented because they were caused by
        events that are too rare to merit additional steps. For example, if a tornado
        hit your facility and you are located in an area where tornados are very rare,
        it may not be reasonable to design a "tornado proof" process even if it is
        technically feasible.

## 3.3    WHEN MUST ACCIDENTS BE REPORTED?

When an RMP is first submitted to EPA, it must contain a five-year accident history
including all of the accidents that meet the reporting criteria discussed above and that
occurred within five years of the date of the RMP is submitted. When an RMP is
updated as required by section 68.190 of the rule, it must contain an updated five-
year accident history including all of the accidents that meet the reporting criteria
and that occurred within five years of the date on which the updated RMP is
submitted. In addition, on April 9, 2004, EPA published a final rule that amended
the accident history reporting requirement (and certain other provisions of the risk
management program). Beginning on that date, if an accident occurs that meets the
reporting criteria, it must be reported in the RMP five-year accident history within
six months of the accident, as required by section 68.195 of the rule, unless it is
included in an RMP update prior to that time. EPA took this action to require more
timely reporting of significant accidents in RMPs so that government, industry and
the public would be more quickly alerted to the possibility of similar accidents
occurring elsewhere.

## 3.4    OTHER ACCIDENT REPORTING REQUIREMENTS

You should already have much of the data required for the five-year accident history
because of the reporting requirements under the Comprehensive Emergency
Response, Compensation, and Liability Act (CERCLA), EPCRA, and OSHA (e.g.,
log of work-related injuries and illnesses). This information should minimize the
effort necessary to complete the accident history.

At the same time, some of the information originally reported to response agencies
may have been incomplete or inaccurate because it was reported during the release
when a full assessment was not possible. It is imperative that you include the most
accurate, up-to-date information possible in the five-year accident history. This
information may not always match the original estimates from the initial reporting of
the accident's effects.

**CERCLA** Section 103(a) requires you to immediately notify the National Response
Center if your facility releases a hazardous substance to the environment in greater
than a reportable quantity (see 40 CFR part 302). Toxic substances regulated under

part 68 are also CERCLA hazardous substances, but most of the flammable substances regulated under part 68 are not subject to CERCLA reporting. Notice required under CERCLA includes the following information:

◆　　　The chemical name or identity of any substance involved in the release

◆　　　An indication of whether the substance is on the list referred to in Section 302(a)

◆　　　An estimate of the quantity of substance that was released into the environment

◆　　　The time and duration of the release

◆　　　The medium or media into which the release occurred.

**EPCRA** Section 304 requires facilities to report to the community emergency coordinator of the appropriate local emergency planning committee (LEPC) and state emergency response commission (SERC) releases of extremely hazardous substances to the environment in excess of reportable quantities (as set forth in 40 CFR part 302). All toxic substances regulated under part 68 are subject to EPCRA reporting; flammables regulated under part 68 are generally not subject to EPCRA reporting. The report required by EPCRA is to include:

◆　　　Chemical name or identity of all substances involved in the accident

◆　　　An estimate of the quantity of substances released to the environment

◆　　　The time and duration of the release.

The owner or operator is also required to release a Follow-up Emergency Notice as soon as possible after a release which requires notification. This notice should update the previously released information and include additional information regarding actions taken to respond to the release, any known or anticipated acute or chronic health risks associated with the release, and where appropriate, advice regarding medical attention necessary for exposed individuals.

OSHA's Log of Work-Related Injuries and Illnesses, OSHA Form 300, is used for recording and classifying recordable occupational injuries and illnesses, and for noting the extent and outcome of each case. The log shows when the occupational injury or illness occurred, to whom, what the injured or ill person's regular job was at the time of the injury or illness exposure, the kind of injury or illness, how much time was lost, and whether the case resulted in a fatality, etc. The following are the sections of the illness/ injury log that are useful in completing the accident history.

◆　　　**Column B:** Employee's name

◆　　　**Column C:** Job title

◆       **Column D:** Date of injury or onset of illness

◆       **Column  F:** Description of injury or illness

◆       **Columns G, H, I, K, L:** Indicate whether a death occurred, whether injury
        resulted in lost workdays or restricted duty, and number of work days away
        form work or on restricted duty.

◆       **Column  M:** Indicates whether injury occurred or type of illness.

## PART 68 INCIDENT INVESTIGATION

An incident investigation is a requirement of the rule (§68.60 and 68.81).  For
accidents involving processes in Program 2 or Program 3, you must investigate each
incident which resulted in, or could reasonably have resulted in, a catastrophic
release of a regulated substance.  A report, which includes the following information,
should be prepared at the conclusion of the investigation:

◆       Date of incident

◆       Date investigation began

◆       Description of the incident

◆       Factors that contributed to the incident

◆       Any recommendations resulting from the investigation.

Because the incident investigation report must be retained for five years, you will
have a record for completing the five-year accident history for updates of the RMP.

## Qs &As
### ACCIDENT HISTORY

**Q.** When does the five-year period to be reported in the accident history begin?

**A.** The five-year accident history must include information on all accidental releases from covered processes meeting the specified criteria that occurred in the five years preceding the date of submission of your initial RMP or your most recent update required under section 68.190, as well as information provided to revise the accident history for any accidental releases that occur prior to the next required update. For example, if an RMP is updated on June 21, 2009, the five-year accident history must cover the period between June 21, 2004, and June 21, 2009. If a reportable accident occurs two months later, the five-year accident history must cover the period between June 21, 2004, and the date of the accident (see next question for further explanation).

**Q.** I recently submitted my five-year RMP update required by section 68.190 (b)(1) and included my accident history for the previous five years. Two months later, we had another reportable accident. Do I have to do anything to revise my RMP?

**A.** Yes. You must revise your accident history within six months of the date of the new accident to include information about it. You do this by correcting section 6 of your RMP (the accident history section) so that it includes all accident history information reported on the most recent update, as well as the information about the new accident. You should also indicate the reason for your correction (i.e., new accident history information) in the appropriate field in section 1 of RMP*eSubmit. Facilities reporting under Programs 2 and 3 must also revise the incident investigation information in their RMPs (section 7 or 8 of their RMP). Specifically, the date of investigation (40 CFR 68.170(j)) and the expected date of completion of any changes (40 CFR 68.175(l)) should be revised. You do not need to update or correct any other section of the RMP, unless you have taken actions (e.g., as a result of the accident) that trigger an update in accordance with section 68.190.

**Q.** If a facility has recently changed ownership, is the new facility owner required to include accidents which occurred prior to the transfer of ownership in the accident history portion of the RMP submitted for the facility?

**A.** Yes, accidents involving covered processes that occurred prior to the transfer of ownership should be included in the five-year accident history. You may want to explain that the ownership has changed in your Executive Summary.

**Q.** If I have a large on-site incident, but no offsite impact, would I have to report it in the five-year accident history?

**A.** It would depend on whether you have onsite deaths, injuries, or significant property damage. You could have a large accident without any of these consequences (e.g., a large spill that was contained); this type of release would not have to be included in the five-year accident history.

**Q.** I had a release where several people were treated at the hospital and released; they attributed their symptoms to exposure. We do not believe that their symptoms were in fact the result of exposure to the released substance. Do we have to report these as offsite impacts?

**A.** Yes, you should report them in your five-year accident history. You may want to use the executive summary to state that you do not believe that the impacts can be legitimately attributed to the release and explain why.

# CHAPTER 4: OFFSITE CONSEQUENCE ANALYSIS

You are required to conduct an offsite consequence analysis to provide information to the government and the public about the potential consequences of an accidental chemical release. The offsite consequence analysis (OCA) consists of two elements:

- • A worst-case release scenario and
- • Alternative release scenarios.

To simplify the analysis and ensure a common basis for comparisons, EPA has defined the worst-case scenario as the release of the largest quantity of a regulated substance from a single vessel or process line failure that results in the greatest distance to an endpoint. In broad terms, the distance to the endpoint is the distance a toxic vapor cloud, heat from a fire, or blast waves from an explosion will travel before dissipating to the point that serious injuries from short-term exposures are no longer likely.

The purpose of this chapter is to give guidance on how to perform the OCA for regulated substances at warehouses.

Section 68.130 lists 77 toxic substances and 63 flammable substances that are subject to regulation; however, it is unlikely that they will all be handled at commercial warehouses. Therefore, during the development of this chapter, EPA consulted representatives of the American Warehouse Association (AWA) and obtained a list of regulated chemicals that are commonly handled at warehouses. These substances are listed in Exhibit 4-1 (toxic substances) and 4-2 (flammable substances). In addition, generic guidance is given for substances that are not listed in Exhibit 4-1 or Exhibit 4-2.

This guidance is based on EPA's *RMP Offsite Consequence Analysis Guidance* (OCAG). Those parts of the OCAG that you need to use are included in Appendix 4A (which is located at the end of this chapter). See the OCAG (available from EPA) for more information on the methodology presented here.

---

### RMP*Comp™

To assist those using this guidance, the National Oceanic and Atmospheric Administration (NOAA) and EPA have developed a software program, RMP*Comp™, that performs the calculations described in this document. This software can be downloaded from the EPA Internet website at http://www.epa.gov/swercepp/tools/rmp-comp/rmp-comp.html.

---

## EXHIBIT 4-1
## EXAMPLES OF REGULATED TOXIC SUBSTANCES IN WAREHOUSES

| Chemical | Toxic Endpoint (mg/L) | Typical Container | Potential Quantity in Warehouse | Comments | Buoyant (B) or Dense (D) |
|---|---|---|---|---|---|
| Boron Trifluoride Compound with Methyl Ether (1:1) | 0.023 | 55-gallon drum | Truckload[a] | Pure liquid | D/B[b] |
| Cyclohexylamine | 0.16 | 55-gallon drum | 1-3 truckloads | Pure liquid | D/B |
| Diborane | 0.0011 | 150-lb cylinder | 10 cylinders | 30% conc. in hydrogen | B/B |
| Epichlorohydrin | 0.076 | 55-gallon drum | Truckload | Pure liquid | D/B |
| Ethylenediamine | 0.49 | 55-gallon drum | 2 truckloads | 80% with other solvent-type materials, alcohol and ethyl acetate | D/B |
| Ethylene Oxide | 0.090 | 150-lb cylinder | 100 cylinders | Pressurized gas | D/D |
| Formaldehyde | 0.012 | 55-gallon drum and smaller containers | 2 truckloads | 10%-50% solutions in water[c] | B/B |
| Hydrazine | 0.011 | 55-gallon drum | 3 truckloads | 35%, 50%, 85% solutions in water[c] | B/B |
| Hydrochloric Acid | 0.030 | 55-gallon drum | 2 truckloads | 30-38% solutions in water[e] | D/B |
| Methyl Chloride | 0.82 | 150-lb cylinder | 100-200 cylinders | Pressurized gas | D/D |
| Nitric Acid | 0.026 | 55-gallon drum | 2 truckloads | 80-90% solutions in water | D/B |
| Propylene Oxide | 0.59 | 55-gallon drum | 2-3 truckloads | Pure liquid | D/D |
| Sulfur Dioxide | 0.0078 | 150-lb cylinder | 100-200 cylinders | Pressurized gas | D/D |
| Titanium Tetrachloride | 0.02 | 55-gallon drum | ** | Pure liquid | D/B |
| Toluene 2,4- and 2,6-diisocyanate, plus unspecified mixtures | 0.007 | 55-gallon drum | 1 or more truckloads | Pure liquid[d] | B/B |

See next page for footnotes.

Footnotes for Exhibit 4-1:

[a]A truckload typically contains 78 55-gallon drums
[b]D/B indicates that the material behaves as a dense gas in worst-case weather conditions and as a neutrally buoyant (passive) gas in the case of the alternative scenarios, etc.
[c]For these solutions in water, the vapor pressures of formaldehyde and hydrazine over the solutions are less than 10 mm Hg. Such mixtures do not fall under the requirements of 40 CFR Part 68 and are not considered further in this guidance.
[d]Toluene diisocyanate can also be present as • 1 to ~ 10 wt% in a mixture of resins, plastics, etc., in 5-gallon pails up to 350-gallon totes. It is also used as a catalyst in the production of polyurethane foam and may be present in small quantities in, for example, furniture. These forms of toluene diisocyanate are not considered further in this guidance.
[e]Hydrochloric acid in concentrations below 37% is not regulated.

## EXHIBIT 4-2
## EXAMPLES OF REGULATED FLAMMABLE SUBSTANCES IN WAREHOUSES

| Chemical | Typical Container | Potential Quantity in Warehouse | Comments |
|---|---|---|---|
| Acetylene | 150-lb cylinder | One or two cylinders | Generally not enough to be covered by the RMP[a] |
| Dimethylamine[b] | 55-gallon drum | 1-2 truckloads | 40% solution in water |
| Isopropyl Chloride | 55-gallon drum | 2 truckloads | Pure liquid |
| Methylamine[b] | 55-gallon drum | 2+ truckloads | 40% solution in water |
| Pentane | - | - | Residual pentane in packages of pellets[a,c] |
| Propane | 33.3-lb steel kegs or 5,000-7,000-lb tanks | Potentially more than 10,000 lb | Used for fueling forklifts and other purposes; not covered if used as a fuel on site |
| Trimethylamine[b] | 55-gallon drum | 3 truckloads | 40% solution in water |
| Vinyl Ethyl Ether | 55-gallon drum | 1 truckload | Pure liquid |

Footnotes for Table 4-2:
[a]Not considered further in this model guidance.
[b]According to Ullman's *Encyclopedia of Industrial Chemistry*, the boiling point of 40% monomethylamine is 121 °F and that of 40% dimethylamine is 125 °F. Therefore, these mixtures are not regulated under 40 CFR Part 68, which places an upper limit of 100 °F on the boiling point of regulated flammable substances. However, 40% trimethylamine has a boiling point of 87 °F and is covered.
[c]Some studies indicate that there is no residual pentane.

**The methodology and reference tables of distances presented here are optional. You are not required to use this guidance.** You may use publicly available or proprietary air dispersion models to do your offsite consequence analysis, subject to certain conditions. If you choose to use other models, you should review the rule and Chapter 4 of the *General Guidance for Risk Management Programs*, which outline required conditions for use of other models.

The results obtained using the methods in this document may be conservative (i.e., they may overestimate the distance to endpoints). Complex models that can account for many site-specific factors may give less conservative estimates of offsite consequences than the simple methods in this guidance. This is particularly true for alternative scenarios, for which EPA has not specified many assumptions. However, complex models may be expensive and require considerable expertise to use; this guidance is designed to be simple and straightforward. You will need to consider these tradeoffs in deciding how to carry out your required consequence analyses.

This chapter presents discussions and tables for the worst-case scenario for warehouses in section 4.1, followed by discussions and tables for alternative scenarios for warehouses in section 4.2. Mitigation provided by buildings is discussed in section 4.3. Section 4.4 provides information on estimating offsite receptors, and section 4.5 discusses required documentation.

## 4.1   WORST-CASE RELEASE SCENARIOS

This section provides guidance on how to analyze worst-case scenarios. Information is provided on the general requirements of the regulations, followed by specific guidance relevant to warehouses. Exhibit 4-3 presents the parameters that must be used in worst-case and alternative release scenarios.

### GENERAL REQUIREMENTS FOR TOXIC SUBSTANCES

The following input is required for toxic substances:

- The *worst-case release quantity* Q (lb) is the greater of the following:

  - For substances in a vessel, the greatest amount held in that vessel, taking into account administrative controls that limit the maximum quantity; or

  - For substances in pipes, the greatest amount in a pipe, taking into account administrative controls that limit the maximum quantity.

- For a release from a vessel, you need only consider the largest amount in the vessel. For the specific case of a warehouse, the largest vessels are 350-gallon totes, 55-gallon drums, 10-gallon pails, 150-lb cylinders, and other containers that are small relative to typical vessels in a chemical plant. Therefore, the spillage of the contents of one of these containers constitutes the worst-case scenario, although you may well be able to think of scenarios

## EXHIBIT 4-3
## REQUIRED PARAMETERS FOR MODELING (40 CFR 68.22)

| WORST CASE | ALTERNATIVE SCENARIO |
|---|---|
| **Endpoints (§68.22(a))** | |
| Toxic endpoints are listed in part 68 Appendix A. | Toxic endpoints are listed in part 68 Appendix A. |
| For flammable substances, endpoint is overpressure of 1 pound per square inch (psi) for vapor cloud explosions. | • For flammable substances, endpoint is overpressure of 1 psi for vapor cloud explosions<br>• Radiant heat level of 5 kilowatts per square meter ($kW/m^2$) for 40 seconds for heat from fires (or equivalent dose)<br>• Lower flammability limit (LFL) as specified in NFPA documents or other generally recognized sources. |
| **Wind speed/stability (§68.22(b))** | |
| This guidance assumes 1.5 meters per second and F stability. For other models, use wind speed of 1.5 meters per second and F stability class unless you can demonstrate that local meteorological data applicable to the site show a higher minimum wind speed or less stable atmosphere at all times during the previous three years. If you can so demonstrate, these minimums may be used for site-specific modeling. | This guidance assumes wind speed of 3 meters per second and D stability. For other models, you must use typical meteorological conditions for your site. |
| **Ambient temperature/humidity (§68.22(c))** | |
| This guidance assumes 25•C (77•F) and 50 percent humidity. For other models for toxic substances, you must use the highest daily maximum temperature and average humidity for the site during the past three years. | This guidance assumes 25•C and 50 percent humidity. For other models, you may use average temperature/humidity data gathered at the site or at a local meteorological station. |
| **Height of release (§68.22(d))** | |
| For toxic substances, you must assume a ground level release. | This guidance assumes a ground-level release. For other models, release height may be determined by the release scenario. |
| **Surface roughness (§68.22(e))** | |
| Use urban (obstructed terrain) or rural (flat terrain) topography, as appropriate. | Use urban (obstructed terrain) or rural (flat terrain) topography, as appropriate. |
| **Dense or neutrally buoyant gases (§68.22(f))** | |
| Tables or models used for dispersion of regulated toxic substances must appropriately account for gas density. | Tables or models used for dispersion must appropriately account for gas density. |
| **Temperature of released substance (§68.22(g))** | |
| You must consider liquids (other than gases liquefied by refrigeration) to be released at the highest daily maximum temperature, from data for the previous three years, or at process temperature, whichever is higher. Assume gases liquefied by refrigeration at atmospheric pressure to be released at their boiling points. | Substances may be considered to be released at a process or ambient temperature that is appropriate for the scenario. |

in which a quantity greater than Q as defined above can be released. Other credible scenarios could involve simultaneous damage to more than one vessel. EPA recommends that you consider multiple-vessel release scenarios, if credible and appropriate, as alternative release scenarios (see Section 4.2).

- *Weather conditions.* The rule allows anyone who conducts their OCA based on this guidance to use specific default weather conditions for wind speed, stability class, average temperature, and humidity. Liquids other than gases liquefied by refrigeration should be considered to be released at the highest daily maximum temperature, based on local data for the previous three years, or at process temperature, whichever is the higher. For warehouses, the liquids are assumed to be stored at ambient temperature. You can obtain weather data from local weather stations. You can also obtain temperature and wind speed data from the National Climatic Data Center at (828) 271-4800.

- For the worst-case scenario, the release must be assumed to take place at *ground level.*

- *The toxic endpoints* for toxic regulated substances are listed 40 CFR Part 68, Appendix A and in Appendix A of this document. Many of these endpoints (which are airborne concentrations) have been published by the American Industrial Hygiene Association (AIHA) as the second level of the Emergency Response Planning Guidelines (ERPG-2) and are the maximum airborne concentrations below which it is believed that nearly all individuals can be exposed for up to one hour without experiencing or developing irreversible or other serious health effects or symptoms which could impair an individual's ability to take protective action. These endpoints should be applied independent of the exposure time.

- *Rural vs. urban sites.* The regulations require you to take account of whether your site is rural or urban. To decide whether the site is rural or urban, the rule offers the following: "Urban means that there are many obstacles in the immediate area; obstacles include buildings or trees. Rural means that there are no buildings in the immediate area and the terrain is generally flat or unobstructed." Some areas outside of cities may still be considered urban if they are forested.

  The distinction between urban and rural sites is important because the atmosphere at urban sites is generally more turbulent than at rural sites, causing more rapid dilution of the cloud as it travels downwind. Therefore, for ground-level releases, predicted distances to toxic endpoints are always smaller at urban sites than at rural sites.

- *Gas density.* The regulations require you to use tables or models that appropriately account for gas density. This guidance provides lookup tables for dense or neutrally buoyant gases or vapors (i.e., for gases that are denser-than-air or for gases that have the same density as air, respectively).

- *Mitigation.* You are only allowed to take account of passive mitigation systems. Passive mitigation systems could include diked areas and buildings (see Section 4.3 for more information on buildings). You are not allowed to consider active mitigation systems such as sprinkler systems or remotely operated valves.

- The predicted frequency of occurrence of the worst-case scenario is not an allowable consideration. You are not required to determine a possible cause of the failure of the vessel.

## TOXIC LIQUIDS

The worst-case scenario for toxic liquids is a spill of the total quantity in the largest vessel. The quantity spilled is assumed to spread instantaneously to a depth of one centimeter in an undiked area or to cover a diked area instantaneously. (This guidance does not consider diked areas.) The distance to the endpoint is estimated based on evaporation from the pool and downwind dispersion of the vapor.

For this guidance, the basic assumption is that the container spills and forms an unconfined pool with a depth of one centimeter. The spill is assumed to be outside, on or near the loading dock. For discussion of spills inside buildings, see Section 4.3. No credit is taken for drains or other features that might contain the spilled liquid. For liquids listed in Exhibit 4-1, Exhibit 4-4 provides the distance to the endpoint for spills from vessels usually found at warehouses. The procedure for calculating the worst-case distance for liquids is as follows:

1. For a specific toxic material, identify the largest container and the quantity Q (lb) in it.

2. Use the following equation to calculate the rate of evaporation QR (lb/min) from the pool:

$$QR = 1.4 \, Q \times LFA \times DF = \bullet \, Q \tag{1}$$

where LFA is the "Liquid Factor Ambient," DF is the "Density Factor," and $\bullet$ = 1.4 $x$ LFA $x$ DF. LFA and DF have values that depend on the specific liquid that has been spilled. Their values at 25$\bullet$C[1] are tabulated in Exhibits B-2 (toxic liquids) and B-3 (water solutions) in Appendix 4A at the end of this chapter. For the convenience of the reader, the values of $\bullet$ for the toxic liquids and solutions that are listed in Exhibit 4-1 are provided in Exhibit 4-4.

---

[1]For spills at a temperature greater than 25$\bullet$C, see the footnote to Exhibit B-4 in Appendix 4A. Unless otherwise stated, all spills are assumed to be at 25$\bullet$C.

## EXHIBIT 4-4

## PREDICTED DISTANCES TO TOXIC ENDPOINTS FOR REGULATED TOXIC MATERIALS

### Worst Case Scenario, Stability Class F, Wind Speed 1.5 m/s

| Chemical/Solution Name | • | Container Size | Density (lb/gallon)[b] | Quantity Q (lb) | Rate of Release (lb/min) | Distance Rural Site (mi) | Distance Urban Site (mi) |
|---|---|---|---|---|---|---|---|
| Boron Trifluoride Compound with Methyl Ether (1:1) | 0.002 | 55 gallon | 8.29 | 456 | 0.91 | 0.3 | 0.2 |
| Cyclohexylamine | 0.002 | 55 gallon | 7.0 | 385 | 0.780 | <0.1 | <0.1 |
| Diborane | NA[a] | 150 lb | NA[a] | 50 | 5 | 2.5 | 1.3 |
| Epichlorohydrin | 0.0024 | 55 gallon | 10.0 | 550 | 1.32 | 0.2 | 0.10 |
| Ethylenediamine | 0.0017 | 55 gallon | 8.0 | 440 | 0.748 | <0.1 | <0.1 |
| Ethylene Oxide | NA[a] | 150 lb | NA[a] | 150 | 15 | 0.5 | 0.3 |
| Hydrochloric Acid 30%[c] | 0.0009 | 55 gallon | 9.64 | 530 | 0.47 | 0.1 | <0.1 |
| Hydrochloric Acid 34%[c] | 0.0028 | 55 gallon | 9.67 | 532 | 1.49 | 0.3 | 0.2 |
| Hydrochloric Acid 36%[c] | 0.0042 | 55 gallon | 9.69 | 533 | 2.23 | 0.3 | 0.2 |
| Hydrochloric Acid 37% | 0.0050 | 55 gallon | 9.69 | 533 | 2.66 | 0.4 | 0.2 |
| Hydrochloric Acid 38% | 0.0060 | 55 gallon | 9.69 | 533 | 3.19 | 0.5 | 0.2 |
| Methyl Chloride | NA[a] | 150 lb | NA[a] | 150 | 15 | 0.1 | <0.1 |
| Nitric Acid 80% | 0.0009 | 55 gallon | 11.13 | 612 | 0.55 | 0.1 | 0.1 |
| Nitric Acid 85% | 0.0015 | 55 gallon | 11.69 | 643 | 0.96 | 0.3 | 0.2 |
| Nitric Acid 90% | 0.0021 | 55 gallon | 12.55 | 690 | 1.45 | 0.4 | 0.2 |
| Propylene Oxide | 0.0770 | 55 gallon | 7.44 | 409 | 31.5 | 0.3 | <0.1 |
| Sulfur Dioxide | NA[a] | 150 lb | NA[a] | 150 | 15 | 0.7 | 0.3 |
| Titanium Tetrachloride | 0.0020 | 55 gallon | 14.42 | 793 | 1.586 | 0.4 | 0.2 |
| Toluene 2,4-diisocyanate | $3.36 \times 10^{-6}$ | 55 gallon | 10.0 | 550 | 0.0018 | <0.1 | <0.1 |
| Toluene 2,6-diisocyanate | $1.0 \times 10^{-5}$ | 55 gallon | 10.0 | 550 | 0.0055 | <0.1 | <0.1 |

Footnotes:

[a]Not applicable—chemical is a gas

[b]The density can be obtained from the density factor DF, which is given in Exhibits B-2 and B-3 of Appendix 4A. The density is $1/(DF \times 0.033)$ lb/ft$^3$. Thus, for epichlorohydrin, DF = 0.42, so that the density is $1/(0.42 \times 0.033) = 72$ lb/ft$^3$. There are 0.134 gallons/ft$^3$, so that the density becomes $0.134 \times 72 \sim 10$ lb/gallon.

[c]Hydrochloric acid in concentrations below 37% is not regulated.

3.      Obtain the toxic endpoint from Exhibit 4-1, together with information on whether the vapor cloud should be regarded as neutrally buoyant or dense. For liquids not included in Exhibit 4-1, you can find this information in Exhibit B-2 (pure liquids) or B-3 (solutions) in Appendix 4A at the end of this chapter.

4.      Determine whether your site is rural or urban.

5.      Take the release rate QR and read the predicted distances to the toxic endpoint from the following tables in Appendix 4A[2]:

    ••      Reference Table 1 (rural site, 10-minute release, neutrally buoyant vapor cloud)
    ••      Reference Table 2 (rural site, 60-minute release, neutrally buoyant vapor cloud)
    ••      Reference Table 3 (urban site, 10-minute release, neutrally buoyant vapor cloud)
    ••      Reference Table 4 (urban site, 60-minute release, neutrally buoyant vapor cloud)
    ••      Reference Table 5 (rural site, 10-minute release, dense vapor cloud)
    ••      Reference Table 6 (rural site, 60-minute release, dense vapor cloud)
    ••      Reference Table 7 (urban site, 10-minute release, dense vapor cloud)
    ••      Reference Table 8 (urban site, 60-minute release, dense vapor cloud).
    ••      Reference Table 10 (specifically for aqueous ammonia).

For the specific case of the toxic liquids and solutions listed on Table 4-1, predicted distances for the failure of 55-gallon drums are given on Table 4-4 and can simply be quoted in your Risk Management Plan if the scenario matches your own worst-case scenario.

**Example 1.** You have epichlorohydrin above the threshold quantity (TQ) in your warehouse. The largest container is a 55-gallon drum. You determine that your site is rural. From Table 4-4, the distance D to the toxic endpoint is 0.2 mi.

**Example 2.** You have epichlorohydrin in 10-gallon pails, which is a quantity that is not represented on Exhibit 4-4. Therefore, you need to use both Exhibit 4-4 and the tables in Appendix 4A, as follows. The density of epichlorohydrin is 10 lb/gallon (Exhibit 4-4), so a pail contains 100 lb. From Exhibit 4-4, • – 0.0024 so that, from Equation 1, QR = (0.0024)(100) = 0.24 lb/min. From Exhibit 4-1, the toxic endpoint of epichlorohydrin is 0.076 mg/L, and it should be treated as a dense vapor. You

---

[2]The 60-minute release tables are generally appropriate for spills of toxic liquids because the materials kept in warehouses usually have low vapor pressures and, hence, low rates of evaporation should they be spilled. For the examples given in Exhibit 4-4, the time to complete evaporation is greater than 10 minutes for all the liquids. The time would be several hours for all the liquids except propylene oxide. For solutions of toxic substances in water, the 10-minute tables should be used.

determine that your site is rural. Therefore, refer to Reference Table 6 in Appendix 4A.

The predicted release rate of 0.24 lb/min is below the lowest release rate of 1 lb/min on Reference Table 6. The closest toxic endpoint to 0.076 mg/L is 0.075 mg/L. The corresponding distance is 0.1 mi[3].

**Example 3.** You have dimethyldichlorosilane in your warehouse in 55-gallon drums at a rural site. This is an example of a material that is <u>not</u> listed in Exhibit 4-1.

From Exhibit B-2, Appendix 4A, the toxic endpoint of dimethyldichlorosilane is 0.026 mg/L, the LFA is 0.042, and the DF is 0.46. The density • of liquid dimethyldichlorosilane is 65.9 lb/ft$^3$ (the density can be calculated from the DF by using the formula • = 1/(0.033DF), as described in the footnote to Exhibit 4-4). This converts to 8.83 lb/gallon (1 gallon is 0.134 ft$^3$). Therefore, a 55-gallon drum contains Q = 55 × 8.76 = 486 lb.

The quantity • in Equation 1 is 1.4 × LFA × DF = 0.027. The rate of evaporation is • Q = 0.027 × 486 = 13 lb/min.

According to Exhibit B-2 of Appendix 4A, dimethyldichlorosilane should be modeled as a dense vapor. Turning to Reference Table 6 in Appendix 4A, for a rural site, look for the closest evaporation rate and toxic endpoint to 13 lb/min and 0.026 mg/L. The closest release rate is 10 lb/min, and the closest endpoint is 0.02 mg/L, giving a distance of 1.4 mi.

## TOXIC GASES

For toxic gases, the worst-case scenario is release of the contents of the largest vessel over 10 minutes. For toxic gases listed in Exhibit 4-1, Exhibit 4-4 provides the distance to the endpoint for the release of a toxic gas from the largest vessel usually found at a warehouse. The procedure to use for analysis of the worst-case scenario for toxic gases is as follows.

1.    Estimate the quantity Q (lb) in the largest container

2.    Assume that the quantity Q is released over 10 minutes, so that the release rate is (Q/10) lb/min.

3.    Obtain the toxic endpoint from Exhibit B-1 of Appendix 4A, together with information on whether the vapor cloud should be regarded as neutrally buoyant or dense.

4.    Determine whether your site is rural or urban.

---

[3]In Reference Table 6 and other tables in Appendix 4-A, results are rounded to the nearest tenth of a mile for distances under ten miles, and to the nearest mile for distances over ten miles. This is a reminder that the results of atmospheric dispersion modeling are uncertain and that more accurate predictions are not warranted.

5.    Determine the distance to the toxic endpoint by using one of the following
      tables:

      •      Reference Table 1 (rural site, 10-minute release, neutrally buoyant
             vapor cloud)
      •      Reference Table 3 (urban site, 10-minute release, neutrally buoyant
             vapor cloud)
      •      Reference Table 5 (rural site, 10-minute release, dense vapor cloud),
             or
      •      Reference Table 7 (urban site, 10-minute release, dense vapor
             cloud).
      •      Reference Table 9 - specifically for anhydrous ammonia
      •      Reference Table 11 - specifically for chlorine
      •      Reference Table 12 - specifically for sulfur dioxide

      Only 10-minute tables are relevant because the rule mandates that for the
      worst-case scenario, gases are released over 10 minutes.

**Example 4.** Diborane is one of the gaseous materials listed on Exhibit 4-1. The
steps listed above have already been carried out for this material, and the results are
displayed on Exhibit 4-4 —3 mi at a rural site and 2.1 mi at an urban site. (Note that
the diborane is assumed to be in a 150-lb cylinder, but as only about 30 wt% in
hydrogen. Therefore, only 50 lb of diborane is released and the predicted release
rate is 5 lb/min, not 15 lb/min.)

**Example 5.** You have arsine in 150-lb cylinders at a rural site. This is another
example of a material that is not on Exhibit 4-1. The worst-case release rate is 15
lb/min. From Exhibit B-1 of Appendix 4A, the toxic endpoint is 0.0019 mg/L and
the release should be treated as a dense vapor cloud. Turning to reference Table 5 of
Appendix 4A, the closest tabulated toxic endpoint is 0.002 mg/L. The closest
release rate is 10 lb/min, for a predicted distance of 3 mi.

## FLAMMABLE SUBSTANCES

For the worst-case scenario involving a release of a regulated flammable substance
(a flammable gas or volatile flammable liquid), you must assume that the quantity of
the flammable substance is released into a vapor cloud. A vapor cloud explosion is
assumed to result from the release. You must estimate the distance D (mi) to an
endpoint to an overpressure level of 1 pound per square inch (psi) from the explosion
of the vapor cloud.

      •      If the flammable substance is normally a gas at ambient temperature and
             handled as gas or liquid under pressure or if the flammable substance is a gas
             handled as a refrigerated liquid and is not contained when released or the
             contained pool is one centimeter or less deep, you must assume the total
             quantity is released as a gas and is involved in a vapor cloud explosion.
      •      If the flammable substance is a liquid or a refrigerated gas released into a
             containment area with a depth greater than one centimeter, you may assume

that the quantity that volatilizes in 10 minutes is involved in a vapor cloud
explosion.

A simple method of obtaining an approximate answer is to use the TNT equivalency
method, which states that:

$$D = 0.0037(Q \times H/H_{TNT})^{1/3} \tag{2}$$

where Q (lb) is the quantity of flammable material released, H is the heat of
combustion of the flammable substance and $H_{TNT}$ is the heat of combustion of
trinitrotoluene (TNT). Implicit in Equation 2 is the assumption that the yield factor
is 10%, as required by the rule. The yield factor is the fraction of the material in the
vessel that effectively participates in the explosion. Equation 2 can be rewritten as

$$D = \bullet \; (Q)^{1/3} \text{ miles} \tag{3}$$

The values of • for the flammable materials listed in Table 4-2 are given below:

|  | •• |
|---|---|
| Ethyl Mercaptan | 0.0067 |
| Isopropyl Chloride | 0.0063 |
| Propane | 0.0080 |
| Trimethylamine | 0.0074 |
| Vinyl Ethyl Ether | 0.0070 |

If you prefer, you can find the distance to 1 psi overpressure for the quantity released
from Reference Table 13 in Appendix 4A, instead of using Equation 3. Reference
Table 13 also includes flammable substances that are not in the above list.

**Example 6**. If 5,000 lb of propane explodes, Equation 3 gives:

$$D = 0.0080 \, (5,000)^{1/3} = 0.14 \text{ mi}$$

For reporting, you would round the distance to 0.1 mile.

**Example 7**. If a 55-gallon drum of vinyl ethyl ether spills and forms a vapor cloud,
this vapor cloud will contain (55 gallons)(6.36 lb/gallon) = 350 lb of vinyl ethyl
ether. If it explodes, then:

$$D = (0.0070)(350)^{1/3} = 0.05 \text{ mi}$$

**Example 8**. A 55-gallon drum containing a 40 wt% solution of trimethylamine
spills. The density of this solution is 6 lb/gallon, so there is a total of (0.4)(55)(6) =
132 lb of trimethylamine. From Equation 3, $D = 0.0074(132)^{1/3} = 0.038$ mi, which
you would round to 0.04 mi.

Vapor cloud explosions involving small quantities generally are considered very
unlikely. If you have a flammable chemical above the threshold quantity, however,

you must assume a vapor cloud explosion for your worst-case scenario, even if your largest vessel contains a relatively small quantity.

The explosions discussed above are assumed to take place outside. Many warehouses keep flammable materials in a room that is specially designed with explosion venting per NFPA requirements. However, the intention of this section is to provide information on the worst-case scenario. There are probably times when you handle the containers of flammable substances outside. Nevertheless, in discussions with local agencies and local communities you may want to explain how your facility is designed to ensure that worst-case explosions do not occur or are effectively mitigated.

## 4.2   ALTERNATIVE SCENARIOS

The purpose of this section is to give guidance on how to model alternative scenarios.

### GENERAL REQUIREMENTS

The requirements that differ from those for the worst-case scenarios are as follows:

- You can take into account active as well as passive mitigation systems, as long as these systems are expected to withstand the causes of the accident. For warehouses, the building itself could function as a passive system and the fire sprinklers could be regarded as an active system.

- The alternative scenario should reach an endpoint offsite, unless no such scenario exists.

- If you are doing your own modeling, you should use "typical meteorological conditions for the stationary source." You may obtain these data from local weather stations. You can obtain wind speed and temperature data from the National Climatic Data Center at (828) 271-4800. This guidance uses an "average" weather condition of wind speed 3 m/s and D stability class with an ambient temperature of 25•C.

- The number of alternative scenarios you are required to develop is as follows:

  - At least one scenario for each regulated toxic substance held in Program 2 and Program 3 processes.
  - At least one scenario to represent all flammables held in Program 2 and Program 3 processes.

- The release is not necessarily restricted to ground level. It can be elevated if appropriate. An elevated release might be appropriate for a warehouse with several floors (analysis of elevated releases would be site-specific and is not considered in this guidance).

## CHOICE OF ALTERNATIVE SCENARIOS FOR TOXIC LIQUIDS

There are some significant issues when it comes to choosing alternative scenarios for warehouses:

- • As already noted, plausible alternative scenarios could well be larger than the single-container worst-case scenarios;

- • The alternative scenario should be more probable than the worst-case scenario, yet spillage from a single container is among the more probable scenarios.

The following subsections contain some suggested alternative scenarios.

### SINGLE-CONTAINER ALTERNATIVE SCENARIO

One possibility is to take the same scenario as the worst-case (i.e., a spill of the contents of a single vessel), but to assume that it takes place in typical weather conditions. Exhibit 4-5 provides distances to the endpoint for this scenario for the substances listed in Exhibit 4-1. The procedure for calculating the alternative scenario distance is very similar to that for the worst-case scenario:

- For a specific toxic material, identify the largest vessel and the quantity Q (lb) in it.

- Use Equation 4 below to calculate the rate of evaporation QR (lb/min.) from the pool formed by the spill of quantity Q in alternative weather conditions:

$$QR = 2.4 \, Q \times LFA \times DF = \bullet \, Q \qquad (4)$$

where LFA is the "Liquid Factor Ambient," DF is the "Density Factor," and • = 2.4 × LFA × DF. LFA and DF have values that depend on the specific liquid that has been spilled. Their values are tabulated in Exhibits B-2 (toxic liquids) and B-3 (water solutions) of Appendix 4A. For the convenience of the reader, the values of • that are appropriate for alternative scenarios are provided in Exhibit 4-5.

- Determine whether your site is rural or urban.

- Read the toxic endpoint from Exhibit B-2 or Exhibit B-3 of Appendix 4A, together with whether the vapor should be modeled as neutrally buoyant or dense.

- Take the release rate QR and read the predicted distances to the toxic endpoint from the following tables in Appendix 4A:

    - Reference Table 14 (rural site, 10-minute release, neutrally buoyant vapor cloud)

- •  Reference Table 15 (rural site, 60-minute release, neutrally buoyant vapor cloud)
- •  Reference Table 16 (urban site, 10-minute release, neutrally buoyant vapor cloud)
- •  Reference Table 17 (urban site, 60-minute release, neutrally buoyant vapor cloud)
- •  Reference Table 18 (rural site, 10-minute release, dense vapor cloud)
- •  Reference Table 19 (rural site, 60-minute release, dense vapor cloud)
- •  Reference Table 20 (urban site, 60-minute release, dense vapor cloud)
- •  Reference Table 21 (urban site, 60-minute release, dense vapor cloud)
- •  Reference Table 23 - specifically for aqueous ammonia

For spilled water solutions, you should assume a 10-minute duration of release because LFA is calculated as a 10-minute average.  For the other toxic liquids, the predicted time to total evaporation of the pool at the rate given by Equation 4 is generally much more than 10 minutes, so you will need to use the 60-minute tables unless your alternative scenario has been terminated in 10 minutes or less (e.g., by some emergency countermeasure).

For the specific case of the toxic liquids and solutions listed on Exhibit 4-1, predicted distances are given on Exhibit 4-5 and can simply be quoted in your Risk Management Plan if the scenario matches one of your own alternative scenarios.

**Example 9**.  You have ethylenediamine above the TQ in your warehouse.  The largest container is a 55-gallon drum.  You determine that your site is rural.  Ethylenediamine is on Exhibit 4-5, and the corresponding distance to the toxic endpoint is < 0.1 mi.

**Example 10**.  You have epichlorohydrin in 10-gallon pails, which are not represented on Exhibit 4-5.  Therefore, you need to use the tables in Appendix 4A, as follows.  The density of epichlorohydrin is 10 lb/gallon, so a pail contains Q = 100 lb.  From Exhibit 4-5, • = 0.0040 so that, from Equation 4, QR = (0.0040)(100) = 0.40 lb/min.  From Exhibit 4-1, the toxic endpoint of epichlorohydrin is 0.076 mg/L, and it should be modeled as a buoyant vapor.  You determine that your site is rural.  Turning to Reference Table 15 in Appendix 4A, you need to calculate the ratio between the release rate and the toxic endpoint, which, for convenience, we call •.  In this case, • = 0.38/0.076 = 5, which lies in the lowest tabulated range for • on Reference Table 15, corresponding to a distance to the toxic endpoint of 0.1 mi.

## *MULTIPLE CONTAINER ALTERNATIVE SCENARIO*

A plausible scenario is that two or more containers could be punctured by a fork lift truck.  The procedure to follow for this scenario for toxic liquids is the same procedure presented in the previous subsection; only the quantity spilled is different.  The evaporation rate displayed in Exhibit 4-5 should then be multiplied by the number of ruptured containers. The predicted distances to the toxic endpoint are given on Exhibit 4-6 for the example of two ruptured containers.

**Example 11**. You have an accident in which four containers of 37% hydrochloric acid are ruptured in typical weather conditions at a rural site. The predicted rate of evaporation is then four times that on Exhibit 4-5, i.e., $4 \times 4.58 = 18.32$ lb/min. From Exhibit 4-1, 37% hydrochloric acid is neutrally buoyant in alternative scenarios, and its toxic endpoint is 0.03 mg/L. Hence, • = 18.32/0.03 = 610. Turning to Reference Table 1 in Appendix 4A, this value of • is close to the tabulated value of 630, corresponding to 0.2 mi.

## ALTERNATIVE SCENARIO - TOXIC GASES

For alternative scenarios for toxic gases, you may consider the worst-case release under typical weather conditions or a release involving multiple containers, as discussed above for liquids. Exhibits 4-5 and 4-6 provide distances to the endpoint for these alternative scenarios for toxic gases listed in Exhibit 4-1. The procedure to use is as follows.

1.      Estimate the release rate in lb/min for the alternative scenario.

2.      Obtain the toxic endpoint from Exhibit B-1 of Appendix 4A, together with information on whether the vapor cloud should be regarded as neutrally buoyant or dense.

3.      Determine whether your site is rural or urban.

4.      Determine the distance to the toxic endpoint by using one of the following tables:

- Reference Table 14 (rural site, 10-minute release, neutrally buoyant vapor cloud)
- Reference Table 15 (rural site, 60-minute release, neutrally buoyant vapor cloud)
- Reference Table 16 (urban site, 10-minute release, neutrally buoyant vapor cloud)
- Reference Table 17 (urban site, 60-minute release, neutrally buoyant vapor cloud)
- Reference Table 18 (rural site, 10-minute release, dense vapor cloud)
- Reference Table 19 (rural site, 60-minute release, dense vapor cloud)
- Reference Table 20 (urban site, 10-minute release, dense vapor cloud)

## EXHIBIT 4-5

### PREDICTED DISTANCES TO TOXIC ENDPOINTS FOR REGULATED TOXIC MATERIALS
### Alternative Case Scenario No. 1 - Release from Single Container, Stability Class D, Wind Speed 3.0 m/s

| Chemical/Solution Name | • | Container Size | Quantity Q (lb) | Rate of Release (lb/min) | Distance Rural Site (mi) | Distance Urban Site (mi) |
|---|---|---|---|---|---|---|
| Boron Trifluoride Compound with Methyl Ether (1:1) | 0.0035 | 55 gallon | 570 | 1.93 | 0.2 | <0.1 |
| Cyclohexylamine | 0.0034 | 55 gallon | 385 | 1.30 | <0.1 | <0.1 |
| Diborane | NA[a] | 150 lb | 50 | 5 | 0. | 0.4 |
| Epichlorohydrin | 0.0040 | 55 gallon | 550 | 2.20 | <0.1 | <0.1 |
| Ethylenediamine | 0.0029 | 55 gallon | 440 | 1.28 | <0.1 | <0.1 |
| Ethylene Oxide | NA[a] | 150 lb | 150 | 15 | 0.1 | <0.1 |
| Hydrochloric Acid 30%[b] | 0.0016 | 55 gallon | 530 | 0.85 | <0.1 | <0.1 |
| Hydrochloric Acid 34%[b] | 0.0048 | 55 gallon | 532 | 2.55 | <0.1 | <0.1 |
| Hydrochloric Acid 36%[b] | 0.0073 | 55 gallon | 533 | 3.89 | 0.1 | <0.1 |
| Hydrochloric Acid 37% | 0.0086 | 55 gallon | 533 | 4.58 | 0.1 | <0.1 |
| Hydrochloric Acid 38% | 0.0098 | 55 gallon | 533 | 5.22 | 0.1 | <0.1 |
| Methyl Chloride | NA[a] | 150 lb | 150 | 15 | <0.1 | <0.1 |
| Nitric Acid 80% | 0.0015 | 55 gallon | 612 | 0.92 | <0.1 | <0.1 |
| Nitric Acid 85% | 0.0025 | 55 gallon | 643 | 1.61 | <0.1 | <0.1 |
| Nitric Acid 90% | 0.0036 | 55 gallon | 690 | 2.48 | 0.1 | <0.1 |
| Propylene Oxide | 0.13 | 55 gallon | 409 | 53.17 | 0.1 | <0.1 |
| Sulfur Dioxide | NA[a] | 150 lb | 150 | 15 | 0.4 | 0.3 |
| Titanium Tetrachloride | 0.0032 | 55 gallon | 793 | 2.54 | 0.1 | <0.1 |
| Toluene 2,4-diisocyanate | 5.76x10[-6] | 55 gallon | 550 | 0.0032 | <0.1 | <0.1 |
| Toluene 2,5-diisocyanate | 1.7x10[-5] | 55 gallon | 550 | 0.0095 | <0.1 | <0.1 |

Footnotes:

[a]Not applicable—chemical is a gas

[b]Hydrochloric acid in concentrations below 37% is not regulated.

- Reference Table 21 (urban site, 60-minute release, dense vapor cloud)
- Reference Table 22 - specifically for anhydrous ammonia
- Reference Table 24 - specifically for chlorine
- Reference Table 25 - specifically for sulfur dioxide

**Example 12**. Diborane is one of the gaseous materials listed on Exhibit 4-1. Assuming that the alternative scenario is the same as the worst-case scenario, Example 4, except that the weather conditions have changed from worst-case to typical, the steps listed above have already been carried out for this material. The results are listed on Exhibit 4-5: ~ 0.8 mi at a rural site and ~ 0.4 mi at an urban site.

For an alternative scenario that consists of the release of two cylinders of diborane over a period of 10 minutes, the results may be read from Exhibit 4-6 and are • 1.0 mi at a rural site and • 0.6 mi at an urban site.

**Example 13**. You have arsine in 150-lb cylinders at a rural site. Arsine is not on Exhibit 4-5. The worst-case release rate is 15 lb/min for 10 min. Assume that this release rate is the same for the alternative scenario and that the only difference is in the weather conditions. From Exhibit B-1 of Appendix 4A, the toxic endpoint is 0.0019 mg/L and the release should be treated as a dense vapor cloud. Turning to Reference Table 18 of Appendix 4A, the closest tabulated toxic endpoint is 0.002 mg/L. The closest release rate is 10 lb/min, for a predicted distance of 0.9 mi.

## ALTERNATIVE SCENARIOS FOR FLAMMABLE SUBSTANCES

For many owners of warehouses, a fire is potentially the most damaging scenario. However, there are several considerations involved in modeling scenarios for flammable substances within the context of the RMP.

- Toxic materials in plumes generated by fires do not have to be considered under the RMP rule, although toxic combustion products might be generated.

- Under the RMP rule, endpoints for regulated flammable substances are specified for explosions, radiant heat, and dispersion to the lower flammability limit (LFL).

- Effects such as heat released by unregulated materials are not considered under the RMP rule; such effects might be a major problem in warehouse fires.

Alternative scenarios for regulated flammable substances at warehouses could include pool fires and vapor cloud fires (assuming dispersion to the LFL), as discussed below.

## EXHIBIT 4-6
## PREDICTED DISTANCES TO TOXIC ENDPOINTS FOR REGULATED TOXIC MATERIALS
### Alternative Case Scenario No. 2 - Release from Two Containers, Stability Class D, Wind Speed 3.0 m/s

| Chemical/Solution Name | • | Container Size | Quantity Q (lb) | Rate of Release (lb/min) | Distance Rural Site (mi) | Distance Urban Site (mi) |
|---|---|---|---|---|---|---|
| Boron Trifluoride Compound with Methyl Ether (1:1) | 0.0035 | 55 gallon | 1140 | 3.87 | <0.1 | <0.1 |
| Cyclohexylamine | 0.0034 | 55 gallon | 770 | 2.61 | <0.1 | <0.1 |
| Diborane | NA[a] | 150 lb | 100 | 10 | 1.0 | 0.6 |
| Epichlorohydrin | 0.0040 | 55 gallon | 1100 | 4.40 | <0.1 | <0.1 |
| Ethylenediamine | 0.0029 | 55 gallon | 880 | 2.55 | <0.1 | <0.1 |
| Ethylene Oxide | NA[a] | 150 lb | 300 | 30 | 0.1 | <0.1 |
| Hydrochloric Acid 30%[b] | 0.0016 | 55 gallon | 1060 | 1.7 | <0.1 | <0.1 |
| Hydrochloric Acid 34%[b] | 0.0048 | 55 gallon | 1064 | 5.1 | <0.1 | <0.1 |
| Hydrochloric Acid 36%[b] | 0.0073 | 55 gallon | 1066 | 7.78 | 0.1 | <0.1 |
| Hydrochloric Acid 37% | 0.0086 | 55 gallon | 1066 | 9.16 | 0.1 | <0.1 |
| Hydrochloric Acid 38% | 0.0098 | 55 gallon | 1066 | 10.44 | 0.1 | 0.1 |
| Methyl Chloride | NA[a] | 150 lb | 300 | 30 | <0.1 | <0.1 |
| Nitric Acid 80% | 0.0015 | 55 gallon | 1224 | 1.84 | <0.1 | <0.1 |
| Nitric Acid 85% | 0.0025 | 55 gallon | 1286 | 3.22 | <0.1 | <0.1 |
| Nitric Acid 90% | 0.0036 | 55 gallon | 1380 | 4.96 | 0.1 | <0.1 |
| Propylene Oxide | 0.13 | 55 gallon | 818 | 106 | 0.1 | <0.1 |
| Sulfur Dioxide | NA[a] | 150 lb | 300 | 30 | 0.6 | 0.3 |
| Titanium Tetrachloride | 0.0032 | 55 gallon | 1586 | 5.08 | 0.1 | <0.1 |
| Toluene 2,4-diisocyanate | $5.79 \times 10^{-6}$ | 55 gallon | 1100 | 0.0064 | <0.1 | <0.1 |
| Toluene 2,6-diisocyanate | $1.7 \times 10^{-5}$ | 55 gallon | 1100 | 0.019 | <0.1 | <0.1 |

Footnotes:
[a]Not applicable—chemical is a gas.
[b]Hydrochloric acid in concentrations below 37% is not regulated.

For <u>pool fires</u> involving spillages of flammable liquids, the following equation gives an estimate for the distance D (ft) from a pool fire at which people could potentially receive a second-degree burn after 40 seconds:

$$D = PFF\ A^{0.5} \tag{5}$$

where PFF is the "Pool Fire Factor", and A is the area of the pool in $ft^2$. The PFF is tabulated in Exhibits C-2 and C-3 of Appendix 4A.

**Example 14.** You have vinyl ethyl ether in a 55-gallon drum (350 lb). It spreads to form a pool of depth 1 cm. The area is given by $A = 350 \times DF$, where DF is tabulated in Exhibit C-3 and is 0.65 for vinyl ethyl ether: $A = 350 \times 0.65 = 228\ ft^2$ so that $A^{0.5} = 15$ ft. From Exhibit C-3 of the OCAG, PFF = 4.2 so that $D = 4.2 \times 15$ • 60 ft. (You also can consider spills from multiple drums. Multiply A for one drum by the number of drums involved.)

Another alternative flammable scenario is a vapor cloud fire. You would calculate the <u>distance to the LFL</u>, that is, the distance to which the cloud propagates before diluting below the lower flammable limit, assuming the vapor cloud then ignites. You need to determine the release rate, as described for toxic substances. The data you need for flammable substances, including the LFL, are in Exhibits C-2 (for flammable gases) and C-3 (for flammable liquids) in Appendix 4A.

The appropriate reference tables for determining the predicted distance to the LFL are as follows:

- • Neutrally buoyant plume, rural site, Reference Table 26
- • Neutrally buoyant plume, urban site, Reference Table 27
- • Dense plume, rural site, Reference Table 28
- • Dense plume, urban site, Reference Table 29

**Example 15.** Vinyl ethyl ether is spilled from a 55-gallon drum (containing 350 lb) at an urban site. As shown in Example 14, it forms a pool of depth 1 cm and area 228 $ft^2$. The rate of evaporation is given by $QR = 2.4Q \times LFA \times DF$. LFA is 0.10 and DF is 0.65, from Exhibit C-3 of Appendix 4A. Hence, QR = (2.4)(350)(0.1)(0.65) = 55 lb/min. From Exhibit C-3 of Appendix 4A, the LFL for vinyl ethyl ether is 50 mg/L, and it should be treated as a dense gas. From Reference Table 28, the predicted distance to the LFL at a rural site is < 0.1 mi for any release rate < 1,500 lb/min.

## 4.3   BUILDINGS

Buildings may be considered provide passive mitigation in some cases. Unless your containers of regulated substances are delivered directly into the building (i.e., they are not unloaded outdoors and moved inside later), you should not consider buildings in your worst-case scenario, because there will be some time when the vessels are outdoors. If your containers are delivered indoors or if your largest vessel is indoors, you may want to analyze the mitigating effects of the building when you do your worst-case analysis. You may also want to consider alternative scenarios that consider buildings as mitigation systems. However, warehouses vary over a wide range in their strength of construction, the surface area of ventilation outlets, and their purpose. In addition, warehouse doors are often left open for considerable periods.

For toxic liquids, EPA has provided simple building release rate reduction factors for indoor releases of 10% for worst-case scenarios and 5% for alternative scenarios (i.e., the predicted rate of release is 10% or 5% of that for the same accident if it should occur outdoors). These factors are based on data for a building with a ventilation rate of 0.5 air changes per hour. The factors are applicable to releases in a fully enclosed, non-airtight space that is directly adjacent to the outside air. They do not apply to a space that has doors or windows that could be open during a release. (See Appendix D of the OCAG for more discussion of the mitigation factors.)

For toxic gases, the EPA's reduction factor is 55%, for both worst-case and alternative scenarios. This factor also is based on 0.5 air changes per hour. It is applicable to releases in the same type of enclosure as the factors for liquids. (See Appendix D of the OCAG for more discussion.)

You are at liberty to provide building-specific models of the mitigating effects of structures. Generally, such modeling requires the use of numerical simulation and specialized computer programs.

**Example 16**. Returning to Example 3, you have dimethyldichlorosilane in your warehouse in 55-gallon drums at a rural site. Assume that the release takes place inside the warehouse in worst-case weather conditions. The effective release rate is then reduced by a factor of 10 from 13 lb/min to 1.3 lb/min. From Exhibit B-2 of Appendix 4A, the toxic endpoint of dimethyldichlorosilane is 0.026 mg/L, and it should be modeled as a dense vapor. Turning to Reference Table 6 of Appendix 4A, for a rural site, look for the evaporation rate closest to 1.3 lb/min and the toxic endpoint closest to 0.026 mg/L. This is 1 lb/min and 0.02 mg/L, giving a distance of 0.3 mi.

## 4.4    ESTIMATING OFFSITE RECEPTORS

The rule requires that you estimate in the RMP residential populations within the circle defined by the endpoint for your worst-case and alternative release scenarios (i.e., the center of the circle is the point of release and the radius is the distance to the endpoint). In addition, you must report in the RMP whether certain types of public receptors and environmental receptors are within the circles.

### RESIDENTIAL POPULATIONS

To estimate residential populations, you may use the most recent Census data or any other source of data that you believe is more accurate. You are not required to update Census data or conduct any surveys to develop your estimates. Census data are available in public libraries and in the LandView system, which is available on CD-ROM (see box below). The rule requires that you estimate populations to two-significant digits. For example, if there are 1,260 people within the circle, you may report 1,300 people. If the number of people is between 10 and 100, estimate to the nearest 10. If the number of people is less than 10, provide the actual number. Census data are presented by Census tract. If your circle covers only a portion of the tract, you should develop an estimate for that portion. The easiest way to do this is to determine the population density per square mile (total population of the Census tract divided by the number of square miles in the tract) and apply that density figure to the number of square miles within your circle. Because there is likely to be considerable variation in actual densities within a Census tract, this number will be approximate. The rule, however, does not require you to correct the number.

## OTHER PUBLIC RECEPTORS

Other public receptors must be noted in the RMP (see the discussion of public receptors in Chapter 2). If there are any schools, residences, hospitals, prisons, public recreational areas or arenas, or commercial or industrial areas within the circle, you must report that. You are not required to develop a list of all public receptors; you must simply check off that one or more such areas is within the circle. Most receptors can be identified from local street maps.

## ENVIRONMENTAL RECEPTORS

Environmental receptors are defined as natural areas such as national or state parks, forests, or monuments; officially designated wildlife sanctuaries, preserves, refuges, or areas; and Federal wilderness areas. Only environmental receptors that can be identified on local U.S. Geological Survey (USGS) maps (see box below) need to be considered. You are not required to locate each of these specifically. You are only required to check off in the RMP which specific types of areas are within the circle. If any part of one of these receptors is within your circle, you must note that in the RMP.

**Important:** The rule does not require you to assess the likelihood, type, or severity of potential impacts on either public or environmental receptors. Identifying them as within the circle simply indicates that they could be adversely affected by the release.

Besides the results you are required to report in the RMP, you may want to consider submitting to EPA or providing your local community with a map showing the distances to the endpoint. Figure 4-1 is one suggested example of how the consequences of worst-case and alternative scenarios might be presented. It is a simplified map that shows the radius to which the vapor cloud might extend, given the worst-case release in worst-case weather conditions (the owner or operator should use a real map of the area surrounding the site).

Organizations that have already begun to prepare Risk Management Programs and Plans have used this form of presentation (for example, in the Kanawha Valley or in Tampa Bay).

---

### HOW TO OBTAIN CENSUS DATA AND LANDVIEW®

Census data can be found in publications of the Bureau of the Census, available in public libraries, including *County and City Data Book*.

LandView ®III is a desktop mapping system that includes database extracts from EPA, the Bureau of the Census, the U.S. Geological Survey, the Nuclear Regulatory Commission, the Department of Transportation, and the Federal Emergency Management Agency. These databases are presented in a geographic context on maps that show jurisdictional boundaries, detailed networks of roads, rivers, and railroads, census block group and tract polygons, schools, hospitals, churches, cemeteries, airports, dams, and other landmark features.

CD-ROM for IBM-compatible PCS
CD-TGR95-LV3-KIT $99 per disc (by region) or $549 for 11 disc set

U.S. Department of Commerce
Bureau of the Census
P.O. Box 277943
Atlanta, GA 30384-7943
Phone: 301-457-4100 (Customer Services — orders)
Fax: (888) 249-7295 (toll-free)
Fax: (301) 457-3842 (local)
Phone: (301) 457-1128 (Geography Staff — content)
http://www.census.gov/ftp/pub/geo/www/tiger/

Further information on LandView and other sources of Census data is available at the Bureau of the Census web site at www.census.gov.

---

---

### HOW TO OBTAIN USGS MAPS

The production of digital cartographic data and graphic maps comprises the largest component of the USGS National Mapping Program. The USGS's most familiar product is the 1:24,000-scale Topographic Quadrangle Map. This is the primary scale of data produced, and depicts greater detail for a smaller area than intermediate-scale (1:50,000 and 1:100,000) and small-scale (1:250,000, 1:2,000,000 or smaller) products, which show selectively less detail for larger areas.

U.S. Geological Survey
508 National Center
12201 Sunrise Valley Drive
Reston, VA 20192
http://mapping.usgs.gov/

To order USGS maps by fax, select, print, and complete one of the online forms and fax to 303-202-4693. A list of commercial dealers also is available at http://mapping.usgs.gov/esic/usimage/dealers.html/. For more information or ordering assistance, call 1-800-HELP-MAP, or write:

USGS Information Services
Box 25286
Denver, CO 80225

For additional information, contact any USGS Earth Science Information Center or call 1-800-USA-MAPS.

---

## 4.5   DOCUMENTATION

You need to maintain onsite the following records on the offsite consequence analyses:

For the worst-case scenario, a description of the vessel or pipeline selected as worst-case, assumptions and parameters used and the rationale for selection; assumptions include use of any administrative controls and any passive mitigation systems that you assumed to limit the quantity that could be released.

For alternative release scenarios, a description of the scenarios identified, assumptions and parameters used and the rationale for the selection of specific scenarios; assumptions include use of any administrative controls and any mitigation that were assumed to limit the quantity that could be released. Documentation includes the effect of the controls and mitigation on the release quantity and rate.

Other data that you should provide includes:

- Documentation of estimated quantity released, release rate and duration of release.

- • Methodology used to determine distance to endpoints (it will be sufficient to reference this guidance).
- • Data used to identify potentially affected population and environmental receptors.

# Figure 4-1 Simplified Presentation of Worst-Case and Alternative Scenario on a Local Map

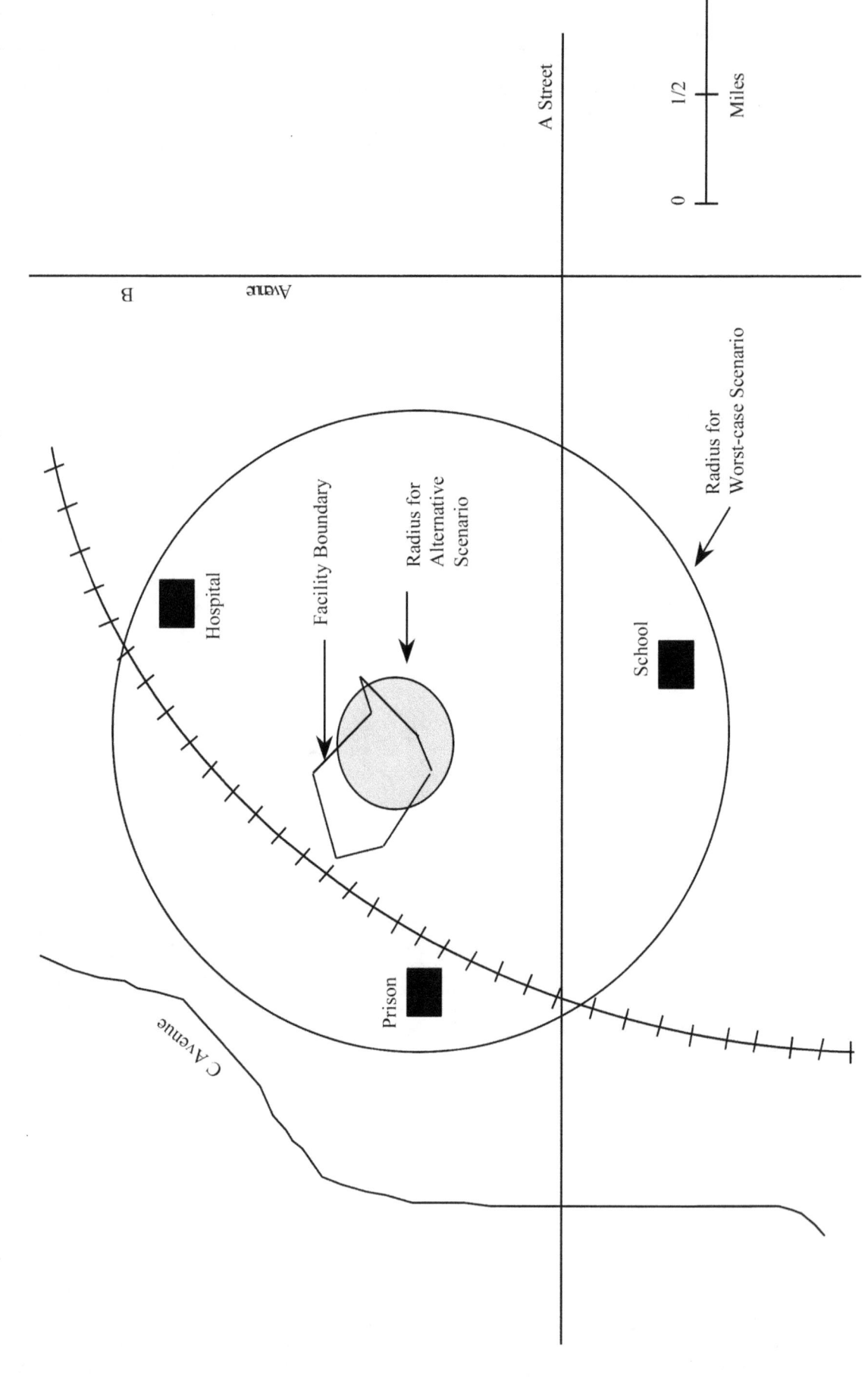

# APPENDIX 4A

## REFERENCE TABLES OF DISTANCES AND TABLES OF DATA FROM THE OFFSITE CONSEQUENCE ANALYSIS GUIDANCE

The following tables from the *RMP Offsite Consequence Analysis Guidance* are reproduced here for the convenience of the reader.

### Reference Tables of Distances

**Reference Table 1**—Neutrally Buoyant Plume Distances to Toxic Endpoint for Release Rate Divided by Endpoint, 10-Minute Release, Rural Conditions, F Stability, Wind Speed 1.5 Meters per Second
**Reference Table 2**—Neutrally Buoyant Plume Distances to Toxic Endpoint for Release Rate Divided by Endpoint, 60-Minute Release, Rural Conditions, F Stability, Wind Speed 1.5 Meters per Second
**Reference Table 3**—Neutrally Buoyant Plume Distances to Toxic Endpoint for Release Rate Divided by Endpoint, 10-Minute Release, Urban Conditions, F Stability, Wind Speed 1.5 Meters per Second
**Reference Table 4**—Neutrally Buoyant Plume Distances to Toxic Endpoint for Release Rate Divided by Endpoint, 60-Minute Release, Urban Conditions, F Stability, Wind Speed 1.5 Meters per Second
**Reference Table 5**—Dense Gas Distances to Toxic Endpoint, 10-Minute Release, Rural Conditions, F Stability, Wind Speed 1.5 Meters per Second
**Reference Table 6**—Dense Gas Distances to Toxic Endpoint, 60-Minute Release, Rural Conditions, F Stability, Wind Speed 1.5 Meters per Second
**Reference Table 7**—Dense Gas Distances to Toxic Endpoint, 10-Minute Release, Urban Conditions, F Stability, Wind Speed 1.5 Meters per Second
**Reference Table 8**—Dense Gas Distances to Toxic Endpoint, 60-Minute Release, Urban Conditions, F Stability, Wind Speed 1.5 Meters per Second
**Reference Table 9**—Distances to Toxic Endpoint for Anhydrous Ammonia Liquefied Under Pressure, F Stability, Wind Speed 1.5 Meters per Second
**Reference Table 10**—Distances to Toxic Endpoint for Aqueous Ammonia, F Stability, Wind Speed 1.5 Meters per Second
**Reference Table 11**—Distances to Toxic Endpoint for Chlorine, F Stability, Wind Speed 1.5 Meters per Second
**Reference Table 12**—Distances to Toxic Endpoint for Sulfur Dioxide, F Stability, Wind Speed 1.5 Meters per Second
**Reference Table 13**—Distance to Overpressure of 1.0 psi for Vapor Cloud Explosions of 500 - 2,000,000 Pounds of Regulated Flammable Substances Based on TNT Equivalent Method, 10 Percent Yield Factor
**Reference Table 14**—Neutrally Buoyant Plume Distances to Toxic Endpoint for Release Rate Divided by Endpoint, 10-Minute Release, Rural Conditions, D Stability, Wind Speed 3.0 Meters per Second
**Reference Table 15**—Neutrally Buoyant Plume Distances to Toxic Endpoint for Release Rate Divided by Endpoint, 60-Minute Release, Rural Conditions, D Stability, Wind Speed 3.0 Meters per Second
**Reference Table 16**—Neutrally Buoyant Plume Distances to Toxic Endpoint for Release Rate Divided by Endpoint, 10-Minute Release, Urban Conditions, D Stability, Wind Speed 3.0 Meters per Second
**Reference Table 17**—Neutrally Buoyant Plume Distances to Toxic Endpoint for Release Rate Divided by Endpoint, 60-Minute Release, Urban Conditions, D Stability, Wind Speed 3.0 Meters per Second
**Reference Table 18**—Dense Gas Distances to Toxic Endpoint, 10-Minute Release, Rural Conditions, D Stability, Wind Speed 3.0 Meters per Second

**Reference Table 19**—Dense Gas Distances to Toxic Endpoint, 60-Minute Release, Rural Conditions, D Stability, Wind Speed 3.0 Meters per Second

**Reference Table 20**—Dense Gas Distances to Toxic Endpoint, 10-Minute Release, Urban Conditions, D Stability, Wind Speed 3.0 Meters per Second

**Reference Table 21**—Dense Gas Distances to Toxic Endpoint, 60-Minute Release, Urban Conditions, D Stability, Wind Speed 3.0 Meters per Second

**Reference Table 22**—Distances to Toxic Endpoint for Anhydrous Ammonia, D Stability, Wind Speed 3.0 Meters per Second

**Reference Table 23**—Distances to Toxic Endpoint for Aqueous Ammonia, D Stability, Wind Speed 3.0 Meters per Second

**Reference Table 24**—Distances to Toxic Endpoint for Chlorine, D Stability, Wind Speed 3.0 Meters per Second

**Reference Table 25**—Distances to Toxic Endpoint for Sulfur Dioxide, D Stability, Wind Speed 3.0 Meters per Second

**Reference Table 26**—Neutrally Buoyant Plume Distances to Lower Flammability Limit (LFL) for Release Rate Divided by LFL, Rural Conditions, D Stability, Wind Speed 3.0 Meters per Second

**Reference Table 27**—Neutrally Buoyant Plume Distances to Lower Flammability Limit (LFL) for Release Rate Divided by LFL, Urban Conditions, D Stability, Wind Speed 3.0 Meters per Second

**Reference Table 28**—Dense Gas Distances to Lower Flammability Limit Rural Conditions, D Stability, Wind Speed 3.0 Meters per Second

**Reference Table 29**—Dense Gas Distances to Lower Flammability Limit Urban Conditions, D Stability, Wind Speed 3.0 Meters per Second

## Tables of Data

**Exhibit B-1**—Data for Toxic Gases

**Exhibit B-2**—Data for Toxic Liquids

**Exhibit B-3**—Data for Water Solutions of Toxic Substances and for Oleum, Average Vapor Pressure and Liquid Factors over 10 Minutes for Wind Speeds of 1.5 and 3.0 Meters per Second (m/s)

**Exhibit B-4**—Temperature Correction Factors for Liquids Evaporating from Pools at Temperatures Between 25•C and 50•C (77•F and 122•F)

**Exhibit C-1**—Heats of Combustion for Flammable Substances

**Exhibit C-2**—Data for Flammable Gases

**Exhibit C-3**—Data for Flammable Liquids

## Reference Table 1
### Neutrally Buoyant Plume Distances to Toxic Endpoint for Release Rate Divided by Endpoint
### 10-Minute Release, Rural Conditions, F Stability, Wind Speed 1.5 Meters per Second

| Release Rate/Endpoint [(lbs/min)/(mg/L)] (•) | Distance to Endpoint (miles) | Release Rate/Endpoint [(lbs/min)/(mg/L)] (•) | Distance to Endpoint (miles) |
|---|---|---|---|
| 0 - 4.4 | 0.1 | 16,000 - 18,000 | 4.8 |
| 4.4 - 37 | 0.2 | 18,000 - 19,000 | 5.0 |
| 37 - 97 | 0.3 | 19,000 - 21,000 | 5.2 |
| 97 - 180 | 0.4 | 21,000 - 23,000 | 5.4 |
| 180 - 340 | 0.6 | 23,000 - 24,000 | 5.6 |
| 340 - 530 | 0.8 | 24,000 - 26,000 | 5.8 |
| 530 - 760 | 1.0 | 26,000 - 28,000 | 6.0 |
| 760 - 1,000 | 1.2 | 28,000 - 29,600 | 6.2 |
| 1,000 - 1,500 | 1.4 | 29,600 - 35,600 | 6.8 |
| 1,500 - 1,900 | 1.6 | 35,600 - 42,000 | 7.5 |
| 1,900 - 2,400 | 1.8 | 42,000 - 48,800 | 8.1 |
| 2,400 - 2,900 | 2.0 | 48,800 - 56,000 | 8.7 |
| 2,900 - 3,500 | 2.2 | 56,000 - 63,600 | 9.3 |
| 3,500 - 4,400 | 2.4 | 63,600 - 71,500 | 9.9 |
| 4,400 - 5,100 | 2.6 | 71,500 - 88,500 | 11 |
| 5,100 - 5,900 | 2.8 | 88,500 - 107,000 | 12 |
| 5,900 - 6,800 | 3.0 | 107,000 - 126,000 | 14 |
| 6,800 - 7,700 | 3.2 | 126,000 - 147,000 | 15 |
| 7,700 - 9,000 | 3.4 | 147,000 - 169,000 | 16 |
| 9,000 - 10,000 | 3.6 | 169,000 - 191,000 | 17 |
| 10,000 - 11,000 | 3.8 | 191,000 - 215,000 | 19 |
| 11,000 - 12,000 | 4.0 | 215,000 - 279,000 | 22 |
| 12,000 - 14,000 | 4.2 | 279,000 - 347,000 | 25 |
| 14,000 - 15,000 | 4.4 | >347,000 | >25* |
| 15,000 - 16,000 | 4.6 | | |

\* Report distance as 25 miles

## Reference Table 2
### Neutrally Buoyant Plume Distances to Toxic Endpoint for Release Rate Divided by Endpoint
### 60-Minute Release, Rural Conditions, F Stability, Wind Speed 1.5 Meters per Second

| Release Rate/Endpoint [(lbs/min)/(mg/L)] (•) | Distance to Endpoint (miles) | Release Rate/Endpoint [(lbs/min)/(mg/L)] (•) | Distance to Endpoint (miles) |
|---|---|---|---|
| 0 - 5.5 | 0.1 | 7,400 - 7,700 | 4.8 |
| 5.5 - 46 | 0.2 | 7,700 - 8,100 | 5.0 |
| 46 - 120 | 0.3 | 8,100 - 8,500 | 5.2 |
| 120 - 220 | 0.4 | 8,500 - 8,900 | 5.4 |
| 220 - 420 | 0.6 | 8,900 - 9,200 | 5.6 |
| 420 - 650 | 0.8 | 9,200 - 9,600 | 5.8 |
| 650 - 910 | 1.0 | 9,600 - 10,000 | 6.0 |
| 910 - 1,200 | 1.2 | 10,000 - 10,400 | 6.2 |
| 1,200 - 1,600 | 1.4 | 10,400 - 11,700 | 6.8 |
| 1,600 - 1,900 | 1.6 | 11,700 - 13,100 | 7.5 |
| 1,900 - 2,300 | 1.8 | 13,100 - 14,500 | 8.1 |
| 2,300 - 2,600 | 2.0 | 14,500 - 15,900 | 8.7 |
| 2,600 - 2,900 | 2.2 | 15,900 - 17,500 | 9.3 |
| 2,900 - 3,400 | 2.4 | 17,500 - 19,100 | 9.9 |
| 3,400 - 3,700 | 2.6 | 19,100 - 22,600 | 11 |
| 3,700 - 4,100 | 2.8 | 22,600 - 26,300 | 12 |
| 4,100 - 4,400 | 3.0 | 26,300 - 30,300 | 14 |
| 4,400 - 4,800 | 3.2 | 30,300 - 34,500 | 15 |
| 4,800 - 5,200 | 3.4 | 34,500 - 38,900 | 16 |
| 5,200 - 5,600 | 3.6 | 38,900 - 43,600 | 17 |
| 5,600 - 5,900 | 3.8 | 43,600 - 48,400 | 19 |
| 5,900 - 6,200 | 4.0 | 48,400 - 61,500 | 22 |
| 6,200 - 6,700 | 4.2 | 61,500 - 75,600 | 25 |
| 6,700 - 7,000 | 4.4 | >75,600 | >25* |
| 7,000 - 7,400 | 4.6 | | |

* Report distance as 25 miles

## Reference Table 3
### Neutrally Buoyant Plume Distances to Toxic Endpoint for Release Rate Divided by Endpoint
### 10-minute Release, Urban Conditions, F Stability, Wind Speed 1.5 Meters per Second

| Release Rate/Endpoint [(lbs/min)/(mg/L)] (•) | Distance to Endpoint (miles) | Release Rate/Endpoint [(lbs/min)/(mg/L) (•) | Distance to Endpoint (miles) |
|---|---|---|---|
| 0 - 21 | 0.1 | 76,000 - 83,000 | 4.8 |
| 21 - 170 | 0.2 | 83,000 - 90,000 | 5.0 |
| 170 - 420 | 0.3 | 90,000 - 100,000 | 5.2 |
| 420 - 760 | 0.4 | 100,000 - 110,000 | 5.4 |
| 760 - 1,400 | 0.6 | 110,000 - 120,000 | 5.6 |
| 1,400 - 2,100 | 0.8 | 120,000 - 130,000 | 5.8 |
| 2,100 - 3,100 | 1.0 | 130,000 - 140,000 | 6.0 |
| 3,100 - 4,200 | 1.2 | 140,000 - 148,000 | 6.2 |
| 4,200 - 6,100 | 1.4 | 148,000 - 183,000 | 6.8 |
| 6,100 - 7,800 | 1.6 | 183,000 - 221,000 | 7.5 |
| 7,800 - 9,700 | 1.8 | 221,000 - 264,000 | 8.1 |
| 9,700 - 12,000 | 2.0 | 264,000 - 310,000 | 8.7 |
| 12,000 - 14,000 | 2.2 | 310,000 - 361,000 | 9.3 |
| 14,000 - 18,000 | 2.4 | 361,000 - 415,000 | 9.9 |
| 18,000 - 22,000 | 2.6 | 415,000 - 535,000 | 11 |
| 22,000 - 25,000 | 2.8 | 535,000 - 671,000 | 12 |
| 25,000 - 29,000 | 3.0 | 671,000 - 822,000 | 14 |
| 29,000 - 33,000 | 3.2 | 822,000 - 990,000 | 15 |
| 33,000 - 39,000 | 3.4 | 990,000 - 1,170,000 | 16 |
| 39,000 - 44,000 | 3.6 | 1,170,000 - 1,370,000 | 17 |
| 44,000 - 49,000 | 3.8 | 1,370,000 - 1,590,000 | 19 |
| 49,000 - 55,000 | 4.0 | 1,590,000 - 2,190,000 | 22 |
| 55,000 - 63,000 | 4.2 | 2,190,000 - 2,890,000 | 25 |
| 63,000 - 69,000 | 4.4 | >2,890,000 | >25* |
| 69,000 - 76,000 | 4.6 | | |

* Report distance as 25 miles

## Reference Table 4
### Neutrally Buoyant Plume Distances to Toxic Endpoint for Release Rate Divided by Endpoint
### 60-Minute Release, Urban Conditions, F Stability, Wind Speed 1.5 Meters per Second

| Release Rate/Endpoint [(lbs/min)/(mg/L)] (•) | Distance to Endpoint (miles) | Release Rate/Endpoint [(lbs/min)/(mg/L)] (•) | Distance to Endpoint (miles) |
|---|---|---|---|
| 0 - 26 | 0.1 | 34,000 - 36,000 | 4.8 |
| 26 - 210 | 0.2 | 36,000 - 38,000 | 5.0 |
| 210 - 530 | 0.3 | 38,000 - 41,000 | 5.2 |
| 530 - 940 | 0.4 | 41,000 - 43,000 | 5.4 |
| 940 - 1,700 | 0.6 | 43,000 - 45,000 | 5.6 |
| 1,700 - 2,600 | 0.8 | 45,000 - 47,000 | 5.8 |
| 2,600 - 3,700 | 1.0 | 47,000 - 50,000 | 6.0 |
| 3,700 - 4,800 | 1.2 | 50,000 - 52,200 | 6.2 |
| 4,800 - 6,400 | 1.4 | 52,200 - 60,200 | 6.8 |
| 6,400 - 7,700 | 1.6 | 60,200 - 68,900 | 7.5 |
| 7,700 - 9,100 | 1.8 | 68,900 - 78,300 | 8.1 |
| 9,100 - 11,000 | 2.0 | 78,300 - 88,400 | 8.7 |
| 11,000 - 12,000 | 2.2 | 88,400 - 99,300 | 9.3 |
| 12,000 - 14,000 | 2.4 | 99,300 - 111,000 | 9.9 |
| 14,000 - 16,000 | 2.6 | 111,000 - 137,000 | 11 |
| 16,000 - 17,000 | 2.8 | 137,000 - 165,000 | 12 |
| 17,000 - 19,000 | 3.0 | 165,000 - 197,000 | 14 |
| 19,000 - 21,000 | 3.2 | 197,000 - 232,000 | 15 |
| 21,000 - 23,000 | 3.4 | 232,000 - 271,000 | 16 |
| 23,000 - 24,000 | 3.6 | 271,000 - 312,000 | 17 |
| 24,000 - 26,000 | 3.8 | 312,000 - 357,000 | 19 |
| 26,000 - 28,000 | 4.0 | 357,000 - 483,000 | 22 |
| 28,000 - 30,000 | 4.2 | 483,000 - 629,000 | 25 |
| 30,000 - 32,000 | 4.4 | >629,000 | >25* |
| 32,000 - 34,000 | 4.6 | | |

* Report distance as 25 miles

## Reference Table 5
## Dense Gas Distances to Toxic Endpoint, 10-minute Release, Rural Conditions, F Stability, Wind Speed 1.5 Meters per Second

| Release Rate (lbs/min) | Toxic Endpoint (mg/L) Distance (Miles) | | | | | | | | | | | | | | | |
|---|---|---|---|---|---|---|---|---|---|---|---|---|---|---|---|---|
| | 0.0004 | 0.0007 | 0.001 | 0.002 | 0.0035 | 0.005 | 0.0075 | 0.01 | 0.02 | 0.035 | 0.05 | 0.075 | 0.1 | 0.25 | 0.5 | 0.75 |
| 1 | > | > | > | > | 0.8 | 0.7 | 0.5 | 0.5 | 0.3 | 0.2 | 0.2 | 0.2 | 0.1 | 0.1 | > | # |
| 2 | > | > | > | > | > | 0.9 | 0.7 | 0.7 | 0.4 | 0.3 | 0.3 | 0.2 | 0.2 | 0.1 | <0.1 | <0.1 |
| 5 | > | > | > | > | > | > | > | 1.0 | 0.7 | 0.5 | 0.4 | 0.3 | 0.3 | 0.2 | 0.1 | 0.1 |
| 10 | > | > | > | > | > | > | > | > | 1.0 | 0.7 | 0.6 | 0.5 | 0.4 | 0.2 | 0.2 | 0.1 |
| 30 | > | > | > | > | > | > | > | > | > | > | > | 0.9 | 0.7 | 0.4 | 0.3 | 0.2 |
| 50 | > | > | > | > | > | > | > | > | > | > | > | > | 0.9 | 0.6 | 0.4 | 0.3 |
| 100 | > | > | > | > | > | > | > | > | > | > | > | > | > | 0.8 | 0.5 | 0.4 |
| 150 | > | > | > | > | > | > | > | > | > | > | > | > | > | 0.9 | 0.6 | 0.5 |
| 250 | > | > | > | > | > | > | > | > | > | > | > | > | > | > | 0.8 | 0.6 |
| 500 | > | > | > | > | > | > | > | > | > | > | > | > | > | > | > | 0.9 |
| 750 | * | * | > | > | > | > | > | > | > | > | > | > | > | > | > | 1.0 |
| 1,000 | * | * | * | > | > | > | > | > | > | > | > | > | > | > | > | > |
| 1,500 | * | * | * | * | > | > | > | > | > | > | > | > | > | > | > | > |
| 2,000 | * | * | * | * | * | > | > | > | > | > | > | > | > | > | > | > |
| 2,500 | * | * | * | * | * | * | > | > | > | > | > | > | > | > | > | > |
| 3,000 | * | * | * | * | * | * | > | > | > | > | > | > | > | > | > | > |
| 4,000 | * | * | * | * | * | * | * | > | > | > | > | > | > | > | > | > |
| 5,000 | * | * | * | * | * | * | * | * | > | > | > | > | > | > | > | > |
| 7,500 | * | * | * | * | * | * | * | * | > | > | > | > | > | > | > | > |
| 10,000 | * | * | * | * | * | * | * | * | > | > | > | > | > | > | > | > |
| 15,000 | * | * | * | * | * | * | * | * | > | > | > | > | > | > | > | > |
| 20,000 | * | * | * | * | * | * | * | * | > | > | > | > | > | > | > | > |
| 50,000 | * | * | * | * | * | * | * | * | > | > | > | > | > | > | > | > |
| 75,000 | * | * | * | * | * | * | * | * | > | > | > | > | > | > | > | > |
| 100,000 | * | * | * | * | * | * | * | * | > | > | > | > | > | > | > | > |
| 150,000 | * | * | * | * | * | * | * | * | > | > | > | > | > | > | > | > |
| 200,000 | * | * | * | * | * | * | * | * | > | > | > | > | > | > | > | > |

\* > 25 miles (report distance as 25 miles)  
\# <0.1 mile (report distance as 0.1 mile)

## Reference Table 6

### Dense Gas Distances to Toxic Endpoint, 60-minute Release, Rural Conditions, F Stability, Wind Speed 1.5 Meters per Second

| Release Rate (lbs/min) | Toxic Endpoint (mg/L) — Distance (Miles) | | | | | | | | | | | | | | | |
|---|---|---|---|---|---|---|---|---|---|---|---|---|---|---|---|---|
| | 0.0004 | 0.0007 | 0.001 | 0.002 | 0.0035 | 0.005 | 0.0075 | 0.01 | 0.02 | 0.035 | 0.05 | 0.075 | 0.1 | 0.25 | 0.5 | 0.75 |
| 1 | ?? | 2.7 | ?? | ?? | 1.0 | 0.8 | 0.6 | 0.5 | 0.3 | 0.2 | 0.2 | 0.1 | 0.1 | <0.1 | ?? | # |
| 2 | ?? | ?? | ?? | ?? | ?? | ?? | 1.0 | 0.8 | 0.5 | 0.4 | 0.3 | 0.2 | 0.2 | 0.1 | <01 | <0.1 |
| 5 | ?? | ?? | ?? | ?? | ?? | ?? | ?? | ?? | 0.9 | 0.6 | 0.5 | 0.4 | 0.3 | 0.2 | 0.1 | 0.1 |
| 10 | ?? | ?? | ?? | ?? | ?? | ?? | ?? | ?? | ?? | 1.0 | 0.8 | 0.6 | 0.5 | 0.3 | 0.2 | 0.1 |
| 30 | ?? | ?? | ?? | ?? | ?? | ?? | ?? | ?? | ?? | ?? | ?? | ?? | 1.0 | 0.5 | 0.3 | 0.2 |
| 50 | ?? | ?? | ?? | ?? | ?? | ?? | ?? | ?? | ?? | ?? | ?? | ?? | ?? | 0.7 | 0.4 | 0.3 |
| 100 | ?? | ?? | ?? | ?? | ?? | ?? | ?? | ?? | ?? | ?? | ?? | ?? | ?? | ?? | 0.7 | 0.5 |
| 150 | * | * | ?? | ?? | ?? | ?? | ?? | ?? | ?? | ?? | ?? | ?? | ?? | ?? | 0.9 | 0.6 |
| 250 | * | * | * | ?? | ?? | ?? | ?? | ?? | ?? | ?? | ?? | ?? | ?? | ?? | ?? | 0.9 |
| 500 | * | * | * | * | ?? | ?? | ?? | ?? | ?? | ?? | ?? | ?? | ?? | ?? | ?? | ?? |
| 750 | * | * | * | * | * | ?? | ?? | ?? | ?? | ?? | ?? | ?? | ?? | ?? | ?? | ?? |
| 1,000 | * | * | * | * | * | * | ?? | ?? | ?? | ?? | ?? | ?? | ?? | ?? | ?? | ?? |
| 1,500 | * | * | * | * | * | * | * | ?? | ?? | ?? | ?? | ?? | ?? | ?? | ?? | ?? |
| 2,000 | * | * | * | * | * | * | * | ?? | ?? | ?? | ?? | ?? | ?? | ?? | ?? | ?? |
| 2,500 | * | * | * | * | * | * | * | * | ?? | ?? | ?? | ?? | ?? | ?? | ?? | ?? |
| 3,000 | * | * | * | * | * | * | * | * | ?? | ?? | ?? | ?? | ?? | ?? | ?? | ?? |
| 4,000 | * | * | * | * | * | * | * | * | * | ?? | ?? | ?? | ?? | ?? | ?? | ?? |
| 5,000 | * | * | * | * | * | * | * | * | * | ?? | ?? | ?? | ?? | ?? | ?? | ?? |
| 7,500 | * | * | * | * | * | * | * | * | * | ?? | ?? | ?? | ?? | ?? | ?? | ?? |
| 10,000 | * | * | * | * | * | * | * | * | * | * | ?? | ?? | ?? | ?? | ?? | ?? |
| 15,000 | * | * | * | * | * | * | * | * | * | * | ?? | ?? | ?? | ?? | ?? | ?? |
| 20,000 | * | * | * | * | * | * | * | * | * | * | * | ?? | ?? | ?? | ?? | ?? |
| 50,000 | * | * | * | * | * | * | * | * | * | * | * | ?? | ?? | ?? | ?? | ?? |
| 75,000 | * | * | * | * | * | * | * | * | * | * | * | ?? | ?? | ?? | ?? | ?? |
| 100,000 | * | * | * | * | * | * | * | * | * | * | * | ?? | ?? | ?? | ?? | ?? |
| 150,000 | * | * | * | * | * | * | * | * | * | * | * | ?? | ?? | ?? | ?? | ?? |
| 200,000 | * | * | * | * | * | * | * | * | * | * | * | ?? | ?? | ?? | ?? | ?? |

\* > 25 miles (report distance as 25 miles)          # <0.1 mile (report distance as 0.1 mile)

## Reference Table 7

### Dense Gas Distances to Toxic Endpoint, 10-minute Release, Urban Conditions, F Stability, Wind Speed 1.5 Meters per Second

| Release Rate (lbs/min) | Toxic Endpoint (mg/L) Distance (Miles) | | | | | | | | | | | | | | | |
|---|---|---|---|---|---|---|---|---|---|---|---|---|---|---|---|---|
| | 0.0004 | 0.0007 | 0.001 | 0.002 | 0.0035 | 0.005 | 0.0075 | 0.01 | 0.02 | 0.035 | 0.05 | 0.075 | 0.1 | 0.25 | 0.5 | 0.75 |
| 1 | ?? | ?? | ?? | 0.7 | 0.6 | 0.4 | 0.4 | 0.3 | 0.2 | 0.2 | 0.1 | 0.1 | 0.1 | ?? | # | # |
| 2 | ?? | ?? | ?? | ?? | 0.8 | 0.6 | 0.5 | 0.4 | 0.3 | 0.2 | 0.2 | 0.1 | 0.1 | <0.1 | # | # |
| 5 | ?? | ?? | ?? | ?? | ?? | 1.0 | 0.8 | 0.7 | 0.5 | 0.4 | 0.3 | 0.2 | 0.2 | 0.1 | <0.1 | # |
| 10 | ?? | ?? | ?? | ?? | ?? | ?? | ?? | 1.0 | 0.7 | 0.5 | 0.4 | 0.3 | 0.2 | 0.1 | 0.1 | <0.1 |
| 30 | ?? | ?? | ?? | ?? | ?? | ?? | ?? | ?? | ?? | 0.9 | 0.7 | 0.6 | 0.4 | 0.2 | 0.1 | 0.1 |
| 50 | ?? | ?? | ?? | ?? | ?? | ?? | ?? | ?? | ?? | ?? | 0.9 | 0.7 | 0.6 | 0.3 | 0.2 | 0.1 |
| 100 | ?? | ?? | ?? | ?? | ?? | ?? | ?? | ?? | ?? | ?? | ?? | 1.0 | 0.9 | 0.5 | 0.3 | 0.2 |
| 150 | ?? | ?? | ?? | ?? | ?? | ?? | ?? | ?? | ?? | ?? | ?? | ?? | ?? | 0.6 | 0.4 | 0.2 |
| 250 | ?? | ?? | ?? | ?? | ?? | ?? | ?? | ?? | ?? | ?? | ?? | ?? | ?? | 0.7 | 0.5 | 0.3 |
| 500 | ?? | ?? | ?? | ?? | ?? | ?? | ?? | ?? | ?? | ?? | ?? | ?? | ?? | ?? | 0.7 | 0.5 |
| 750 | ?? | ?? | ?? | ?? | ?? | ?? | ?? | ?? | ?? | ?? | ?? | ?? | ?? | ?? | 0.8 | 0.6 |
| 1,000 | * | * | ?? | ?? | ?? | ?? | ?? | ?? | ?? | ?? | ?? | ?? | ?? | ?? | 0.9 | 0.7 |
| 1,500 | * | * | ?? | ?? | ?? | ?? | ?? | ?? | ?? | ?? | ?? | ?? | ?? | ?? | ?? | 0.8 |
| 2,000 | * | * | * | ?? | ?? | ?? | ?? | ?? | ?? | ?? | ?? | ?? | ?? | ?? | ?? | 0.9 |
| 2,500 | * | * | * | * | ?? | ?? | ?? | ?? | ?? | ?? | ?? | ?? | ?? | ?? | ?? | ?? |
| 3,000 | * | * | * | * | ?? | ?? | ?? | ?? | ?? | ?? | ?? | ?? | ?? | ?? | ?? | ?? |
| 4,000 | * | * | * | * | * | ?? | ?? | ?? | ?? | ?? | ?? | ?? | ?? | ?? | ?? | ?? |
| 5,000 | * | * | * | * | * | * | * | ?? | ?? | ?? | ?? | ?? | ?? | ?? | ?? | ?? |
| 7,500 | * | * | * | * | * | * | * | ?? | ?? | ?? | ?? | ?? | ?? | ?? | ?? | ?? |
| 10,000 | * | * | * | * | * | * | * | ?? | ?? | ?? | ?? | ?? | ?? | ?? | ?? | ?? |
| 15,000 | * | * | * | * | * | * | * | ?? | ?? | ?? | ?? | ?? | ?? | ?? | ?? | ?? |
| 20,000 | * | * | * | * | * | * | * | ?? | ?? | ?? | ?? | ?? | ?? | ?? | ?? | ?? |
| 50,000 | * | * | * | * | * | * | * | ?? | ?? | ?? | ?? | ?? | ?? | ?? | ?? | ?? |
| 75,000 | * | * | * | * | * | * | * | ?? | ?? | ?? | ?? | ?? | ?? | ?? | ?? | ?? |
| 100,000 | * | * | * | * | * | * | * | ?? | ?? | ?? | ?? | ?? | ?? | ?? | ?? | ?? |
| 150,000 | * | * | * | * | * | * | * | ?? | ?? | ?? | ?? | ?? | >25 | ?? | ?? | ?? |
| 200,000 | * | * | * | * | * | * | * | ?? | ?? | ?? | ?? | ?? | ?? | ?? | ?? | ?? |

\* > 25 miles (report distance as 25 miles)   # <0.1 mile (report distance as 0.1 mile)

# Reference Table 8

## Dense Gas Distances to Toxic Endpoint, 60-minute Release, Urban Conditions, F Stability, Wind Speed 1.5 Meters per Second

| Release Rate (lbs/min) | Toxic Endpoint (mg/L) — Distance (Miles) | | | | | | | | | | | | | | | |
|---|---|---|---|---|---|---|---|---|---|---|---|---|---|---|---|---|
| | 0.0004 | 0.0007 | 0.001 | 0.002 | 0.0035 | 0.005 | 0.0075 | 0.01 | 0.02 | 0.035 | 0.05 | 0.075 | 0.1 | 0.25 | 0.5 | 0.75 |
| 1 | ?? | ?? | ?? | ?? | 0.7 | 0.6 | 0.4 | 0.4 | 0.2 | 0.2 | 0.1 | 0.1 | 0.1 | ?? | # | # |
| 2 | ?? | ?? | ?? | ?? | ?? | 0.9 | 0.7 | 0.6 | 0.4 | 0.2 | 0.2 | 0.1 | 0.1 | <0.1 | # | # |
| 5 | ?? | ?? | ?? | ?? | ?? | ?? | ?? | 0.9 | 0.6 | 0.4 | 0.3 | 0.2 | 0.2 | 0.1 | <0.1 | # |
| 10 | ?? | ?? | ?? | ?? | ?? | ?? | ?? | ?? | 0.9 | 0.7 | 0.5 | 0.4 | 0.3 | 0.2 | 0.1 | <0.1 |
| 30 | ?? | ?? | ?? | ?? | ?? | ?? | ?? | ?? | ?? | ?? | 1.0 | 0.7 | 0.6 | 0.3 | 0.2 | 0.1 |
| 50 | ?? | ?? | ?? | ?? | ?? | ?? | ?? | ?? | ?? | ?? | ?? | ?? | 0.9 | 0.4 | 0.2 | 0.2 |
| 100 | ?? | ?? | ?? | ?? | ?? | ?? | ?? | ?? | ?? | ?? | ?? | ?? | ?? | 0.7 | 0.4 | 0.3 |
| 150 | ?? | ?? | ?? | ?? | ?? | ?? | ?? | ?? | ?? | ?? | ?? | ?? | ?? | 0.9 | 0.5 | 0.3 |
| 250 | * | * | * | ?? | ?? | ?? | ?? | ?? | ?? | ?? | ?? | ?? | ?? | ?? | 0.7 | 0.5 |
| 500 | * | * | * | * | ?? | ?? | ?? | ?? | ?? | ?? | ?? | ?? | ?? | ?? | ?? | 0.7 |
| 750 | * | * | * | * | * | ?? | ?? | ?? | ?? | ?? | ?? | ?? | ?? | ?? | ?? | 1.0 |
| 1,000 | * | * | * | * | * | * | ?? | ?? | ?? | ?? | ?? | ?? | ?? | ?? | ?? | ?? |
| 1,500 | * | * | * | * | * | * | * | ?? | ?? | ?? | ?? | ?? | ?? | ?? | ?? | ?? |
| 2,000 | * | * | * | * | * | * | * | * | ?? | ?? | ?? | ?? | ?? | ?? | ?? | ?? |
| 2,500 | * | * | * | * | * | * | * | * | * | ?? | ?? | ?? | ?? | ?? | ?? | ?? |
| 3,000 | * | * | * | * | * | * | * | * | * | ?? | ?? | ?? | ?? | ?? | ?? | ?? |
| 4,000 | * | * | * | * | * | * | * | * | * | ?? | ?? | ?? | ?? | ?? | ?? | ?? |
| 5,000 | * | * | * | * | * | * | * | * | * | ?? | ?? | ?? | ?? | ?? | ?? | ?? |
| 7,500 | * | * | * | * | * | * | * | * | * | ?? | ?? | ?? | ?? | ?? | ?? | ?? |
| 10,000 | * | * | * | * | * | * | * | * | * | * | ?? | ?? | ?? | ?? | ?? | ?? |
| 15,000 | * | * | * | * | * | * | * | * | * | * | ?? | ?? | ?? | ?? | ?? | ?? |
| 20,000 | * | * | * | * | * | * | * | * | * | * | ?? | ?? | ?? | ?? | ?? | ?? |
| 50,000 | * | * | * | * | * | * | * | * | * | * | ?? | ?? | ?? | ?? | ?? | ?? |
| 75,000 | * | * | * | * | * | * | * | * | * | * | ?? | ?? | ?? | ?? | ?? | ?? |
| 100,000 | * | * | * | * | * | * | * | * | * | * | ?? | ?? | ?? | ?? | ?? | ?? |
| 150,000 | * | * | * | * | * | * | * | * | * | * | ?? | ?? | ?? | ?? | ?? | ?? |
| 200,000 | * | * | * | * | * | * | * | * | * | * | ?? | ?? | ?? | ?? | ?? | ?? |

\* > 25 miles (report distance as 25 miles)    # <0.1 mile (report distance as 0.1 mile)

## Reference Table 9
### Distances to Toxic Endpoint for Anhydrous Ammonia Liquefied Under Pressure
### F Stability, Wind Speed 1.5 Meters per Second

| Release Rate (lbs/min) | Distance to Endpoint (miles) | | Release Rate (lbs/min) | Distance to Endpoint (miles) | |
|---|---|---|---|---|---|
| | Rural | Urban | | Rural | Urban |
| 1 | 0.1 | <0.1* | 1,000 | 1.8 | 1.2 |
| 2 | 0.1 | 0.1 | 1,500 | 2.2 | 1.5 |
| 5 | 0.1 | 0.1 | 2,000 | 2.6 | 1.7 |
| 10 | 0.2 | 0.1 | 2,500 | 2.9 | 1.9 |
| 15 | 0.2 | 0.2 | 3,000 | 3.1 | 2.0 |
| 20 | 0.3 | 0.2 | 4,000 | 3.6 | 2.3 |
| 30 | 0.3 | 0.2 | 5,000 | 4.0 | 2.6 |
| 40 | 0.4 | 0.3 | 6,000 | 4.4 | 2.8 |
| 50 | 0.4 | 0.3 | 7,000 | 4.7 | 3.1 |
| 60 | 0.5 | 0.3 | 7,500 | 4.9 | 3.2 |
| 70 | 0.5 | 0.3 | 8,000 | 5.1 | 3.3 |
| 80 | 0.5 | 0.4 | 9,000 | 5.4 | 3.4 |
| 90 | 0.6 | 0.4 | 10,000 | 5.6 | 3.6 |
| 100 | 0.6 | 0.4 | 15,000 | 6.9 | 4.4 |
| 150 | 0.7 | 0.5 | 20,000 | 8.0 | 5.0 |
| 200 | 0.8 | 0.6 | 25,000 | 8.9 | 5.6 |
| 250 | 0.9 | 0.6 | 30,000 | 9.7 | 6.1 |
| 300 | 1.0 | 0.7 | 40,000 | 11 | 7.0 |
| 400 | 1.2 | 0.8 | 50,000 | 12 | 7.8 |
| 500 | 1.3 | 0.9 | 75,000 | 15 | 9.5 |
| 600 | 1.4 | 0.9 | 100,000 | 18 | 10 |
| 700 | 1.5 | 1.0 | 150,000 | 22 | 13 |
| 750 | 1.6 | 1.0 | 200,000 | ** | 15 |
| 800 | 1.6 | 1.1 | 250,000 | ** | 17 |
| 900 | 1.7 | 1.2 | 750,000 | ** | ** |

*Report distance as 0.1 mile

** More than 25 miles (report distance as 25 miles)

## Reference Table 10
### Distances to Toxic Endpoint for Aqueous Ammonia
### F Stability, Wind Speed 1.5 Meters per Second

| Release Rate (lbs/min) | Distance to Endpoint (miles) | | Release Rate (lbs/min) | Distance to Endpoint (miles) | |
|---|---|---|---|---|---|
| | Rural | Urban | | Rural | Urban |
| 1 | 0.1 | <0.1* | 1,000 | 1.6 | 0.6 |
| 2 | 0.1 | | 1,500 | 2.0 | 0.7 |
| 5 | 0.1 | | 2,000 | 2.2 | 0.8 |
| 10 | 0.2 | 0.1 | 2,500 | 2.5 | 0.9 |
| 15 | 0.2 | 0.1 | 3,000 | 2.7 | 1.0 |
| 20 | 0.3 | 0.1 | 4,000 | 3.1 | 1.1 |
| 30 | 0.3 | 0.1 | 5,000 | 3.4 | 1.2 |
| 40 | 0.4 | 0.1 | 6,000 | 3.7 | 1.3 |
| 50 | 0.4 | 0.1 | 7,000 | 4.0 | 1.4 |
| 60 | 0.4 | 0.2 | 7,500 | 4.1 | 1.5 |
| 70 | 0.5 | 0.2 | 8,000 | 4.2 | 1.5 |
| 80 | 0.5 | 0.2 | 9,000 | 4.5 | 1.6 |
| 90 | 0.5 | 0.2 | 10,000 | 4.7 | 1.7 |
| 100 | 0.6 | 0.2 | 15,000 | 5.6 | 2.0 |
| 150 | 0.7 | 0.2 | 20,000 | 6.5 | 2.4 |
| 200 | 0.8 | 0.3 | 25,000 | 7.2 | 2.6 |
| 250 | 0.8 | 0.3 | 30,000 | 7.8 | 2.8 |
| 300 | 0.9 | 0.3 | 40,000 | 8.9 | 3.3 |
| 400 | 1.1 | 0.4 | 50,000 | 9.8 | 3.6 |
| 500 | 1.2 | 0.4 | 75,000 | 12 | 4.4 |
| 600 | 1.3 | 0.4 | 100,000 | 14 | 5.0 |
| 700 | 1.4 | 0.5 | 150,000 | 16 | 6.1 |
| 750 | 1.4 | 0.5 | 200,000 | 19 | 7.0 |
| 800 | 1.5 | 0.5 | 250,000 | 21 | 7.8 |
| 900 | 1.5 | 0.6 | 750,000 | ** | 13 |

*Report distance as 0.1 mile

** More than 25 miles (report distance as 25 miles)

**Reference Table 11**
**Distances to Toxic Endpoint for Chlorine**
**F Stability, Wind Speed 1.5 Meters per Second**

| Release Rate (lbs/min) | Distance to Endpoint (miles) | |
|---|---|---|
| | Rural | Urban |
| 1 | 0.2 | 0.1 |
| 2 | 0.3 | 0.1 |
| 5 | 0.5 | 0.2 |
| 10 | 0.7 | 0.3 |
| 15 | 0.8 | 0.4 |
| 20 | 1.0 | 0.4 |
| 30 | 1.2 | 0.5 |
| 40 | 1.4 | 0.6 |
| 50 | 1.5 | 0.6 |
| 60 | 1.7 | 0.7 |
| 70 | 1.8 | 0.8 |
| 80 | 1.9 | 0.8 |
| 90 | 2.0 | 0.9 |
| 100 | 2.2 | 0.9 |
| 150 | 2.6 | 1.2 |
| 200 | 3.0 | 1.3 |
| 250 | 3.4 | 1.5 |
| 300 | 3.7 | 1.6 |
| 400 | 4.2 | 1.9 |
| 500 | 4.7 | 2.1 |
| 600 | 5.2 | 2.3 |
| 700 | 5.6 | 2.5 |

| Release Rate (lbs/min) | Distance to Endpoint (miles) | |
|---|---|---|
| | Rural | Urban |
| 750 | 5.8 | 2.6 |
| 800 | 5.9 | 2.7 |
| 900 | 6.3 | 2.9 |
| 1,000 | 6.6 | 3.0 |
| 1,500 | 8.1 | 3.8 |
| 2,000 | 9.3 | 4.4 |
| 2,500 | 10 | 4.9 |
| 3,000 | 11 | 5.4 |
| 4,000 | 13 | 6.2 |
| 5,000 | 14 | 7.0 |
| 6,000 | 16 | 7.6 |
| 7,000 | 17 | 8.3 |
| 7,500 | 18 | 8.6 |
| 8,000 | 18 | 8.9 |
| 9,000 | 19 | 9.4 |
| 10,000 | 20 | 9.9 |
| 15,000 | 25 | 12 |
| 20,000 | * | 14 |
| 25,000 | * | 16 |
| 30,000 | * | 18 |
| 40,000 | * | 20 |
| 50,000 | * | * |

* More than 25 miles (report distance as 25 miles)

## Reference Table 12
## Distances to Toxic Endpoint for Anhydrous Sulfur Dioxide
## F Stability, Wind Speed 1.5 Meters per Second

| Release Rate (lbs/min) | Distance to Endpoint (miles) | | Release Rate (lbs/min) | Distance to Endpoint (miles) | |
|---|---|---|---|---|---|
| | Rural | Urban | | Rural | Urban |
| 1 | 0.2 | 0.1 | 750 | 6.6 | 2.6 |
| 2 | 0.2 | 0.1 | 800 | 6.8 | 2.7 |
| 5 | 0.4 | 0.2 | 900 | 7.2 | 2.9 |
| 10 | 0.6 | 0.2 | 1,000 | 7.7 | 3.1 |
| 15 | 0.7 | 0.3 | 1,500 | 9.6 | 3.8 |
| 20 | 0.9 | 0.4 | 2,000 | 11 | 4.5 |
| 30 | 1.1 | 0.5 | 2,500 | 13 | 5.0 |
| 40 | 1.3 | 0.5 | 3,000 | 14 | 5.6 |
| 50 | 1.4 | 0.6 | 4,000 | 17 | 6.5 |
| 60 | 1.6 | 0.7 | 5,000 | 19 | 7.3 |
| 70 | 1.8 | 0.7 | 6,000 | 21 | 8.1 |
| 80 | 1.9 | 0.8 | 7,000 | 23 | 8.8 |
| 90 | 2.0 | 0.8 | 7,500 | 24 | 9.1 |
| 100 | 2.1 | 0.9 | 8,000 | 25 | 9.5 |
| 150 | 2.7 | 1.1 | 9,000 | * | 10 |
| 200 | 3.1 | 1.3 | 10,000 | * | 11 |
| 250 | 3.6 | 1.4 | 15,000 | * | 13 |
| 300 | 3.9 | 1.6 | 20,000 | * | 16 |
| 400 | 4.6 | 1.9 | 25,000 | * | 18 |
| 500 | 5.2 | 2.1 | 30,000 | * | 19 |
| 600 | 5.8 | 2.3 | 40,000 | * | 23 |
| 700 | 6.3 | 2.5 | 50,000 | * | * |

* More than 25 miles (report distance as 25 miles)

Reference Table 13

Distance to Overpressure of 1.0 psi for Vapor Cloud Explosions of 500 - 2,000,000 Pounds of Regulated Flammable Substances Based on TNT Equivalent Method, 10 Percent Yield Factor

| CAS No. | Quantity in Cloud (pounds) Chemical Name | 500 | 2,000 | 5,000 | 10,000 | 20,000 | 50,000 | 100,000 | 200,000 | 500,000 | 1,000,000 | 2,000,000 |
|---|---|---|---|---|---|---|---|---|---|---|---|---|
| | | | | | | | | Distance (Miles) to 1 psi Overpressure | | | | |
| 75-07-0 | Acetaldehyde | 0.05 | 0.08 | 0.1 | 0.1 | 0.2 | 0.2 | 0.3 | 0.4 | 0.5 | 0.7 | 0.8 |
| 74-86-2 | Acetylene | 0.07 | 0.1 | 0.1 | 0.2 | 0.2 | 0.3 | 0.4 | 0.5 | 0.7 | 0.8 | 1.0 |
| 598-73-2 | Bromotrifluoroethylene | 0.02 | 0.04 | 0.05 | 0.06 | 0.08 | 0.1 | 0.1 | 0.2 | 0.2 | 0.3 | 0.4 |
| 106-99-0 | 1,3-Butadiene | 0.06 | 0.1 | 0.1 | 0.2 | 0.2 | 0.3 | 0.4 | 0.5 | 0.6 | 0.8 | 1.0 |
| 106-97-8 | Butane | 0.06 | 0.1 | 0.1 | 0.2 | 0.2 | 0.3 | 0.4 | 0.5 | 0.6 | 0.8 | 1.0 |
| 25167-67-3 | Butene | 0.06 | 0.1 | 0.1 | 0.2 | 0.2 | 0.3 | 0.4 | 0.5 | 0.6 | 0.8 | 1.0 |
| 590-18-1 | 2-Butene-cis | 0.06 | 0.1 | 0.1 | 0.2 | 0.2 | 0.3 | 0.4 | 0.5 | 0.6 | 0.8 | 1.0 |
| 624-64-6 | 2-Butene-trans | 0.06 | 0.1 | 0.1 | 0.2 | 0.2 | 0.3 | 0.4 | 0.5 | 0.6 | 0.8 | 1.0 |
| 106-98-9 | 1-Butene | 0.06 | 0.1 | 0.1 | 0.2 | 0.2 | 0.3 | 0.4 | 0.5 | 0.6 | 0.8 | 1.0 |
| 107-01-7 | 2-Butene | 0.06 | 0.1 | 0.1 | 0.2 | 0.2 | 0.3 | 0.4 | 0.5 | 0.6 | 0.8 | 1.0 |
| 463-58-1 | Carbon oxysulfide | 0.04 | 0.06 | 0.08 | 0.1 | 0.1 | 0.2 | 0.2 | 0.3 | 0.4 | 0.5 | 0.6 |
| 7791-21-1 | Chlorine monoxide | 0.02 | 0.03 | 0.04 | 0.05 | 0.06 | 0.08 | 0.1 | 0.1 | 0.2 | 0.2 | 0.3 |
| 590-21-6 | 1-Chloropropylene | 0.05 | 0.08 | 0.1 | 0.1 | 0.2 | 0.2 | 0.3 | 0.4 | 0.5 | 0.6 | 0.8 |
| 557-98-2 | 2-Chloropropylene | 0.05 | 0.08 | 0.1 | 0.1 | 0.2 | 0.2 | 0.3 | 0.4 | 0.5 | 0.6 | 0.8 |
| 460-19-5 | Cyanogen | 0.05 | 0.08 | 0.1 | 0.1 | 0.2 | 0.2 | 0.3 | 0.4 | 0.5 | 0.6 | 0.8 |
| 75-19-4 | Cyclopropane | 0.06 | 0.1 | 0.1 | 0.2 | 0.2 | 0.3 | 0.4 | 0.5 | 0.6 | 0.8 | 1.0 |
| 4109-96-0 | Dichlorosilane | 0.04 | 0.06 | 0.08 | 0.1 | 0.1 | 0.2 | 0.2 | 0.3 | 0.4 | 0.5 | 0.6 |
| 75-37-6 | Difluoroethane | 0.04 | 0.06 | 0.09 | 0.1 | 0.1 | 0.2 | 0.2 | 0.3 | 0.4 | 0.5 | 0.6 |
| 124-40-3 | Dimethylamine | 0.06 | 0.09 | 0.1 | 0.2 | 0.2 | 0.3 | 0.3 | 0.4 | 0.6 | 0.7 | 0.9 |
| 463-82-1 | 2,2-Dimethylpropane | 0.06 | 0.1 | 0.1 | 0.2 | 0.2 | 0.3 | 0.4 | 0.5 | 0.6 | 0.8 | 1.0 |

| CAS No. | Chemical Name | Quantity in Cloud (pounds) Distance (Miles) to 1 psi Overpressure | | | | | | | | | | |
| | | 500 | 2,000 | 5,000 | 10,000 | 20,000 | 50,000 | 100,000 | 200,000 | 500,000 | 1,000,000 | 2,000,000 |
|---|---|---|---|---|---|---|---|---|---|---|---|---|
| 74-84-0 | Ethane | 0.06 | 0.1 | 0.1 | 0.2 | 0.2 | 0.3 | 0.4 | 0.5 | 0.6 | 0.8 | 1.0 |
| 107-00-6 | Ethyl acetylene | 0.06 | 0.1 | 0.1 | 0.2 | 0.2 | 0.3 | 0.4 | 0.5 | 0.6 | 0.8 | 1.0 |
| 75-04-7 | Ethylamine | 0.06 | 0.09 | 0.1 | 0.2 | 0.2 | 0.3 | 0.3 | 0.4 | 0.6 | 0.7 | 0.9 |
| 75-00-3 | Ethyl chloride | 0.05 | 0.08 | 0.1 | 0.1 | 0.2 | 0.2 | 0.3 | 0.4 | 0.5 | 0.6 | 0.8 |
| 74-85-1 | Ethylene | 0.06 | 0.1 | 0.1 | 0.2 | 0.2 | 0.3 | 0.4 | 0.5 | 0.7 | 0.8 | 1.0 |
| 60-29-7 | Ethyl ether | 0.06 | 0.09 | 0.1 | 0.2 | 0.2 | 0.3 | 0.3 | 0.4 | 0.6 | 0.7 | 0.9 |
| 75-08-1 | Ethyl mercaptan | 0.05 | 0.09 | 0.1 | 0.2 | 0.2 | 0.2 | 0.3 | 0.4 | 0.5 | 0.7 | 0.9 |
| 109-95-5 | Ethyl nitrite | 0.05 | 0.07 | 0.1 | 0.1 | 0.2 | 0.2 | 0.3 | 0.3 | 0.5 | 0.6 | 0.7 |
| 1333-74-0 | Hydrogen | 0.09 | 0.1 | 0.2 | 0.2 | 0.3 | 0.4 | 0.5 | 0.6 | 0.9 | 1.1 | 1.4 |
| 75-28-5 | Isobutane | 0.06 | 0.1 | 0.1 | 0.2 | 0.2 | 0.3 | 0.4 | 0.5 | 0.6 | 0.8 | 1.0 |
| 78-78-4 | Isopentane | 0.06 | 0.1 | 0.1 | 0.2 | 0.2 | 0.3 | 0.4 | 0.5 | 0.6 | 0.8 | 1.0 |
| 78-79-5 | Isoprene | 0.06 | 0.1 | 0.1 | 0.2 | 0.2 | 0.3 | 0.4 | 0.5 | 0.6 | 0.8 | 1.0 |
| 75-31-0 | Isopropylamine | 0.06 | 0.09 | 0.1 | 0.2 | 0.2 | 0.3 | 0.3 | 0.4 | 0.6 | 0.7 | 0.9 |
| 75-29-6 | Isopropyl chloride | 0.05 | 0.08 | 0.1 | 0.1 | 0.2 | 0.2 | 0.3 | 0.4 | 0.5 | 0.6 | 0.8 |
| 74-82-8 | Methane | 0.07 | 0.1 | 0.1 | 0.2 | 0.2 | 0.3 | 0.4 | 0.5 | 0.7 | 0.8 | 1.0 |
| 74-89-5 | Methylamine | 0.06 | 0.09 | 0.1 | 0.2 | 0.2 | 0.3 | 0.3 | 0.4 | 0.6 | 0.7 | 0.9 |
| 563-45-1 | 3-Methyl-1-butene | 0.06 | 0.1 | 0.1 | 0.2 | 0.2 | 0.3 | 0.4 | 0.5 | 0.6 | 0.8 | 1.0 |
| 563-46-2 | 2-Methyl-1-butene | 0.06 | 0.1 | 0.1 | 0.2 | 0.2 | 0.3 | 0.4 | 0.5 | 0.6 | 0.8 | 1.0 |
| 115-10-6 | Methyl ether | 0.05 | 0.09 | 0.1 | 0.1 | 0.2 | 0.3 | 0.3 | 0.4 | 0.5 | 0.7 | 0.9 |
| 107-31-3 | Methyl formate | 0.04 | 0.07 | 0.1 | 0.1 | 0.2 | 0.2 | 0.3 | 0.3 | 0.4 | 0.6 | 0.7 |
| 115-11-7 | 2-Methylpropene | 0.06 | 0.1 | 0.1 | 0.2 | 0.2 | 0.3 | 0.4 | 0.5 | 0.6 | 0.8 | 1.0 |
| 504-60-9 | 1,3-Pentadiene | 0.06 | 0.1 | 0.1 | 0.2 | 0.2 | 0.3 | 0.4 | 0.5 | 0.6 | 0.8 | 1.0 |
| 109-66-0 | Pentane | 0.06 | 0.1 | 0.1 | 0.2 | 0.2 | 0.3 | 0.4 | 0.5 | 0.6 | 0.8 | 1.0 |
| 109-67-1 | 1-Pentene | 0.06 | 0.1 | 0.1 | 0.2 | 0.2 | 0.3 | 0.4 | 0.5 | 0.6 | 0.8 | 1.0 |

| CAS No. | Chemical Name | Quantity in Cloud (pounds) | | | | | | | | | | |
|---|---|---|---|---|---|---|---|---|---|---|---|---|
| | | 500 | 2,000 | 5,000 | 10,000 | 20,000 | 50,000 | 100,000 | 200,000 | 500,000 | 1,000,000 | 2,000,000 |
| | | Distance (Miles) to 1 psi Overpressure | | | | | | | | | | |
| 646-04-8 | 2-Pentene, (E)- | 0.06 | 0.1 | 0.1 | 0.2 | 0.2 | 0.3 | 0.4 | 0.5 | 0.6 | 0.8 | 1.0 |
| 627-20-3 | 2-Pentene, (Z)- | 0.06 | 0.1 | 0.1 | 0.2 | 0.2 | 0.3 | 0.4 | 0.5 | 0.6 | 0.8 | 1.0 |
| 463-49-0 | Propadiene | 0.06 | 0.1 | 0.1 | 0.2 | 0.2 | 0.3 | 0.4 | 0.5 | 0.6 | 0.8 | 1.0 |
| 74-98-6 | Propane | 0.06 | 0.1 | 0.1 | 0.2 | 0.2 | 0.3 | 0.4 | 0.5 | 0.6 | 0.8 | 1.0 |
| 115-07-1 | Propylene | 0.06 | 0.1 | 0.1 | 0.2 | 0.2 | 0.3 | 0.4 | 0.5 | 0.6 | 0.8 | 1.0 |
| 74-99-7 | Propyne | 0.06 | 0.1 | 0.1 | 0.2 | 0.2 | 0.3 | 0.4 | 0.5 | 0.6 | 0.8 | 1.0 |
| 7803-62-5 | Silane | 0.06 | 0.1 | 0.1 | 0.2 | 0.2 | 0.3 | 0.4 | 0.5 | 0.6 | 0.8 | 1.0 |
| 116-14-3 | Tetrafluoroethylene | 0.02 | 0.03 | 0.04 | 0.05 | 0.07 | 0.09 | 0.1 | 0.1 | 0.2 | 0.2 | 0.3 |
| 75-76-3 | Tetramethylsilane | 0.06 | 0.1 | 0.1 | 0.2 | 0.2 | 0.3 | 0.4 | 0.5 | 0.6 | 0.8 | 1.0 |
| 10025-78-2 | Trichlorosilane | 0.03 | 0.04 | 0.06 | 0.08 | 0.1 | 0.1 | 0.2 | 0.2 | 0.3 | 0.4 | 0.4 |
| 79-38-9 | Trifluorochloroethylene | 0.02 | 0.03 | 0.05 | 0.06 | 0.07 | 0.1 | 0.1 | 0.2 | 0.2 | 0.3 | 0.3 |
| 75-50-3 | Trimethylamine | 0.06 | 0.1 | 0.1 | 0.2 | 0.2 | 0.3 | 0.4 | 0.4 | 0.6 | 0.8 | 1.0 |
| 689-97-4 | Vinyl acetylene | 0.06 | 0.1 | 0.1 | 0.2 | 0.2 | 0.3 | 0.4 | 0.5 | 0.6 | 0.8 | 1.0 |
| 75-01-4 | Vinyl chloride | 0.05 | 0.08 | 0.1 | 0.1 | 0.2 | 0.2 | 0.3 | 0.4 | 0.5 | 0.6 | 0.8 |
| 109-92-2 | Vinyl ethyl ether | 0.06 | 0.09 | 0.1 | 0.2 | 0.2 | 0.3 | 0.3 | 0.4 | 0.6 | 0.7 | 0.9 |
| 75-02-5 | Vinyl fluoride | 0.02 | 0.04 | 0.05 | 0.06 | 0.08 | 0.1 | 0.1 | 0.2 | 0.2 | 0.3 | 0.4 |
| 75-35-4 | Vinylidene chloride | 0.04 | 0.06 | 0.08 | 0.1 | 0.1 | 0.2 | 0.2 | 0.3 | 0.4 | 0.5 | 0.6 |
| 75-38-7 | Vinylidene fluoride | 0.04 | 0.06 | 0.09 | 0.1 | 0.1 | 0.2 | 0.2 | 0.3 | 0.4 | 0.5 | 0.6 |
| 107-25-5 | Vinyl methyl ether | 0.06 | 0.09 | 0.1 | 0.2 | 0.2 | 0.3 | 0.3 | 0.4 | 0.6 | 0.7 | 0.9 |

**Reference Table 14**
**Neutrally Buoyant Plume Distances to Toxic Endpoint for Release Rate Divided by Endpoint**
**10-Minute Release, Rural Conditions, D Stability, Wind Speed 3.0 Meters per Second**

| Release Rate/Endpoint [(lbs/min)/(mg/L)] (•) | Distance to Endpoint (miles) | Release Rate/Endpoint [(lbs/min)/(mg/L)] (•) | Distance to Endpoint (miles) |
|---|---|---|---|
| 0 - 64 | 0.1 | 130,000 - 140,000 | 4.8 |
| 64 - 510 | 0.2 | 140,000 - 160,000 | 5.0 |
| 510 - 1,300 | 0.3 | 160,000 - 180,000 | 5.2 |
| 1,300 - 2,300 | 0.4 | 180,000 - 190,000 | 5.4 |
| 2,300 - 4,100 | 0.6 | 190,000 - 210,000 | 5.6 |
| 4,100 - 6,300 | 0.8 | 210,000 - 220,000 | 5.8 |
| 6,300 - 8,800 | 1.0 | 220,000 - 240,000 | 6.0 |
| 8,800 - 12,000 | 1.2 | 240,000 - 261,000 | 6.2 |
| 12,000 - 16,000 | 1.4 | 261,000 - 325,000 | 6.8 |
| 16,000 - 19,000 | 1.6 | 325,000 - 397,000 | 7.5 |
| 19,000 - 22,000 | 1.8 | 397,000 - 477,000 | 8.1 |
| 22,000 - 26,000 | 2.0 | 477,000 - 566,000 | 8.7 |
| 26,000 - 30,000 | 2.2 | 566,000 - 663,000 | 9.3 |
| 30,000 - 36,000 | 2.4 | 663,000 - 769,000 | 9.9 |
| 36,000 - 42,000 | 2.6 | 769,000 - 1,010,000 | 11 |
| 42,000 - 47,000 | 2.8 | 1,010,000 - 1,280,000 | 12 |
| 47,000 - 54,000 | 3.0 | 1,280,000 - 1,600,000 | 14 |
| 54,000 - 60,000 | 3.2 | 1,600,000 - 1,950,000 | 15 |
| 60,000 - 70,000 | 3.4 | 1,950,000 - 2,340,000 | 16 |
| 70,000 - 78,000 | 3.6 | 2,340,000 - 2,770,000 | 17 |
| 78,000 - 87,000 | 3.8 | 2,770,000 - 3,240,000 | 19 |
| 87,000 - 97,000 | 4.0 | 3,240,000 - 4,590,000 | 22 |
| 97,000 - 110,000 | 4.2 | 4,590,000 - 6,190,000 | 25 |
| 110,000 - 120,000 | 4.4 | >6,190,000 | >25* |
| 120,000 - 130,000 | 4.6 | | |

* Report distance as 25 miles

**Reference Table 15**
**Neutrally Buoyant Plume Distances to Toxic Endpoint for Release Rate Divided by Endpoint**
**60-Minute Release, Rural Conditions, D Stability, Wind Speed 3.0 Meters per Second**

| Release Rate/Endpoint [(lbs/min)/(mg/L)] (•) | Distance to Endpoint (miles) | Release Rate/Endpoint [(lbs/min)/(mg/L)] (•) | Distance to Endpoint (miles) |
|---|---|---|---|
| 0 - 79 | 0.1 | 100,000 - 108,000 | 4.8 |
| 79 - 630 | 0.2 | 108,000 - 113,000 | 5.0 |
| 630 - 1,600 | 0.3 | 113,000 - 120,000 | 5.2 |
| 1,600 - 2,800 | 0.4 | 120,000 - 126,000 | 5.4 |
| 2,800 - 5,200 | 0.6 | 126,000 - 132,000 | 5.6 |
| 5,200 - 7,900 | 0.8 | 132,000 - 140,000 | 5.8 |
| 7,900 - 11,000 | 1.0 | 140,000 - 150,000 | 6.0 |
| 11,000 - 14,000 | 1.2 | 150,000 - 151,000 | 6.2 |
| 14,000 - 19,000 | 1.4 | 151,000 - 171,000 | 6.8 |
| 19,000 - 23,000 | 1.6 | 171,000 - 191,000 | 7.5 |
| 23,000 - 27,000 | 1.8 | 191,000 - 212,000 | 8.1 |
| 27,000 - 32,000 | 2.0 | 212,000 - 233,000 | 8.7 |
| 32,000 - 36,000 | 2.2 | 233,000 - 256,000 | 9.3 |
| 36,000 - 42,000 | 2.4 | 256,000 - 280,000 | 9.9 |
| 42,000 - 47,000 | 2.6 | 280,000 - 332,000 | 11 |
| 47,000 - 52,000 | 2.8 | 332,000 - 390,000 | 12 |
| 52,000 - 57,000 | 3.0 | 390,000 - 456,000 | 14 |
| 57,000 - 61,000 | 3.2 | 456,000 - 529,000 | 15 |
| 61,000 - 68,000 | 3.4 | 529,000 - 610,000 | 16 |
| 68,000 - 73,000 | 3.6 | 610,000 - 699,000 | 17 |
| 73,000 - 79,000 | 3.8 | 699,000 - 796,000 | 19 |
| 79,000 - 84,000 | 4.0 | 796,000 - 1,080,000 | 22 |
| 84,000 - 91,000 | 4.2 | 1,080,000 - 1,410,000 | 25 |
| 91,000 - 97,000 | 4.4 | >1,410,000 | >25* |
| 97,000 - 100,000 | 4.6 | | |

* Report distance as 25 miles

## Reference Table 16
### Neutrally Buoyant Plume Distances to Toxic Endpoint for Release Rate Divided by Endpoint
### 10-Minute Release, Urban Conditions, D Stability, Wind Speed 3.0 Meters per Second

| Release Rate/Endpoint [(lbs/min)/(mg/L)] (•) | Distance to Endpoint (miles) | Release Rate/Endpoint [(lbs/min)/(mg/L)] (•) | Distance to Endpoint (miles) |
|---|---|---|---|
| 0 - 160 | 0.1 | 600,000 - 660,000 | 4.8 |
| 160 - 1,400 | 0.2 | 660,000 - 720,000 | 5.0 |
| 1,400 - 3,600 | 0.3 | 720,000 - 810,000 | 5.2 |
| 3,600 - 6,900 | 0.4 | 810,000 - 880,000 | 5.4 |
| 6,900 - 13,000 | 0.6 | 880,000 - 950,000 | 5.6 |
| 13,000 - 22,000 | 0.8 | 950,000 - 1,000,000 | 5.8 |
| 22,000 - 31,000 | 1.0 | 1,000,000 - 1,100,000 | 6.0 |
| 31,000 - 42,000 | 1.2 | 1,100,000 - 1,220,000 | 6.2 |
| 42,000 - 59,000 | 1.4 | 1,220,000 - 1,530,000 | 6.8 |
| 59,000 - 73,000 | 1.6 | 1,530,000 - 1,880,000 | 7.5 |
| 73,000 - 88,000 | 1.8 | 1,880,000 - 2,280,000 | 8.1 |
| 88,000 - 100,000 | 2.0 | 2,280,000 - 2,710,000 | 8.7 |
| 100,000 - 120,000 | 2.2 | 2,710,000 - 3,200,000 | 9.3 |
| 120,000 - 150,000 | 2.4 | 3,200,000 - 3,730,000 | 9.9 |
| 150,000 - 170,000 | 2.6 | 3,730,000 - 4,920,000 | 11 |
| 170,000 - 200,000 | 2.8 | 4,920,000 - 6,310,000 | 12 |
| 200,000 - 230,000 | 3.0 | 6,310,000 - 7,890,000 | 14 |
| 230,000 - 260,000 | 3.2 | 7,890,000 - 9,660,000 | 15 |
| 260,000 - 310,000 | 3.4 | 9,660,000 - 11,600,000 | 16 |
| 310,000 - 340,000 | 3.6 | 11,600,000 - 13,800,000 | 17 |
| 340,000 - 390,000 | 3.8 | 13,800,000 - 16,200,000 | 19 |
| 390,000 - 430,000 | 4.0 | 16,200,000 - 23,100,000 | 22 |
| 430,000 - 490,000 | 4.2 | 23,100,000 - 31,300,000 | 25 |
| 490,000 - 540,000 | 4.4 | >31,300,000 | >25* |
| 540,000 - 600,000 | 4.6 | | |

* Report distance as 25 miles

## Reference Table 17
### Neutrally Buoyant Plume Distances to Toxic Endpoint for Release Rate Divided by Endpoint
### 60-Minute Release, Urban Conditions, D Stability, Wind Speed 3.0 Meters per Second

| Release Rate/Endpoint [(lbs/min)/(mg/L)] (•) | Distance to Endpoint (miles) | Release Rate/Endpoint [(lbs/min)/(mg/L)] (•) | Distance to Endpoint (miles) |
|---|---|---|---|
| 0 - 200 | 0.1 | 460,000 - 490,000 | 4.8 |
| 200 - 1,700 | 0.2 | 490,000 - 520,000 | 5.0 |
| 1,700 - 4,500 | 0.3 | 520,000 - 550,000 | 5.2 |
| 4,500 - 8,600 | 0.4 | 550,000 - 580,000 | 5.4 |
| 8,600 - 17,000 | 0.6 | 580,000 - 610,000 | 5.6 |
| 17,000 - 27,000 | 0.8 | 610,000 - 640,000 | 5.8 |
| 27,000 - 39,000 | 1.0 | 640,000 - 680,000 | 6.0 |
| 39,000 - 53,000 | 1.2 | 680,000 - 705,000 | 6.2 |
| 53,000 - 73,000 | 1.4 | 705,000 - 804,000 | 6.8 |
| 73,000 - 90,000 | 1.6 | 804,000 - 905,000 | 7.5 |
| 90,000 - 110,000 | 1.8 | 905,000 - 1,010,000 | 8.1 |
| 110,000 - 130,000 | 2.0 | 1,010,000 - 1,120,000 | 8.7 |
| 130,000 - 150,000 | 2.2 | 1,120,000 - 1,230,000 | 9.3 |
| 150,000 - 170,000 | 2.4 | 1,230,000 - 1,350,000 | 9.9 |
| 170,000 - 200,000 | 2.6 | 1,350,000 - 1,620,000 | 11 |
| 200,000 - 220,000 | 2.8 | 1,620,000 - 1,920,000 | 12 |
| 220,000 - 240,000 | 3.0 | 1,920,000 - 2,250,000 | 14 |
| 240,000 - 270,000 | 3.2 | 2,250,000 - 2,620,000 | 15 |
| 270,000 - 300,000 | 3.4 | 2,620,000 - 3,030,000 | 16 |
| 300,000 - 320,000 | 3.6 | 3,030,000 - 3,490,000 | 17 |
| 320,000 - 350,000 | 3.8 | 3,490,000 - 3,980,000 | 19 |
| 350,000 - 370,000 | 4.0 | 3,980,000 - 5,410,000 | 22 |
| 370,000 - 410,000 | 4.2 | 5,410,000 - 7,120,000 | 25 |
| 410,000 - 430,000 | 4.4 | >7,120,000 | >25* |
| 430,000 - 460,000 | 4.6 | | |

\* Report distance as 25 miles

## Reference Table 18

### Dense Gas Distances to Toxic Endpoint, 10-minute Release, Rural Conditions, D Stability, Wind Speed 3.0 Meters per Second

| Release Rate (lbs/min) | Toxic Endpoint (mg/L) — Distance (Miles) | | | | | | | | | | | | | | | |
|---|---|---|---|---|---|---|---|---|---|---|---|---|---|---|---|---|
| | 0.0004 | 0.0007 | 0.001 | 0.002 | 0.0035 | 0.005 | 0.0075 | 0.01 | 0.02 | 0.035 | 0.05 | 0.075 | 0.1 | 0.25 | 0.5 | 0.75 |
| 1 | 0.6 | 0.4 | 0.4 | 0.2 | 0.2 | 0.1 | 0.1 | 0.1 | <0.1 | <0.1 | ?? | # | # | # | # | # |
| 2 | 0.9 | 0.6 | 0.5 | 0.4 | 0.3 | 0.2 | 0.2 | 0.1 | 0.1 | 0.1 | <0.1 | <0.1 | # | # | # | # |
| 5 | ?? | ?? | 0.9 | 0.6 | 0.4 | 0.4 | 0.3 | 0.2 | 0.2 | 0.1 | 0.1 | 0.1 | <0.1 | # | # | # |
| 10 | ?? | ?? | ?? | 0.9 | 0.6 | 0.5 | 0.4 | 0.4 | 0.2 | 0.2 | 0.1 | 0.1 | 0.1 | <0.1 | <0.1 | # |
| 30 | ?? | ?? | ?? | ?? | ?? | 0.9 | 0.7 | 0.7 | 0.5 | 0.3 | 0.3 | 0.2 | 0.2 | 0.1 | 0.1 | <0.1 |
| 50 | ?? | ?? | ?? | ?? | ?? | ?? | 1.0 | 0.9 | 0.6 | 0.4 | 0.4 | 0.3 | 0.2 | 0.2 | 0.1 | 0.1 |
| 100 | ?? | ?? | ?? | ?? | ?? | ?? | ?? | ?? | 0.9 | 0.6 | 0.6 | 0.4 | 0.4 | 0.2 | 0.2 | 0.1 |
| 150 | ?? | ?? | ?? | ?? | ?? | ?? | ?? | ?? | ?? | 0.8 | 0.7 | 0.6 | 0.5 | 0.3 | 0.2 | 0.2 |
| 250 | ?? | ?? | ?? | ?? | ?? | ?? | ?? | ?? | ?? | ?? | 0.9 | 0.7 | 0.5 | 0.4 | 0.3 | 0.2 |
| 500 | ?? | ?? | ?? | ?? | ?? | ?? | ?? | ?? | ?? | ?? | ?? | ?? | 0.9 | 0.6 | 0.4 | 0.3 |
| 750 | ?? | ?? | ?? | ?? | ?? | ?? | ?? | ?? | ?? | ?? | ?? | ?? | ?? | 0.7 | 0.5 | 0.4 |
| 1,000 | ?? | ?? | ?? | ?? | ?? | ?? | ?? | ?? | ?? | ?? | ?? | ?? | ?? | 0.8 | 0.6 | 0.4 |
| 1,500 | ?? | ?? | ?? | ?? | ?? | ?? | ?? | ?? | ?? | ?? | ?? | ?? | ?? | 1.0 | 0.7 | 0.6 |
| 2,000 | ?? | ?? | ?? | ?? | ?? | ?? | ?? | ?? | ?? | ?? | ?? | ?? | ?? | ?? | 0.8 | 0.6 |
| 2,500 | ?? | ?? | ?? | ?? | ?? | ?? | ?? | ?? | ?? | ?? | ?? | ?? | ?? | ?? | 0.9 | 0.7 |
| 3,000 | * | * | * | * | ?? | ?? | ?? | ?? | ?? | ?? | ?? | ?? | ?? | ?? | 1.0 | 0.8 |
| 4,000 | * | * | * | * | * | ?? | ?? | ?? | ?? | ?? | ?? | ?? | ?? | ?? | ?? | 0.9 |
| 5,000 | * | * | * | * | * | ?? | ?? | ?? | ?? | ?? | ?? | ?? | ?? | ?? | ?? | ?? |
| 7,500 | * | * | * | * | * | ?? | ?? | ?? | ?? | ?? | ?? | ?? | ?? | ?? | ?? | ?? |
| 10,000 | * | * | * | * | * | * | ?? | ?? | ?? | ?? | ?? | ?? | ?? | ?? | ?? | ?? |
| 15,000 | * | * | * | * | * | * | ?? | ?? | ?? | ?? | ?? | ?? | ?? | ?? | ?? | ?? |
| 20,000 | * | * | * | * | * | * | ?? | ?? | ?? | ?? | ?? | ?? | ?? | ?? | ?? | ?? |
| 50,000 | * | * | * | * | * | * | ?? | ?? | ?? | ?? | ?? | ?? | ?? | ?? | ?? | ?? |
| 75,000 | * | * | * | * | * | * | ?? | ?? | ?? | ?? | ?? | ?? | ?? | ?? | ?? | ?? |
| 100,000 | * | * | * | * | * | * | ?? | ?? | ?? | ?? | ?? | ?? | ?? | ?? | ?? | ?? |
| 150,000 | * | * | * | * | * | * | ?? | ?? | ?? | ?? | ?? | ?? | ?? | ?? | ?? | ?? |
| 200,000 | | | * | * | * | * | ?? | ?? | ?? | ?? | ?? | ?? | ?? | ?? | ?? | ?? |

* > 25 miles (report distance as 25 miles)    # <0.1 mile (report distance as 0.1 mile)

## Reference Table 19

### Dense Gas Distances to Toxic Endpoint, 60-minute Release, Rural Conditions, D Stability, Wind Speed 3.0 Meters per Second

| Release Rate (lbs/min) | Toxic Endpoint (mg/L) — Distance (Miles) | | | | | | | | | | | | | | | |
|---|---|---|---|---|---|---|---|---|---|---|---|---|---|---|---|---|
| | 0.0004 | 0.0007 | 0.001 | 0.002 | 0.0035 | 0.005 | 0.0075 | 0.01 | 0.02 | 0.035 | 0.05 | 0.075 | 0.1 | 0.25 | 0.5 | 0.75 |
| 1 | 0.5 | 0.4 | 0.3 | 0.2 | 0.2 | 0.1 | 0.1 | 0.1 | <0.1 | # | # | # | # | # | # | # |
| 2 | 0.8 | 0.6 | 0.5 | 0.3 | 0.2 | 0.2 | 0.2 | 0.1 | 0.1 | <0.1 | <0.1 | ?? | # | # | # | # |
| 5 | ?? | 1.0 | 0.8 | 0.5 | 0.4 | 0.3 | 0.2 | 0.2 | 0.2 | 0.1 | 0.1 | 0.1 | <0.1 | # | # | # |
| 10 | ?? | ?? | ?? | 0.8 | 0.6 | 0.5 | 0.4 | 0.3 | 0.2 | 0.2 | 0.1 | 0.1 | 0.1 | <0.1 | <0.1 | # |
| 30 | ?? | ?? | ?? | ?? | ?? | 0.9 | 0.7 | 0.6 | 0.4 | 0.3 | 0.2 | 0.2 | 0.2 | 0.1 | 0.1 | <0.1 |
| 50 | ?? | ?? | ?? | ?? | ?? | ?? | 1.0 | 0.8 | 0.6 | 0.4 | 0.3 | 0.3 | 0.2 | 0.1 | 0.1 | 0.1 |
| 100 | ?? | ?? | ?? | ?? | ?? | ?? | ?? | ?? | 0.8 | 0.6 | 0.5 | 0.4 | 0.3 | 0.2 | 0.1 | 0.1 |
| 150 | ?? | ?? | ?? | ?? | ?? | ?? | ?? | ?? | ?? | 0.7 | 0.6 | 0.5 | 0.4 | 0.3 | 0.2 | 0.1 |
| 250 | ?? | ?? | ?? | ?? | ?? | ?? | ?? | ?? | ?? | ?? | 0.9 | 0.7 | 0.6 | 0.4 | 0.2 | 0.2 |
| 500 | ?? | ?? | ?? | ?? | ?? | ?? | ?? | ?? | ?? | ?? | ?? | 1.0 | 0.9 | 0.5 | 0.4 | 0.3 |
| 750 | ?? | ?? | ?? | ?? | ?? | ?? | ?? | ?? | ?? | ?? | ?? | ?? | ?? | 0.6 | 0.4 | 0.4 |
| 1,000 | ?? | ?? | ?? | ?? | ?? | ?? | ?? | ?? | ?? | ?? | ?? | ?? | ?? | 0.7 | 0.5 | 0.4 |
| 1,500 | * | * | ?? | ?? | ?? | ?? | ?? | ?? | ?? | ?? | ?? | ?? | ?? | 1.0 | 0.7 | 0.5 |
| 2,000 | * | * | * | * | ?? | ?? | ?? | ?? | ?? | ?? | ?? | ?? | ?? | ?? | 0.7 | 0.6 |
| 2,500 | * | * | * | * | * | ?? | ?? | ?? | ?? | ?? | ?? | ?? | ?? | ?? | 0.9 | 0.7 |
| 3,000 | * | * | * | * | * | * | ?? | ?? | ?? | ?? | ?? | ?? | ?? | ?? | 1.0 | 0.8 |
| 4,000 | * | * | * | * | * | * | ?? | ?? | ?? | ?? | ?? | ?? | ?? | ?? | ?? | 0.9 |
| 5,000 | * | * | * | * | * | * | * | ?? | ?? | ?? | ?? | ?? | ?? | ?? | ?? | ?? |
| 7,500 | * | * | * | * | * | * | * | ?? | ?? | ?? | ?? | ?? | ?? | ?? | ?? | ?? |
| 10,000 | * | * | * | * | * | * | * | ?? | ?? | ?? | ?? | ?? | ?? | ?? | ?? | ?? |
| 15,000 | * | * | * | * | * | * | * | ?? | ?? | ?? | ?? | ?? | ?? | ?? | ?? | ?? |
| 20,000 | * | * | * | * | * | * | * | ?? | ?? | ?? | ?? | ?? | ?? | ?? | ?? | ?? |
| 50,000 | * | * | * | * | * | * | * | ?? | ?? | ?? | ?? | ?? | ?? | ?? | ?? | ?? |
| 75,000 | * | * | * | * | * | * | * | ?? | ?? | ?? | ?? | ?? | ?? | ?? | ?? | ?? |
| 100,000 | * | * | * | * | * | * | * | ?? | ?? | ?? | ?? | ?? | ?? | ?? | ?? | ?? |
| 150,000 | * | * | * | * | * | * | * | ?? | ?? | ?? | ?? | ?? | ?? | ?? | ?? | ?? |
| 200,000 | * | * | * | * | * | * | * | ?? | ?? | ?? | ?? | ?? | ?? | ?? | ?? | ?? |

* > 25 miles (report distance as 25 miles)　　# <0.1 mile (report distance as 0.1 mile)

## Reference Table 20

### Dense Gas Distances to Toxic Endpoint, 10-minute Release, Urban Conditions, D Stability, Wind Speed 3.0 Meters per Second

| Release Rate (lbs/min) | Toxic Endpoint (mg/L) — Distance (Miles) | | | | | | | | | | | | | | | |
|---|---|---|---|---|---|---|---|---|---|---|---|---|---|---|---|---|
| | 0.0004 | 0.0007 | 0.001 | 0.002 | 0.0035 | 0.005 | 0.0075 | 0.01 | 0.02 | 0.035 | 0.05 | 0.075 | 0.1 | 0.25 | 0.5 | 0.75 |
| 1 | 0.5 | 0.3 | 0.2 | 0.2 | 0.1 | 0.1 | 0.1 | 0.1 | <0.1 | # | # | # | # | # | # | # |
| 2 | 0.7 | 0.5 | 0.4 | 0.3 | 0.2 | 0.2 | 0.1 | 0.1 | 0.1 | <0.1 | <0.1 | # | # | # | # | # |
| 5 | ?? | 0.8 | 0.6 | 0.5 | 0.3 | 0.3 | 0.2 | 0.2 | 0.1 | 0.1 | 0.1 | <0.1 | <0.1 | # | # | # |
| 10 | ?? | ?? | 1.0 | 0.7 | 0.5 | 0.4 | 0.3 | 0.3 | 0.2 | 0.1 | 0.1 | 0.1 | 0.1 | <0.1 | # | # |
| 30 | ?? | ?? | ?? | ?? | 0.9 | 0.8 | 0.6 | 0.6 | 0.4 | 0.3 | 0.2 | 0.2 | 0.2 | 0.1 | <0.1 | # |
| 50 | ?? | ?? | ?? | ?? | ?? | 1.0 | 0.8 | 0.7 | 0.5 | 0.3 | 0.3 | 0.2 | 0.2 | 0.1 | 0.1 | <0.1 |
| 100 | ?? | ?? | ?? | ?? | ?? | ?? | ?? | 1.0 | 0.7 | 0.6 | 0.4 | 0.4 | 0.3 | 0.2 | 0.1 | 0.1 |
| 150 | ?? | ?? | ?? | ?? | ?? | ?? | ?? | ?? | 0.9 | 0.7 | 0.6 | 0.4 | 0.4 | 0.2 | 0.2 | 0.1 |
| 250 | ?? | ?? | ?? | ?? | ?? | ?? | ?? | ?? | ?? | 0.9 | 0.7 | 0.6 | 0.5 | 0.3 | 0.2 | 0.1 |
| 500 | ?? | ?? | ?? | ?? | ?? | ?? | ?? | ?? | ?? | ?? | 1.0 | 0.8 | 0.7 | 0.4 | 0.3 | 0.2 |
| 750 | ?? | ?? | ?? | ?? | ?? | ?? | ?? | ?? | ?? | ?? | ?? | 1.0 | 0.9 | 0.5 | 0.4 | 0.3 |
| 1,000 | ?? | ?? | ?? | ?? | ?? | ?? | ?? | ?? | ?? | ?? | ?? | ?? | 1.0 | 0.6 | 0.4 | 0.3 |
| 1,500 | ?? | ?? | ?? | ?? | ?? | ?? | ?? | ?? | ?? | ?? | ?? | ?? | ?? | 0.7 | 0.5 | 0.4 |
| 2,000 | ?? | ?? | ?? | ?? | ?? | ?? | ?? | ?? | ?? | ?? | ?? | ?? | ?? | 0.9 | 0.6 | 0.5 |
| 2,500 | ?? | ?? | ?? | ?? | ?? | ?? | ?? | ?? | ?? | ?? | ?? | ?? | ?? | 1.0 | 0.7 | 0.6 |
| 3,000 | ?? | ?? | ?? | ?? | ?? | ?? | ?? | ?? | ?? | ?? | ?? | ?? | ?? | ?? | 0.7 | 0.6 |
| 4,000 | * | * | ?? | ?? | ?? | ?? | ?? | ?? | ?? | ?? | ?? | ?? | ?? | ?? | 0.9 | 0.7 |
| 5,000 | * | * | * | ?? | ?? | ?? | ?? | ?? | ?? | ?? | ?? | ?? | ?? | ?? | 0.9 | 0.7 |
| 7,500 | * | * | * | ?? | ?? | ?? | ?? | ?? | ?? | ?? | ?? | ?? | ?? | ?? | ?? | 0.9 |
| 10,000 | * | * | * | * | * | ?? | ?? | ?? | ?? | ?? | ?? | ?? | ?? | ?? | ?? | ?? |
| 15,000 | * | * | * | * | * | ?? | ?? | ?? | ?? | ?? | ?? | ?? | ?? | ?? | ?? | ?? |
| 20,000 | * | * | * | * | * | ?? | ?? | ?? | ?? | ?? | ?? | ?? | ?? | ?? | ?? | ?? |
| 50,000 | * | * | * | * | * | ?? | ?? | ?? | ?? | ?? | ?? | ?? | ?? | ?? | ?? | ?? |
| 75,000 | * | * | * | * | * | ?? | ?? | ?? | ?? | ?? | ?? | ?? | ?? | ?? | ?? | ?? |
| 100,000 | * | * | * | * | * | ?? | ?? | ?? | ?? | ?? | ?? | ?? | ?? | ?? | ?? | ?? |
| 150,000 | * | * | * | * | * | ?? | ?? | ?? | ?? | ?? | ?? | ?? | ?? | ?? | ?? | ?? |
| 200,000 | * | * | * | * | * | ?? | ?? | ?? | ?? | ?? | ?? | ?? | ?? | ?? | ?? | ?? |

\* > 25 miles (report distance as 25 miles)   # <0.1 mile (report distance as 0.1 mile)

## Reference Table 21

**Dense Gas Distances to Toxic Endpoint, 60-minute Release, Urban Conditions, D Stability, Wind Speed 3.0 Meters per Second**

Toxic Endpoint (mg/L) — Distance (Miles)

| Release Rate (lbs/min) | 0.0004 | 0.0007 | 0.001 | 0.002 | 0.0035 | 0.005 | 0.0075 | 0.01 | 0.02 | 0.035 | 0.05 | 0.075 | 0.1 | 0.25 | 0.5 | 0.75 |
|---|---|---|---|---|---|---|---|---|---|---|---|---|---|---|---|---|
| 1 | 0.4 | 0.3 | 0.2 | 0.2 | 0.1 | 0.1 | 0.1 | <0.1 | # | # | # | # | # | # | # | # |
| 2 | 0.7 | 0.5 | 0.4 | 0.2 | 0.2 | 0.2 | 0.1 | 0.1 | <0.1 | <0.1 | # | # | # | # | # | # |
| 5 | ?? | 0.8 | 0.7 | 0.4 | 0.3 | 0.2 | 0.2 | 0.2 | 0.1 | 0.1 | <0.1 | <0.1 | <0.1 | # | # | # |
| 10 | ?? | ?? | 1.0 | 0.7 | 0.5 | 0.4 | 0.3 | 0.3 | 0.2 | 0.1 | 0.1 | 0.1 | 0.1 | <0.1 | # | # |
| 30 | ?? | ?? | ?? | ?? | 0.9 | 0.7 | 0.6 | 0.5 | 0.3 | 0.2 | 0.2 | 0.2 | 0.1 | 0.1 | <0.1 | # |
| 50 | ?? | ?? | ?? | ?? | ?? | 1.0 | 0.8 | 0.7 | 0.4 | 0.3 | 0.3 | 0.2 | 0.2 | 0.1 | 0.1 | <0.1 |
| 100 | ?? | ?? | ?? | ?? | ?? | ?? | ?? | 1.0 | 0.7 | 0.5 | 0.4 | 0.3 | 0.3 | 0.2 | 0.1 | 0.1 |
| 150 | ?? | ?? | ?? | ?? | ?? | ?? | ?? | ?? | 0.9 | 0.6 | 0.5 | 0.4 | 0.3 | 0.2 | 0.1 | 0.1 |
| 250 | ?? | ?? | ?? | ?? | ?? | ?? | ?? | ?? | ?? | 0.8 | 0.7 | 0.5 | 0.4 | 0.3 | 0.2 | 0.1 |
| 500 | ?? | ?? | ?? | ?? | ?? | ?? | ?? | ?? | ?? | ?? | 1.0 | 0.8 | 0.7 | 0.4 | 0.2 | 0.2 |
| 750 | ?? | ?? | ?? | ?? | ?? | ?? | ?? | ?? | ?? | ?? | ?? | 1.0 | 0.9 | 0.5 | 0.3 | 0.3 |
| 1,000 | ?? | ?? | ?? | ?? | ?? | ?? | ?? | ?? | ?? | ?? | ?? | ?? | 1.0 | 0.6 | 0.4 | 0.3 |
| 1,500 | ?? | ?? | ?? | ?? | ?? | ?? | ?? | ?? | ?? | ?? | ?? | ?? | ?? | 0.7 | 0.5 | 0.4 |
| 2,000 | " | * | * | * | ?? | ?? | ?? | ?? | ?? | ?? | ?? | ?? | ?? | 0.9 | 0.6 | 0.4 |
| 2,500 | " | * | * | * | ?? | ?? | ?? | ?? | ?? | ?? | ?? | ?? | ?? | 1.0 | 0.6 | 0.5 |
| 3,000 | " | * | * | * | ?? | ?? | ?? | ?? | ?? | ?? | ?? | ?? | ?? | ?? | 0.7 | 0.6 |
| 4,000 | " | * | * | * | ?? | ?? | ?? | ?? | ?? | ?? | ?? | ?? | ?? | ?? | 0.9 | 0.7 |
| 5,000 | " | * | * | * | * | ?? | ?? | ?? | ?? | ?? | ?? | ?? | ?? | ?? | 1.0 | 0.7 |
| 7,500 | " | * | * | * | * | * | ?? | ?? | ?? | ?? | ?? | ?? | ?? | ?? | ?? | 0.9 |
| 10,000 | " | * | * | * | * | * | * | ?? | ?? | ?? | ?? | ?? | ?? | ?? | ?? | ?? |
| 15,000 | " | * | * | * | * | * | * | ?? | ?? | ?? | ?? | ?? | ?? | ?? | ?? | ?? |
| 20,000 | " | * | * | * | * | * | * | ?? | ?? | ?? | ?? | ?? | ?? | ?? | ?? | ?? |
| 50,000 | " | * | * | * | * | * | * | ?? | ?? | ?? | ?? | ?? | ?? | ?? | ?? | ?? |
| 75,000 | " | * | * | * | * | * | * | ?? | ?? | ?? | ?? | ?? | ?? | ?? | ?? | ?? |
| 100,000 | " | * | * | * | * | * | * | ?? | ?? | ?? | ?? | ?? | ?? | ?? | ?? | ?? |
| 150,000 | " | * | * | * | * | * | * | ?? | ?? | ?? | ?? | ?? | ?? | ?? | ?? | ?? |
| 200,000 | " | * | * | * | * | * | * | ?? | ?? | ?? | ?? | ?? | ?? | ?? | ?? | ?? |

\* > 25 miles (report distance as 25 miles)  
\# <0.1 mile (report distance as 0.1 mile)

## Reference Table 22
## Distances to Toxic Endpoint for Anhydrous Ammonia
## D Stability, Wind Speed 3.0 Meters per Second

| Release Rate (lbs/min) | Distance to Endpoint (miles) | | Release Rate (lbs/min) | Distance to Endpoint (miles) | |
|---|---|---|---|---|---|
| | Rural | Urban | | Rural | Urban |
| <10 | <0.1* | | 900 | 0.6 | 0.2 |
| 10 | 0.1 | | 1,000 | 0.6 | 0.2 |
| 15 | 0.1 | | 1,500 | 0.7 | 0.3 |
| 20 | 0.1 | <0.1* | 2,000 | 0.8 | 0.3 |
| 30 | 0.1 | | 2,500 | 0.9 | 0.3 |
| 40 | 0.1 | | 3,000 | 1.0 | 0.4 |
| 50 | 0.1 | | 4,000 | 1.2 | 0.4 |
| 60 | 0.2 | 0.1 | 5,000 | 1.3 | 0.5 |
| 70 | 0.2 | 0.1 | 7,500 | 1.6 | 0.5 |
| 80 | 0.2 | 0.1 | 10,000 | 1.8 | 0.6 |
| 90 | 0.2 | 0.1 | 15,000 | 2.2 | 0.7 |
| 100 | 0.2 | 0.1 | 20,000 | 2.5 | 0.8 |
| 150 | 0.2 | 0.1 | 25,000 | 2.8 | 0.9 |
| 200 | 0.3 | 0.1 | 30,000 | 3.1 | 1.0 |
| 250 | 0.3 | 0.1 | 40,000 | 3.5 | 1.1 |
| 300 | 0.3 | 0.1 | 50,000 | 3.9 | 1.2 |
| 400 | 0.4 | 0.2 | 75,000 | 4.8 | 1.4 |
| 500 | 0.4 | 0.2 | 100,000 | 5.4 | 1.6 |
| 600 | 0.5 | 0.2 | 150,000 | 6.6 | 1.9 |
| 700 | 0.5 | 0.2 | 200,000 | 7.6 | 2.1 |
| 750 | 0.5 | 0.2 | 250,000 | 8.4 | 2.3 |
| 800 | 0.5 | 0.2 | | | |

* Report distance as 0.1 mile

**Reference Table 23**
**Distances to Toxic Endpoint for Aqueous Ammonia**
**D Stability, Wind Speed 3.0 Meters per Second**

| Release Rate (lbs/min) | Distance to Endpoint (miles) | | Release Rate (lbs/min) | Distance to Endpoint (miles) | |
|---|---|---|---|---|---|
| | Rural | Urban | | Rural | Urban |
| 8 | 0.1 | | 800 | 0.7 | 0.2 |
| 10 | 0.1 | | 900 | 0.7 | 0.3 |
| 15 | 0.1 | <0.1* | 1,000 | 0.8 | 0.3 |
| 20 | 0.1 | | 1,500 | 1.0 | 0.4 |
| 30 | 0.1 | | 2,000 | 1.2 | 0.4 |
| 40 | 0.1 | | 2,500 | 1.2 | 0.4 |
| 50 | 0.2 | 0.1 | 3,000 | 1.5 | 0.5 |
| 60 | 0.2 | 0.1 | 4,000 | 1.8 | 0.6 |
| 70 | 0.2 | 0.1 | 5,000 | 2.0 | 0.7 |
| 80 | 0.2 | 0.1 | 7,500 | 2.2 | 0.7 |
| 90 | 0.2 | 0.1 | 10,000 | 2.5 | 0.8 |
| 100 | 0.2 | 0.1 | 15,000 | 3.1 | 1.0 |
| 150 | 0.3 | 0.1 | 20,000 | 3.6 | 1.2 |
| 200 | 0.3 | 0.1 | 25,000 | 4.1 | 1.3 |
| 250 | 0.4 | 0.2 | 30,000 | 4.4 | 1.4 |
| 300 | 0.4 | 0.2 | 40,000 | 5.1 | 1.6 |
| 400 | 0.4 | 0.2 | 50,000 | 5.8 | 1.8 |
| 500 | 0.5 | 0.2 | 75,000 | 7.1 | 2.2 |
| 600 | 0.6 | 0.2 | 100,000 | 8.2 | 2.5 |
| 700 | 0.6 | 0.2 | 150,000 | 10 | 3.1 |
| 750 | 0.6 | 0.2 | 200,000 | 12 | 3.5 |

* Report distance as 0.1 mile

## Reference Table 24
### Distances to Toxic Endpoint for Chlorine
### D Stability, Wind Speed 3.0 Meters per Second

| Release Rate (lbs/min) | Distance to Endpoint (miles) | | Release Rate (lbs/min) | Distance to Endpoint (miles) | |
|---|---|---|---|---|---|
| | Rural | Urban | | Rural | Urban |
| 1 | <0.1* | <0.1* | 750 | 1.2 | 0.4 |
| 2 | 0.1 | | 800 | 1.2 | 0.5 |
| 5 | 0.1 | | 900 | 1.2 | 0.5 |
| 10 | 0.2 | 0.1 | 1,000 | 1.3 | 0.5 |
| 15 | 0.2 | 0.1 | 1,500 | 1.6 | 0.6 |
| 20 | 0.2 | 0.1 | 2,000 | 1.8 | 0.6 |
| 30 | 0.3 | 0.1 | 2,500 | 2.0 | 0.7 |
| 40 | 0.3 | 0.1 | 3,000 | 2.2 | 0.8 |
| 50 | 0.3 | 0.1 | 4,000 | 2.5 | 0.8 |
| 60 | 0.4 | 0.2 | 5,000 | 2.8 | 0.9 |
| 70 | 0.4 | 0.2 | 7,500 | 3.4 | 1.2 |
| 80 | 0.4 | 0.2 | 10,000 | 3.9 | 1.3 |
| 90 | 0.4 | 0.2 | 15,000 | 4.6 | 1.6 |
| 100 | 0.5 | 0.2 | 20,000 | 5.3 | 1.8 |
| 150 | 0.6 | 0.2 | 25,000 | 5.9 | 2.0 |
| 200 | 0.6 | 0.3 | 30,000 | 6.4 | 2.1 |
| 250 | 0.7 | 0.3 | 40,000 | 7.3 | 2.4 |
| 300 | 0.8 | 0.3 | 50,000 | 8.1 | 2.7 |
| 400 | 0.8 | 0.4 | 75,000 | 9.8 | 3.2 |
| 500 | 1.0 | 0.4 | 100,000 | 11 | 3.6 |
| 600 | 1.0 | 0.4 | 150,000 | 13 | 4.2 |
| 700 | 1.1 | 0.4 | 200,000 | 15 | 4.8 |

* Report distance as 0.1 mile

**Reference Table 25**
**Distances to Toxic Endpoint for Sulfur Dioxide**
**D Stability, Wind Speed 3.0 Meters per Second**

| Release Rate (lbs/min) | Distance to Endpoint (miles) | | Release Rate (lbs/min) | Distance to Endpoint (miles) | |
|---|---|---|---|---|---|
| | Rural | Urban | | Rural | Urban |
| 1 | <0.1* | <0.1* | 750 | 1.3 | 0.5 |
| 2 | 0.1 | | 800 | 1.3 | 0.5 |
| 5 | 0.1 | | 900 | 1.4 | 0.5 |
| 10 | 0.2 | 0.1 | 1,000 | 1.5 | 0.5 |
| 15 | 0.2 | 0.1 | 1,500 | 1.9 | 0.6 |
| 20 | 0.2 | 0.1 | 2,000 | 2.2 | 0.7 |
| 30 | 0.2 | 0.1 | 2,500 | 2.3 | 0.8 |
| 40 | 0.3 | 0.1 | 3,000 | 2.7 | 0.8 |
| 50 | 0.3 | 0.1 | 4,000 | 3.1 | 1.0 |
| 60 | 0.4 | 0.2 | 5,000 | 3.3 | 1.1 |
| 70 | 0.4 | 0.2 | 7,500 | 4.0 | 1.3 |
| 80 | 0.4 | 0.2 | 10,000 | 4.6 | 1.4 |
| 90 | 0.4 | 0.2 | 15,000 | 5.6 | 1.7 |
| 100 | 0.5 | 0.2 | 20,000 | 6.5 | 1.9 |
| 150 | 0.6 | 0.2 | 25,000 | 7.3 | 2.1 |
| 200 | 0.6 | 0.2 | 30,000 | 8.0 | 2.3 |
| 250 | 0.7 | 0.3 | 40,000 | 9.2 | 2.6 |
| 300 | 0.8 | 0.3 | 50,000 | 10 | 2.9 |
| 400 | 0.9 | 0.4 | 75,000 | 13 | 3.5 |
| 500 | 1.0 | 0.4 | 100,000 | 14 | 4.0 |
| 600 | 1.1 | 0.4 | 150,000 | 18 | 4.7 |
| 700 | 1.2 | 0.4 | 200,000 | 20 | 5.4 |

* Report distance as 0.1 mile

## Reference Table 26
### Neutrally Buoyant Plume Distances to Lower Flammability Limit (LFL)
### For Release Rate Divided by LFL
### Rural Conditions, D Stability, Wind Speed 3.0 Meters per Second

| Release Rate/Endpoint [(lbs/min)/(mg/L)] | Distance to Endpoint (miles) | Release Rate/Endpoint [(lbs/min)/(mg/L)] | Distance to Endpoint (miles) |
|---|---|---|---|
| 0 - 28 | 0.1 | 2,700 - 3,300 | 0.9 |
| 28 - 40 | 0.1 | 3,300 - 3,900 | 1.0 |
| 40 - 60 | 0.1 | 3,900 - 4,500 | 1.1 |
| 60 - 220 | 0.2 | 4,500 - 5,200 | 1.2 |
| 220 - 530 | 0.3 | 5,200 - 5,800 | 1.3 |
| 530 - 860 | 0.4 | 5,800 - 6,800 | 1.4 |
| 860 - 1,300 | 0.5 | 6,800 - 8,200 | 1.6 |
| 1,300 - 1,700 | 0.6 | 8,200 - 9,700 | 1.8 |
| 1,700 - 2,200 | 0.7 | 9,700 - 11,000 | 2.0 |
| 2,200 - 2,700 | 0.8 | 11,000 - 13,000 | 2.2 |

## Reference Table 27
### Neutrally Buoyant Plume Distances to Lower Flammability Limit (LFL)
### For Release Rate Divided by LFL
### Urban Conditions, D Stability, Wind Speed 3.0 Meters per Second

| Release Rate/Endpoint [(lbs/min)/(mg/L)] | Distance to Endpoint (miles) | Release Rate/Endpoint [(lbs/min)/(mg/L)] | Distance to Endpoint (miles) |
|---|---|---|---|
| 0 - 68 | 0.1 | 5,500 - 7,300 | 0.7 |
| 68 - 100 | 0.1 | 7,300 - 9,200 | 0.8 |
| 100 - 150 | 0.1 | 9,200 - 11,000 | 0.9 |
| 150 - 710 | 0.2 | 11,000 - 14,000 | 1.0 |
| 710 - 1,500 | 0.3 | 14,000 - 18,000 | 1.2 |
| 1,500 - 2,600 | 0.4 | 18,000 - 26,000 | 1.4 |
| 2,600 - 4,000 | 0.5 | 26,000 - 31,000 | 1.6 |
| 4,000 - 5,500 | 0.6 | 31,000 - 38,000 | 1.8 |

## Reference Table 28
## Dense Gas Distances to Lower Flammability Limit
## Rural Conditions, D Stability, Wind Speed 3.0 Meters per Second

| Release Rate (lbs/min) | Lower Flammability Limit (mg/L) | | | | | | | | | | | | | |
|---|---|---|---|---|---|---|---|---|---|---|---|---|---|---|
| | 27 | 30 | 35 | 40 | 45 | 50 | 60 | 70 | 100 | >100 |
| | Distance (Miles) | | | | | | | | | |
| <1,500 | ?? | # | # | # | # | # | # | # | # | # |
| 1,500 | <0.1 | <0.1 | # | # | # | # | # | # | # | # |
| 2,000 | 0.1 | 0.1 | <0.1 | # | # | # | # | # | # | # |
| 2,500 | 0.1 | 0.1 | 0.1 | <0.1 | # | # | # | # | # | # |
| 3,000 | 0.1 | 0.1 | 0.1 | 0.1 | <0.1 | <0.1 | # | # | # | # |
| 4,000 | 0.1 | 0.1 | 0.1 | 0.1 | 0.1 | 0.1 | <0.1 | # | # | # |
| 5,000 | 0.1 | 0.1 | 0.1 | 0.1 | 0.1 | 0.1 | 0.1 | <0.1 | # | # |
| 7,500 | 0.2 | 0.1 | 0.1 | 0.1 | 0.1 | 0.1 | 0.1 | 0.1 | <0.1 | # |
| 10,000 | 0.2 | 0.2 | 0.1 | 0.1 | 0.1 | 0.1 | 0.1 | 0.1 | 0.1 | <0.1 |

# < 0.1 mile (report distance a 0.1 mile)

**Reference Table 29**
**Dense Gas Distances to Lower Flammability Limit**
**Urban Conditions, D Stability, Wind Speed 3.0 Meters per Second**

| Release Rate (lbs/min) | Lower Flammability Limit (mg/L) | | | | |
|---|---|---|---|---|---|
| | 27 | 30 | 35 | 40 | >40 |
| | Distance (Miles) | | | | |
| <5,000 | ?? | # | # | # | # |
| 5,000 | <0.1 | <0.1 | # | # | # |
| 7,500 | 0.1 | 0.1 | <0.1 | # | # |
| 10,000 | 0.1 | 0.1 | 0.1 | <0.1 | # |

#  < 0.1 mile (report distance as 0.1 mile)

**Exhibit B-1**
**Data for Toxic Gases**

| CAS Number | Chemical Name | Molecular Weight | Ratio of Specific Heats | Toxic Endpoint[a] | | | Liquid Factor Boiling (LFB) | Density Factor (DF) (Boiling) | Gas Factor (GF)[k] | Vapor Pressure @25·C (psi) | Reference Table[b] |
|---|---|---|---|---|---|---|---|---|---|---|---|
| | | | | mg/L | ppm | Basis | | | | | |
| 7664-41-7 | Ammonia (anhydrous)[c] | 17.03 | 1.31 | 0.14 | 200 | ERPG-2 | 0.073 | 0.71 | 14 | 145 | Buoyant[d] |
| 7784-42-1 | Arsine | 77.95 | 1.28 | 0.0019 | 0.6 | EHS-LOC (IDLH) | 0.23 | 0.30 | 30 | 239 | Dense |
| 10294-34-5 | Boron trichloride | 117.17 | 1.15 | 0.010 | 2 | EHS-LOC (Tox[e]) | 0.22 | 0.36 | 36 | 22.7 | Dense |
| 7637-07-2 | Boron trifluoride | 67.81 | 1.20 | 0.028 | 10 | EHS-LOC (IDLH) | 0.25 | 0.31 | 28 | f | Dense |
| 7782-50-5 | Chlorine | 70.91 | 1.32 | 0.0087 | 3 | ERPG-2 | 0.9 | 0.31 | 29 | 113 | Dense |
| 10049-04-4 | Chlorine dioxide | 67.45 | 1.25 | 0.0028 | 1 | EHS-LOC equivalent (IDLH)[g] | 0.5 | 0.30 | 28 | 24.3 | Dense |
| 506-77-4 | Cyanogen chloride | 61.47 | 1.22 | 0.030 | 12 | EHS-LOC equivalent (Tox)[h] | 0.4 | 0.41 | 26 | 23.7 | Dense |
| 19287-45-7 | Diborane | 27.67 | 1.17 | 0.0011 | 1 | ERPG-2 | 0.3 | 1.13 | 17 | f | Buoyant[d] |
| 75-21-8 | Ethylene oxide | 44.05 | 1.21 | 0.090 | 50 | ERPG-2 | 0.2 | 0.55 | 22 | 25.4 | Dense |
| 7782-41-4 | Fluorine | 38.00 | 1.36 | 0.0039 | 2.5 | EHS-LOC (IDLH) | 0.35 | 0.32 | 22 | f | Dense |
| 50-00-0 | Formaldehyde (anhydrous)[c] | 30.03 | 1.31 | 0.012 | 10 | ERPG-2 | 0.0 | 0.59 | 19 | 75.2 | Dense |
| 74-90-8 | Hydrocyanic acid | 27.03 | 1.30 | 0.011 | 10 | ERPG-2 | 0.079 | 0.72 | 18 | 14.8 | Buoyant[d] |
| 7647-01-0 | Hydrogen chloride (anhydrous)[c] | 36.46 | 1.40 | 0.030 | 20 | ERPG-2 | 0.5 | 0.41 | 21 | 684 | Dense |
| 7664-39-3 | Hydrogen fluoride (anhydrous)[c] | 20.01 | 1.40 | 0.016 | 20 | ERPG-2 | 0.066 | 0.51 | 16 | 17.7 | Buoyant[i] |
| 7783-07-5 | Hydrogen selenide | 80.98 | 1.32 | 0.00066 | 0.2 | EHS-LOC (IDLH) | 0.21 | 0.25 | 31 | 151 | Dense |
| 7783-06-4 | Hydrogen sulfide | 34.08 | 1.32 | 0.042 | 30 | ERPG-2 | 0.13 | 0.51 | 20 | 302 | Dense |
| 74-87-3 | Methyl chloride | 50.49 | 1.26 | 0.82 | 400 | ERPG-2 | 0.14 | 0.48 | 24 | 83.2 | Dense |
| 74-93-1 | Methyl mercaptan | 48.11 | 1.20 | 0.049 | 25 | ERPG-2 | 0.12 | 0.55 | 23 | 29.2 | Dense |
| 10102-43-9 | Nitric oxide | 30.01 | 1.38 | 0.031 | 25 | EHS-LOC (TLV[j]) | 0.21 | 0.38 | 19 | f | Dense |

| CAS Number | Chemical Name | Molecular Weight | Ratio of Specific Heats | Toxic Endpoint[a] | | | Liquid Factor Boiling (LFB) | Density Factor (DF) (Boiling) | Gas Factor (GF)[k] | Vapor Pressure @25•C (psi) | Reference Table[b] |
|---|---|---|---|---|---|---|---|---|---|---|---|
| | | | | mg/L | ppm | Basis | | | | | |
| 75-44-5 | Phosgene | 98.92 | 1.17 | 0.00081 | 0.2 | ERPG-2 | 0.20 | 0.35 | 33 | 27.4 | Dense |
| 7803-51-2 | Phosphine | 34.00 | 1.29 | 0.0035 | 2.5 | ERPG-2 | 0.15 | 0.66 | 20 | 567 | Dense |
| 7446-09-5 | Sulfur dioxide (anhydrous) | 64.07 | 1.26 | 0.0078 | 3 | ERPG-2 | 0.16 | 0.33 | 27 | 58.0 | Dense |
| 7783-60-0 | Sulfur tetrafluoride | 108.06 | 1.30 | 0.0092 | 2 | EHS-LOC (Tox[c]) | 0.25 | 0.25 (at -73 °C) | 36 | 293 | Dense |

**Notes:**

[a] Toxic endpoints are specified in Appendix A to 40 CFR Part 68 in units of mg/L. To convert from units of mg/L to mg/m$^3$, multiply by 1,000. To convert mg/L to ppm, use the following equation:

$$Endpoint_{ppm} = (Endpoint_{mg/l} \times 1,000 \times 24.5)/Molecular\ Weight$$

[b] "Buoyant" in the Reference Table column refers to the tables for neutrally buoyant gases and vapors; "Dense" refers to the tables for dense gases and vapors. See OCAG, Appendix D, Section D.4.4, for more information on the choice of reference tables.

[c] See Exhibit B-3 of OCAG, Appendix B, for data on water solutions.

[d] Gases that are lighter than air may behave as dense gases upon release if liquefied under pressure or cold; consider the conditions of release when choosing the appropriate table.

[e] LOC is based on the IDLH-equivalent level estimated from toxicity data.

[f] Cannot be liquefied at 25•C.

[g] Not an EHS; LOC-equivalent value was estimated from one-tenth of the IDLH.

[h] Not an EHS; LOC-equivalent value was estimated from one-tenth of the IDLH-equivalent level estimated from toxicity data.

[i] Hydrogen fluoride is lighter than air, but may behave as a dense gas upon release under some circumstances (e.g., release under pressure, high concentration in the released cloud) because of hydrogen bonding; consider the conditions of release when choosing the appropriate table.

[j] LOC based on Threshold Limit Value (TLV) - Time-weighted average (TWA) developed by the American Conference of Governmental Industrial Hygienists (ACGIH).

[k] Use GF for gas leaks under choked (maximum) flow conditions.

**Exhibit B-2**
**Data for Toxic Liquids**

| CAS Number | Chemical Name | Molecular Weight | Vapor Pressure at 25-C (mm Hg) | Toxic Endpoint[a] | | | Liquid Factors | | Density Factor (DF) | Liquid Leak Factor (LLF)[1] | Reference Table[b] | |
|---|---|---|---|---|---|---|---|---|---|---|---|---|
| | | | | mg/L | ppm | Basis | Ambient (LFA) | Boiling (LFB) | | | Worst Case | Alternative Case |
| 107-02-8 | Acrolein | 56.06 | 274 | 0.0011 | 0.5 | ERPG-2 | 0.047 | 0.12 | 0.58 | 40 | Dense | Dense |
| 107-13-1 | Acrylonitrile | 53.06 | 108 | 0.076 | 35 | ERPG-2 | 0.018 | 0.11 | 0.61 | 39 | Dense | Dense |
| 814-68-6 | Acryl chloride | 90.51 | 110 | 0.00090 | 0.2 | EHS-LOC (Tox[c]) | 0.026 | 0.15 | 0.44 | 54 | Dense | Dense |
| 107-18-6 | Allyl alcohol | 58.08 | 26.1 | 0.036 | 15 | EHS-LOC (IDLH) | 0.0046 | 0.11 | 0.58 | 41 | Dense | Buoyant[d] |
| 107-11-9 | Allylamine | 57.10 | 242 | 0.0032 | 1 | EHS-LOC (Tox[c]) | 0.042 | 0.12 | 0.64 | 36 | Dense | Dense |
| 7784-34-1 | Arsenous trichloride | 181.28 | 10 | 0.01 | 1 | EHS-LOC (Tox[c]) | 0.0037 | 0.21 | 0.23 | 100 | Dense | Buoyant[d] |
| 353-42-4 | Boron trifluoride compound with methyl ether (1 1) | 113.89 | 11 | 0.023 | 5 | EHS-LOC (Tox[c]) | 0.0030 | 0.16 | 0.49 | 48 | Dense | Buoyant[d] |
| 7726-95-6 | Bromine | 159.81 | 212 | 0.0065 | 1 | ERPG-2 | 0.073 | 0.23 | 0.16 | 150 | Dense | Dense |
| 75-15-0 | Carbon disulfide | 76.14 | 359 | 0.16 | 50 | ERPG-2 | 0.075 | 0.15 | 0.39 | 60 | Dense | Dense |
| 67-66-3 | Chloroform | 119.38 | 196 | 0.49 | 100 | EHS-LOC (IDLH) | 0.055 | 0.19 | 0.33 | 71 | Dense | Dense |
| 542-88-1 | Chloromethyl ether | 114.96 | 29.4 | 0.00025 | 0.05 | EHS-LOC (Tox[c]) | 0.0080 | 0.17 | 0.37 | 63 | Dense | Dense |
| 107-30-2 | Chloromethyl methyl ether | 80.51 | 199 | 0.0018 | 0.6 | EHS-LOC (Tox[c]) | 0.043 | 0.15 | 0.46 | 51 | Dense | Buoyant[d] |
| 4170-30-3 | Crotonaldehyde | 70.09 | 33.1 | 0.029 | 10 | ERPG-2 | 0.0066 | 0.12 | 0.58 | 41 | Dense | Buoyant[d] |
| 123-73-9 | Crotonaldehyde, (E) | 70.09 | 33.1 | 0.029 | 10 | ERPG-2 | 0.0066 | 0.12 | 0.58 | 41 | Dense | Buoyant[d] |
| 108-91-8 | Cyclohexylamine | 99.18 | 10.1 | 0.16 | 39 | EHS-LOC (Tox[c]) | 0.0025 | 0.14 | 0.56 | 41 | Dense | Buoyant[d] |
| 75-78-5 | Dimethyldichlorosilane | 129.06 | 141 | 0.026 | 5 | ERPG-2 | 0.042 | 0.20 | 0.46 | 51 | Dense | Dense |
| 57-14-7 | 1,1-Dimethylhydrazine | 60.10 | 157 | 0.012 | 5 | EHS-LOC (IDLH) | 0.028 | 0.12 | 0.62 | 38 | Dense | Dense |
| 106-89-8 | Epichlorohydrin | 92.53 | 17.0 | 0.076 | 20 | ERPG-2 | 0.0040 | 0.14 | 0.42 | 57 | Dense | Buoyant[d] |
| 107-15-3 | Ethylenediamine | 60.10 | 12.2 | 0.49 | 200 | EHS-LOC (IDLH) | 0.0022 | 0.10 | 0.54 | 43 | Dense | Buoyant[d] |
| 151-56-4 | Ethyleneimine | 43.07 | 211 | 0.018 | 10 | EHS-LOC (IDLH) | 0.030 | 0.10 | 0.58 | 40 | Dense | Dense |
| 110-00-9 | Furan | 68.08 | 600 | 0.0012 | 0.4 | EHS-LOC (Tox[c]) | 0.12 | 0.14 | 0.52 | 45 | Dense | Dense |
| 302-01-2 | Hydrazine | 32.05 | 14.4 | 0.011 | 8 | EHS-LOC (IDLH) | 0.0017 | 0.069 | 0.48 | 48 | Buoyant[d] | Buoyant[d] |

| CAS Number | Chemical Name | Molecular Weight | Vapor Pressure at 25-C (mm Hg) | Toxic Endpoint[a] | | | Liquid Factors | | Density Factor (DF) | Liquid Leak Factor (LLF)[i] | Reference Table[b] | |
|---|---|---|---|---|---|---|---|---|---|---|---|---|
| | | | | mg/L | ppm | Basis | Ambient (LFA) | Boiling (LFB) | | | Worst Case | Alternative Case |
| 13463-40-6 | Iron, pentacarbonyl- | 195.90 | 40 | 0.00044 | 0.05 | EHS-LOC (Tox[c]) | 0.016 | 0.24 | 0.33 | 70 | Dense | Dense |
| 78-82-0 | Isobutyronitrile | 69.11 | 32.7 | 0.14 | 50 | ERPG-2 | 0.0064 | 0.12 | 0.63 | 37 | Dense | Buoyant[d] |
| 108-23-6 | Isopropyl chloroformate | 122.55 | 28 | 0.10 | 20 | EHS-LOC (Tox[c]) | 0.0080 | 0.17 | 0.45 | 52 | Dense | Dense |
| 126-98-7 | Methacrylonitrile | 67.09 | 71.2 | 0.0027 | 1 | EHS-LOC (TLV[e]) | 0.014 | 0.12 | 0.61 | 38 | Dense | Dense |
| 79-22-1 | Methyl chloroformate | 94.50 | 108 | 0.0019 | 0.5 | EHS-LOC (Tox[c]) | 0.026 | 0.16 | 0.40 | 58 | Dense | Dense |
| 60-34-4 | Methyl hydrazine | 46.07 | 49.6 | 0.0094 | 5 | EHS-LOC (IDLH) | 0.0074 | 0.094 | 0.56 | 42 | Dense | Buoyant[d] |
| 624-83-9 | Methyl isocyanate | 57.05 | 457 | 0.0012 | 0.5 | ERPG-2 | 0.079 | 0.13 | 0.52 | 45 | Dense | Dense |
| 556-64-9 | Methyl thiocyanate | 73.12 | 10 | 0.085 | 29 | EHS-LOC (Tox[c]) | 0.0020 | 0.11 | 0.45 | 51 | Dense | Buoyant[d] |
| 75-79-6 | Methyltrichlorosilane | 149.48 | 173 | 0.018 | 3 | ERPG-2 | 0.057 | 0.22 | 0.38 | 61 | Dense | Dense |
| 13463-39-3 | Nickel carbonyl | 170.73 | 400 | 0.00067 | 0.1 | EHS-LOC (IDLH) | 0.14 | 0.26 | 0.37 | 63 | Dense | Dense |
| 7697-37-2 | Nitric acid (100%)[f] | 63.01 | 63.0 | 0.026 | 10 | EHS-LOC (IDLH) | 0.012 | 0.12 | 0.32 | 73 | Dense | Dense |
| 79-21-0 | Peracetic acid | 76.05 | 13.9 | 0.0045 | 1.5 | EHS-LOC (Tox[c]) | 0.0029 | 0.12 | 0.40 | 58 | Dense | Buoyant[d] |
| 594-42-3 | Perchloromethylmercaptan | 185.87 | 6 | 0.0076 | 1 | EHS-LOC (IDLH) | 0.0023 | 0.20 | 0.29 | 81 | Dense | Buoyant[d] |
| 10025-87-3 | Phosphorus oxychloride | 153.33 | 35.8 | 0.0030 | 0.5 | EHS-LOC (Tox[c]) | 0.012 | 0.20 | 0.29 | 80 | Dense | Dense |
| 7719-12-2 | Phosphorus trichloride | 137.33 | 120 | 0.028 | 5 | EHS-LOC (IDLH) | 0.037 | 0.20 | 0.31 | 75 | Dense | Dense |
| 110-89-4 | Piperidine | 85.15 | 32.1 | 0.022 | 6 | EHS-LOC (Tox[c]) | 0.0072 | 0.13 | 0.57 | 41 | Dense | Buoyant[d] |
| 107-12-0 | Propionitrile | 55.08 | 47.3 | 0.0037 | 1.6 | EHS-LOC (Tox[c]) | 0.0080 | 0.10 | 0.63 | 37 | Dense | Buoyant[d] |
| 109-61-5 | Propyl chloroformate | 122.56 | 20.0 | 0.010 | 2 | EHS-LOC (Tox[c]) | 0.0058 | 0.17 | 0.45 | 52 | Dense | Buoyant[d] |
| 75-55-8 | Propyleneimine | 57.10 | 187 | 0.12 | 50 | EHS-LOC (IDLH) | 0.032 | 0.12 | 0.61 | 39 | Dense | Dense |
| 75-56-9 | Propylene oxide | 58.08 | 533 | 0.59 | 250 | ERPG-2 | 0.093 | 0.13 | 0.59 | 40 | Dense | Dense |
| 7446-11-9 | Sulfur trioxide | 80.06 | 263 | 0.010 | 3 | ERPG-2 | 0.057 | 0.15 | 0.26 | 91 | Dense | Dense |
| 75-74-1 | Tetramethyllead | 267.33 | 22.5 | 0.00040 | 0.4 | EHS-LOC (IDLH) | 0.011 | 0.29 | 0.24 | 96 | Dense | Dense |
| 509-14-8 | Tetranitromethane | 196.04 | 11.4 | 0.00040 | 0.5 | EHS-LOC (IDLH) | 0.0045 | 0.22 | 0.30 | 78 | Dense | Buoyant[d] |
| 7550-45-0 | Titanium tetrachloride | 189.69 | 12.4 | 0.020 | 2.6 | ERPG-2 | 0.0048 | 0.21 | 0.28 | 82 | Dense | Buoyant[d] |

| CAS Number | Chemical Name | Molecular Weight | Vapor Pressure at 25-C (mm Hg) | Toxic Endpoint[a] | | | Liquid Factors | | Density Factor (DF) | Liquid Leak Factor (LLF)[i] | Reference Table[b] | |
| | | | | mg/L | ppm | Basis | Ambient (LFA) | Boiling (LFB) | | | Worst Case | Alternative Case |
|---|---|---|---|---|---|---|---|---|---|---|---|---|
| 584-84-9 | Toluene 2,4-diisocyanate | 174.16 | 0.017 | 0.0070 | 1 | EHS-LOC (IDLH) | 0.000006 | 0.16 | 0.40 | 59 | Buoyant[d] | Buoyant[d] |
| 91-08-7 | Toluene 2,6-diisocyanate | 174.16 | 0.05 | 0.0070 | 1 | EHS-LOC (IDLH[g]) | 0.000018 | 0.16 | 0.40 | 59 | Buoyant[d] | Buoyant[d] |
| 26471-62-5 | Toluene diisocyanate (unspecified isomer) | 174.16 | 0.017 | 0.0070 | 1 | EHS-LOC equivalent (IDLH[h]) | 0.000006 | 0.16 | 0.40 | 59 | Buoyant[d] | Buoyant[d] |
| 75-77-4 | Trimethylchlorosilane | 108.64 | 231 | 0.050 | 11 | EHS-LOC (Tox[c]) | 0.061 | 0.18 | 0.57 | 41 | Dense | Dense |
| 108-05-4 | Vinyl acetate monomer | 86.09 | 113 | 0.26 | 75 | ERPG-2 | 0.026 | 0.15 | 0.53 | 45 | Dense | Dense |

**Notes:**

[a] Toxic endpoints are specified in the Appendix A to 40 CFR Part 68 in units of mg/L. To convert to units of mg/m³, multiply by 1,000. To convert mg/L to ppm, use the following equation:

$$\text{Endpoint}_{ppm} = (\text{Endpoint}_{mg/l} \times 1,000 \times 24.5)/\text{Molecular Weight}$$

[b] "Buoyant" in the Reference Table column refers to the tables for neutrally buoyant gases and vapors; "Dense" refers to the tables for dense gases and vapors. See OCAG, Appendix D, Section D.4.2, for more information on the choice of reference tables.

[c] LOC is based on IDLH-equivalent level estimated from toxicity data.

[d] Use dense gas table if substance is at an elevated temperature.

[e] LOC based on Threshold Limit Value (TLV) – Time-weighted average (TWA) developed by the American Conference of Governmental Industrial Hygienists (ACGIH).

[f] See Exhibit B-3 of OCAG, Appendix B, for data on water solutions.

[g] LOC for this isomer is based on IDLH for toluene 2,4-diisocyanate.

[h] Not an EHS; LOC-equivalent value is based on IDLH for toluene 2,4-diisocyanate.

[i] Use the LLF only for leaks from tanks at atmospheric pressure.

**Exhibit B-3**

**Data for Water Solutions of Toxic Substances and for Oleum**

**Average Vapor Pressure and Liquid Factors Over 10 Minutes for**

**Wind Speeds of 1.5 and 3.0 Meters per Second (m/s)**

| CAS Number | Regulated Substance in Solution | Molecular Weight | Toxic Endpoint[a] | | | Initial Concentration (Wt %) | 10-min. Average Vapor Pressure (mm Hg) | | Liquid Factor at 25°C (LFA) | | Density Factor (DF) | Liquid Leak Factor (LLF) | Reference Table[b] | |
| | | | mg/L | ppm | Basis | | 1.5 m/s | 3.0 m/s | 1.5 m/s | 3.0 m/s | | | Worst | Alternative |
|---|---|---|---|---|---|---|---|---|---|---|---|---|---|---|
| 7664-41-7 | Ammonia | 17.03 | 0.14 | 200 | ERPG-2 | 30 | 332 | 248 | 0.026 | 0.019 | 0.55 | 43 | Buoyant | Buoyant |
| | | | | | | 24 | 241 | 184 | 0.019 | 0.014 | 0.54 | 44 | Buoyant | Buoyant |
| | | | | | | 20 | 190 | 148 | 0.015 | 0.011 | 0.53 | 44 | Buoyant | Buoyant |
| 50-00-0 | Formaldehyde | 30.027 | 0.012 | 10 | ERPG-2 | 37 | 1.5 | 1.4 | 0.0002 | 0.0002 | 0.44 | 53 | Buoyant | Buoyant |
| 7647-01-0 | Hydrochloric acid | 36.46 | 0.030 | 20 | ERPG-2 | 38 | 78 | 55 | 0.010 | 0.0070 | 0.41 | 57 | Dense | Buoyant |
| | | | | | | 37 | 67 | 48 | 0.0085 | 0.0062 | 0.42 | 57 | Dense | Buoyant |
| | | | | | | 36[c] | 56 | 42 | 0.0072 | 0.0053 | 0.42 | 57 | Dense | Buoyant |
| | | | | | | 34[c] | 38 | 29 | 0.0048 | 0.0037 | 0.42 | 56 | Dense | Buoyant |
| | | | | | | 30[c] | 13 | 12 | 0.0016 | 0.0015 | 0.42 | 55 | Buoyant | Buoyant |
| 7664-39-3 | Hydrofluoric acid | 20.01 | 0.016 | 20 | ERPG-2 | 70 | 124 | 107 | 0.011 | 0.010 | 0.39 | 61 | Buoyant | Buoyant |
| | | | | | | 50 | 16 | 15 | 0.0014 | 0.0013 | 0.41 | 58 | Buoyant | Buoyant |
| 7697-37-2 | Nitric acid | 63.01 | 0.026 | 10 | EHS-LOC (IDLH) | 90 | 25 | 22 | 0.0046 | 0.0040 | 0.33 | 71 | Dense | Buoyant |
| | | | | | | 85 | 17 | 16 | 0.0032 | 0.0029 | 0.33 | 70 | Dense | Buoyant |
| | | | | | | 80 | 10.2 | 10 | 0.0019 | 0.0018 | 0.33 | 70 | Dense | Buoyant |
| 8014-95-7 | Oleum - based on SO₃ | 80.06 (SO₃) | 0.010 | 3 | ERPG-2 | 30 (SO₃) | 3.5 (SO₃) | 3.4 (SO₃) | 0.0008 | 0.0007 | 0.25 | 93 | Buoyant | Buoyant |

**Notes:**

[a] Toxic endpoints are specified in the Appendix A to 40 CFR Part 68 in units of mg/L. See Notes to Exhibit B-1 or B-2 for converting to other units.

[b] "Buoyant" in the Reference Table column refers to the tables for neutrally buoyant gases and vapors; "Dense" refers to the tables for dense gases and vapors. See OCAG, Appendix D, Section D.4.4, for more information on the choice of reference tables.

[c] Hydrochloric acid in concentrations below 37 percent is not regulated.

**Exhibit B-4**
**Temperature Correction Factors for Liquids Evaporating from Pools at Temperatures**
**Between 25·C and 50·C (77·F and 122·F)**

| CAS Number | Chemical Name | Boiling Point (·C) | Temperature Correction Factor (TCF) | | | | |
|---|---|---|---|---|---|---|---|
| | | | 30·C (86·F) | 35·C (95·F) | 40·C (104·F) | 45·C (113·F) | 50·C (122·F) |
| 107-02-8 | Acrolein | 52.69 | 1.2 | 1.4 | 1.7 | 2.0 | 2.3 |
| 107-13-1 | Acrylonitrile | 77.35 | 1.2 | 1.5 | 1.8 | 2.1 | 2.5 |
| 814-68-6 | Acrylyl chloride | 75.00 | ND | ND | ND | ND | ND |
| 107-18-6 | Allyl alcohol | 97.08 | 1.3 | 1.7 | 2.2 | 2.9 | 3.6 |
| 107-11-9 | Allylamine | 53.30 | 1.2 | 1.5 | 1.8 | 2.1 | 2.5 |
| 7784-34-1 | Arsenous trichloride | 130.06 | ND | ND | ND | ND | ND |
| 353-42-4 | Boron trifluoride compound with methyl ether (1:1) | 126.85 | ND | ND | ND | ND | ND |
| 7726-95-6 | Bromine | 58.75 | 1.2 | 1.5 | 1.7 | 2.1 | 2.5 |
| 75-15-0 | Carbon disulfide | 46.22 | 1.2 | 1.4 | 1.6 | 1.9 | LFB |
| 67-66-3 | Chloroform | 61.18 | 1.2 | 1.5 | 1.8 | 2.1 | 2.5 |
| 542-88-1 | Chloromethyl ether | 104.85 | 1.3 | 1.6 | 2.0 | 2.5 | 3.1 |
| 107-30-2 | Chloromethyl methyl ether | 59.50 | 1.2 | 1.5 | 1.8 | 2.1 | 2.5 |
| 4170-30-3 | Crotonaldehyde | 104.10 | 1.3 | 1.6 | 2.0 | 2.5 | 3.1 |
| 123-73-9 | Crotonaldehyde, (E)- | 102.22 | 1.3 | 1.6 | 2.0 | 2.5 | 3.1 |
| 108-91-8 | Cyclohexylamine | 134.50 | 1.3 | 1.7 | 2.1 | 2.7 | 3.4 |
| 75-78-5 | Dimethyldichlorosilane | 70.20 | 1.2 | 1.5 | 1.8 | 2.1 | 2.5 |
| 57-14-7 | 1,1-Dimethylhydrazine | 63.90 | ND | ND | ND | ND | ND |
| 106-89-8 | Epichlorohydrin | 118.50 | 1.3 | 1.7 | 2.1 | 2.7 | 3.4 |
| 107-15-3 | Ethylenediamine | 117.40 | 1.3 | 1.8 | 2.3 | 3.0 | 3.8 |
| 151-56-4 | Ethyleneimine | 55.85 | 1.2 | 1.5 | 1.8 | 2.2 | 2.7 |
| 110-00-9 | Furan | 31.35 | 1.2 | LFB | LFB | LFB | LFB |
| 302-01-2 | Hydrazine | 113.50 | 1.3 | 1.7 | 2.2 | 2.9 | 3.6 |
| 13463-40-6 | Iron, pentacarbonyl- | 102.63 | ND | ND | ND | ND | ND |
| 78-82-0 | Isobutyronitrile | 103.61 | 1.3 | 1.6 | 2.0 | 2.5 | 3.1 |
| 108-23-6 | Isopropyl chloroformate | 104.60 | ND | ND | ND | ND | ND |
| 126-98-7 | Methacrylonitrile | 90.30 | 1.2 | 1.5 | 1.8 | 2.2 | 2.6 |
| 79-22-1 | Methyl chloroformate | 70.85 | 1.3 | 1.6 | 1.9 | 2.4 | 2.9 |
| 60-34-4 | Methyl hydrazine | 87.50 | ND | ND | ND | ND | ND |
| 624-83-9 | Methyl isocyanate | 38.85 | 1.2 | 1.4 | LFB | LFB | LFB |
| 556-64-9 | Methyl thiocyanate | 130.00 | ND | ND | ND | ND | ND |
| 75-79-6 | Methyltrichlorosilane | 66.40 | 1.2 | 1.4 | 1.7 | 2.0 | 2.4 |

| CAS Number | Chemical Name | Boiling Point (•C) | Temperature Correction Factor (TCF) | | | | |
|---|---|---|---|---|---|---|---|
| | | | 30•C (86•F) | 35•C (95•F) | 40•C (104•F) | 45•C (113•F) | 50•C (122•F) |
| 13463-39-3 | Nickel carbonyl | 42.85 | ND | ND | ND | ND | ND |
| 7697-37-2 | Nitric acid | 83.00 | 1.3 | 1.6 | 2.0 | 2.5 | 3.1 |
| 79-21-0 | Peracetic acid | 109.85 | 1.3 | 1.8 | 2.3 | 3.0 | 3.8 |
| 594-42-3 | Perchloromethylmercaptan | 147.00 | ND | ND | ND | ND | ND |
| 10025-87-3 | Phosphorus oxychloride | 105.50 | 1.3 | 1.6 | 1.9 | 2.4 | 2.9 |
| 7719-12-2 | Phosphorus trichloride | 76.10 | 1.2 | 1.5 | 1.8 | 2.1 | 2.5 |
| 110-89-4 | Piperidine | 106.40 | 1.3 | 1.6 | 2.0 | 2.4 | 3.0 |
| 107-12-0 | Propionitrile | 97.35 | 1.3 | 1.6 | 1.9 | 2.3 | 2.8 |
| 109-61-5 | Propyl chloroformate | 112.40 | ND | ND | ND | ND | ND |
| 75-55-8 | Propyleneimine | 60.85 | 1.2 | 1.5 | 1.8 | 2.1 | 2.5 |
| 75-56-9 | Propylene oxide | 33.90 | 1.2 | LFB | LFB | LFB | LFB |
| 7446-11-9 | Sulfur trioxide | 44.75 | 1.3 | 1.7 | LFB | LFB | LFB |
| 75-74-1 | Tetramethyllead | 110.00 | ND | ND | ND | ND | ND |
| 509-14-8 | Tetranitromethane | 125.70 | 1.3 | 1.7 | 2.2 | 2.8 | 3.5 |
| 7550-45-0 | Titanium tetrachloride | 135.85 | 1.3 | 1.6 | 2.0 | 2.6 | 3.2 |
| 584-84-9 | Toluene 2,4-diisocyanate | 251.00 | 1.6 | 2.4 | 3.6 | 5.3 | 7.7 |
| 91-08-7 | Toluene 2,6-diisocyanate | 244.85 | ND | ND | ND | ND | ND |
| 26471-62-5 | Toluene diisocyanate (unspecified isomer) | 250.00 | 1.6 | 2.4 | 3.6 | 5.3 | 7.7 |
| 75-77-4 | Trimethylchlorosilane | 57.60 | 1.2 | 1.4 | 1.7 | 2.0 | 2.3 |
| 108-05-4 | Vinyl acetate monomer | 72.50 | 1.2 | 1.5 | 1.9 | 2.3 | 2.7 |

Notes:

ND:  No data available

LFB:  Chemical above boiling point at this temperature; use LFB for analysis

Explanation of Temperature Correction Factors.  Temperature correction factors were developed for toxic liquids released at temperatures above 25•C, the temperature used for development of the LFAs.  The factors are based on vapor pressures calculated from the coefficients provided in *Physical and Thermodynamic Properties of Pure Chemicals, Data Compilation*, developed by the Design Institute for Physical Property Data (DIPPR), American Institute of Chemical Engineers.  The factors are calculated as follows:

$$TCF_T = (VP_T \times 298)/(VP_{298} \times T)$$

where:

| | | |
|---|---|---|
| $TCF_T$ | = | Temperature Correction Factor at temperature T |
| $VP_T$ | = | Vapor pressure at temperature T |
| $VP_{298}$ | = | Vapor pressure at 298K |
| T | = | Temperature (K) |

Factors were developed at intervals of 5•C for temperatures up to 50•C.  The above equation is the same as Equation D-5 in the OCAG.

For temperatures exceeding 25 °C, the value of LFA in Equation 1 or Equation 4 of Chapter 4 should be multiplied by the appropriate temperature correction factor (TCF), or, equivalently, the calculated evaporation rate should be multiplied by TCF. For example, in Example 3 in Chapter 4, a release rate of 13 lb/min has been calculated following an accidental spillage of dimethyldichlorosilane from a 55-gallon drum at 25• C. If the spill were at 45• C, TCF would be 2.1 (from Exhibit B-4), and the predicted rate of evaporation would be 13 x 2.1 = 27.3 lb/min.

## Exhibit C-1
## Heats of Combustion for Flammable Substances

| CAS No. | Chemical Name | Physical State at 25·C | Heat of Combustion (kjoule/kg) |
|---------|---------------|------------------------|-------------------------------|
| 75-07-0 | Acetaldehyde | Gas | 25,072 |
| 74-86-2 | Acetylene [Ethyne] | Gas | 48,222 |
| 598-73-2 | Bromotrifluoroethylene [Ethene, bromotrifluoro-] | Gas | 1,967 |
| 106-99-0 | 1,3-Butadiene | Gas | 44,548 |
| 106-97-8 | Butane | Gas | 45,719 |
| 25167-67-3 | Butene | Gas | 45,200* |
| 590-18-1 | 2-Butene-cis | Gas | 45,171 |
| 624-64-6 | 2-Butene-trans [2-Butene, (E)] | Gas | 45,069 |
| 106-98-9 | 1-Butene | Gas | 45,292 |
| 107-01-7 | 2-Butene | Gas | 45,100* |
| 463-58-1 | Carbon oxysulfide [Carbon oxide sulfide (COS)] | Gas | 9,126 |
| 7791-21-1 | Chlorine monoxide [Chlorine oxide] | Gas | 1,011* |
| 590-21-6 | 1-Chloropropylene [1-Propene, 1-chloro-] | Liquid | 23,000* |
| 557-98-2 | 2-Chloropropylene [1-Propene, 2-chloro-] | Gas | 22,999 |
| 460-19-5 | Cyanogen [Ethanedinitrile] | Gas | 21,064 |
| 75-19-4 | Cyclopropane | Gas | 46,560 |
| 4109-96-0 | Dichlorosilane [Silane, dichloro-] | Gas | 8,225 |
| 75-37-6 | Difluoroethane [Ethane, 1,1-difluoro-] | Gas | 11,484 |
| 124-40-3 | Dimethylamine [Methanamine, N-methyl-] | Gas | 35,813 |
| 463-82-1 | 2,2-Dimethylpropane [Propane, 2,2-dimethyl-] | Gas | 45,051 |
| 74-84-0 | Ethane | Gas | 47,509 |
| 107-00-6 | Ethyl acetylene [1-Butyne] | Gas | 45,565 |
| 75-04-7 | Ethylamine [Ethanamine] | Gas | 35,210 |
| 75-00-3 | Ethyl chloride [Ethane, chloro-] | Gas | 19,917 |
| 74-85-1 | Ethylene [Ethene] | Gas | 47,145 |
| 60-29-7 | Ethyl ether [Ethane, 1,1'-oxybis-] | Liquid | 33,775 |
| 75-08-1 | Ethyl mercaptan [Ethanethiol] | Liquid | 27,948 |

| CAS No. | Chemical Name | Physical State at 25·C | Heat of Combustion (kjoule/kg) |
|---|---|---|---|
| 109-95-5 | Ethyl nitrite [Nitrous acid, ethyl ester] | Gas | 18,000 |
| 1333-74-0 | Hydrogen | Gas | 119,950 |
| 75-28-5 | Isobutane [Propane, 2-methyl] | Gas | 45,576 |
| 78-78-4 | Isopentane [Butane, 2-methyl-] | Liquid | 44,911 |
| 78-79-5 | Isoprene [1,3-Butadiene, 2-methyl-] | Liquid | 43,809 |
| 75-31-0 | Isopropylamine [2-Propanamine] | Liquid | 36,484 |
| 75-29-6 | Isopropyl chloride [Propane, 2-chloro-] | Liquid | 23,720 |
| 74-82-8 | Methane | Gas | 50,029 |
| 74-89-5 | Methylamine [Methanamine] | Gas | 31,396 |
| 563-45-1 | 3-Methyl-1-butene | Gas | 44,559 |
| 563-46-2 | 2-Methyl-1-butene | Liquid | 44,414 |
| 115-10-6 | Methyl ether [Methane, oxybis-] | Gas | 28,835 |
| 107-31-3 | Methyl formate [Formic acid, methyl ester] | Liquid | 15,335 |
| 115-11-7 | 2-Methylpropene [1-Propene, 2-methyl-] | Gas | 44,985 |
| 504-60-9 | 1,3-Pentadiene | Liquid | 43,834 |
| 109-66-0 | Pentane | Liquid | 44,697 |
| 109-67-1 | 1-Pentene | Liquid | 44,625 |
| 646-04-8 | 2-Pentene, (E)- | Liquid | 44,458 |
| 627-20-3 | 2-Pentene, (Z)- | Liquid | 44,520 |
| 463-49-0 | Propadiene [1,2-Propadiene] | Gas | 46,332 |
| 74-98-6 | Propane | Gas | 46,333 |
| 115-07-1 | Propylene [1-Propene] | Gas | 45,762 |
| 74-99-7 | Propyne [1-Propyne] | Gas | 46,165 |
| 7803-62-5 | Silane | Gas | 44,307 |
| 116-14-3 | Tetrafluoroethylene [Ethene, tetrafluoro-] | Gas | 1,284 |
| 75-76-3 | Tetramethylsilane [Silane, tetramethyl-] | Liquid | 41,712 |
| 10025-78-2 | Trichlorosilane [Silane, trichloro-] | Liquid | 3,754 |
| 79-38-9 | Trifluorochloroethylene [Ethene, chlorotrifluoro-] | Gas | 1,837 |
| 75-50-3 | Trimethylamine [Methanamine, N,N-dimethyl-] | Gas | 37,978 |

| CAS No. | Chemical Name | Physical State at 25·C | Heat of Combustion (kjoule/kg) |
|---------|---------------|------------------------|-------------------------------|
| 689-97-4 | Vinyl acetylene  [1-Buten-3-yne] | Gas | 45,357 |
| 75-01-4 | Vinyl chloride  [Ethene, chloro-] | Gas | 18,848 |
| 109-92-2 | Vinyl ethyl ether  [Ethene, ethoxy-] | Liquid | 32,909 |
| 75-02-5 | Vinyl fluoride  [Ethene, fluoro-] | Gas | 2,195 |
| 75-35-4 | Vinylidene chloride  [Ethene, 1,1-dichloro-] | Liquid | 10,354 |
| 75-38-7 | Vinylidene fluoride  [Ethene, 1,1-difluoro-] | Gas | 10,807 |
| 107-25-5 | Vinyl methyl ether  [Ethene, methoxy-] | Gas | 30,549 |

*Estimated heat of combustion

**Exhibit C-2**
**Data for Flammable Gases**

| CAS Number | Chemical Name | Molecular Weight | Ratio of Specific Heats | Flammability Limits (Vol %) | | LFL (mg/L) | Gas Factor (GF)[g] | Liquid Factor Boiling (LFB) | Density Factor (Boiling) (DF) | Reference Table[a] | Pool Fire Factor (PFF) | Flash Fraction Factor (FFF)[f] |
|---|---|---|---|---|---|---|---|---|---|---|---|---|
| | | | | Lower (LFL) | Upper (UFL) | | | | | | | |
| 75-07-0 | Acetaldehyde | 44.05 | 1.18 | 4.0 | 60.0 | 72 | 22 | 0.11 | 0.62 | Dense | 2.7 | 0.018 |
| 74-86-2 | Acetylene | 26.04 | 1.23 | 2.5 | 80.0 | 27 | 17 | 0.12 | 0.78 | Buoyant[b] | 4.8 | 0.23[f] |
| 598-73-2 | Bromotrifluoroethylene | 160.92 | 1.11 | c | 37.0 | c | 41[c] | 0.25[c] | 0.29[c] | Dense | 0.42[c] | 0.15[c] |
| 106-99-0 | 1,3-Butadiene | 54.09 | 1.12 | 2.0 | 11.5 | 44 | 24 | 0.14 | 0.75 | Dense | 5.5 | 0.15 |
| 106-97-8 | Butane | 58.12 | 1.09 | 1.5 | 9.0 | 36 | 25 | 0.14 | 0.81 | Dense | 5.9 | 0.15 |
| 25167-67-3 | Butene | 56.11 | 1.10 | 1.7 | 9.5 | 39 | 24 | 0.14 | 0.77 | Dense | 5.6 | 0.14 |
| 590-18-1 | 2-Butene-cis | 56.11 | 1.12 | 1.6 | 9.7 | 37 | 24 | 0.14 | 0.76 | Dense | 5.6 | 0.11 |
| 624-64-6 | 2-Butene-trans | 56.11 | 1.11 | 1.8 | 9.7 | 41 | 24 | 0.14 | 0.77 | Dense | 5.6 | 0.12 |
| 106-98-9 | 1-Butene | 56.11 | 1.11 | 1.6 | 9.3 | 37 | 24 | 0.14 | 0.78 | Dense | 5.7 | 0.17 |
| 107-01-7 | 2-Butene | 56.11 | 1.10 | 1.7 | 9.7 | 39 | 24 | 0.14 | 0.77 | Dense | 5.6 | 0.12 |
| 463-58-1 | Carbon oxysulfide | 60.08 | 1.25 | 12.0 | 29.0 | 290 | 26 | 0.18 | 0.41 | Dense | 1.3 | 0.29 |
| 7791-21-1 | Chlorine monoxide | 86.91 | 1.21 | 23.5 | NA | 830 | 31 | 0.19 | NA | Dense | 0.15 | NA |
| 557-98-2 | 2-Chloropropylene | 76.53 | 1.12 | 4.5 | 16.0 | 140 | 29 | 0.16 | 0.54 | Dense | 3.3 | 0.011 |
| 460-19-5 | Cyanogen | 52.04 | 1.17 | 6.0 | 32.0 | 130 | 24 | 0.15 | 0.51 | Dense | 2.5 | 0.40 |
| 75-19-4 | Cyclopropane | 42.08 | 1.18 | 2.4 | 10.4 | 41 | 22 | 0.13 | 0.72 | Dense | 5.4 | 0.23 |
| 4109-96-0 | Dichlorosilane | 101.01 | 1.16 | 4.0 | 96.0 | 160 | 33 | 0.20 | 0.40 | Dense | 1.3 | 0.084 |
| 75-37-6 | Difluoroethane | 66.05 | 1.14 | 3.7 | 18.0 | 100 | 27 | 0.17 | 0.48 | Dense | 1.6 | 0.23 |
| 124-40-3 | Dimethylamine | 45.08 | 1.14 | 2.8 | 14.4 | 52 | 22 | 0.12 | 0.73 | Dense | 3.7 | 0.090 |
| 463-82-1 | 2,2-Dimethylpropane | 72.15 | 1.07 | 1.4 | 7.5 | 41 | 27 | 0.16 | 0.80 | Dense | 6.4 | 0.11 |
| 74-84-0 | Ethane | 30.07 | 1.19 | 2.9 | 13.0 | 36 | 18 | 0.14 | 0.89 | Dense | 5.4 | 0.75 |
| 107-00-6 | Ethyl acetylene | 54.09 | 1.11 | 2.0 | 32.9 | 44 | 24 | 0.13 | 0.73 | Dense | 5.4 | 0.091 |
| 75-04-7 | Ethylamine | 45.08 | 1.13 | 3.5 | 14.0 | 64 | 22 | 0.12 | 0.71 | Dense | 3.6 | 0.040 |

| CAS Number | Chemical Name | Molecular Weight | Ratio of Specific Heats | Flammability Limits (Vol %) | | LFL (mg/L) | Gas Factor (GF)[g] | Liquid Factor Boiling (LFB) | Density Factor (Boiling) (DF) | Reference Table[a] | Pool Fire Factor (PFF) | Flash Fraction Factor (FFF)[f] |
| --- | --- | --- | --- | --- | --- | --- | --- | --- | --- | --- | --- | --- |
| | | | | Lower (LFL) | Upper (UFL) | | | | | | | |
| 75-00-3 | Ethyl chloride | 64.51 | 1.15 | 3.8 | 15.4 | 100 | 27 | 0.15 | 0.53 | Dense | 2.6 | 0.053 |
| 74-85-1 | Ethylene | 28.05 | 1.24 | 2.7 | 36.0 | 31 | 18 | 0.14 | 0.85 | Buoyant[b] | 5.4 | 0.63[f] |
| 109-95-5 | Ethyl nitrite | 75.07 | 1.30 | 4.0 | 50.0 | 120 | 30 | 0.16 | 0.54 | Dense | 2.0 | NA |
| 1333-74-0 | Hydrogen | 2.02 | 1.41 | 4.0 | 75.0 | 3.3 | 5.0 | [e] | [e] | [d] | [e] | NA |
| 75-28-5 | Isobutane | 58.12 | 1.09 | 1.8 | 8.4 | 43 | 25 | 0.15 | 0.82 | Dense | 6.0 | 0.23 |
| 74-82-8 | Methane | 16.04 | 1.30 | 5.0 | 15.0 | 33 | 14 | 0.15 | 1.1 | Buoyant | 5.6 | 0.87[f] |
| 74-89-5 | Methylamine | 31.06 | 1.19 | 4.9 | 20.7 | 62 | 19 | 0.10 | 0.70 | Dense | 2.7 | 0.12 |
| 563-45-1 | 3-Methyl-1-butene | 70.13 | 1.08 | 1.5 | 9.1 | 43 | 26 | 0.15 | 0.77 | Dense | 6.0 | 0.030 |
| 115-10-6 | Methyl ether | 46.07 | 1.15 | 3.3 | 27.3 | 64 | 22 | 0.14 | 0.66 | Dense | 3.4 | 0.22 |
| 115-11-7 | 2-Methylpropene | 56.11 | 1.10 | 1.8 | 8.8 | 41 | 24 | 0.14 | 0.77 | Dense | 5.7 | 0.18 |
| 463-49-0 | Propadiene | 40.07 | 1.16 | 2.1 | 2.1 | 34 | 21 | 0.13 | 0.73 | Dense | 5.2 | 0.20 |
| 74-98-6 | Propane | 44.10 | 1.13 | 2.0 | 9.5 | 36 | 22 | 0.14 | 0.83 | Dense | 5.7 | 0.38 |
| 115-07-1 | Propylene | 42.08 | 1.15 | 2.0 | 11.0 | 34 | 21 | 0.14 | 0.79 | Dense | 5.5 | 0.35 |
| 74-99-7 | Propyne | 40.07 | 1.16 | 1.7 | 39.9 | 28 | 21 | 0.12 | 0.72 | Dense | 4.9 | 0.18 |
| 7803-62-5 | Silane | 32.12 | 1.24 | [e] | [e] | [e] | 19[c] | [e] | [e] | Dense | [e] | 0.41[f] |
| 116-14-3 | Tetrafluoroethylene | 100.02 | 1.12 | 11.0 | 60.0 | 450 | 33 | 0.29 | 0.32 | Dense | 0.25 | 0.69 |
| 79-38-9 | Trifluorochloroethylene | 116.47 | 1.11 | 8.4 | 38.7 | 400 | 35 | 0.26 | 0.33 | Dense | 0.34 | 0.27 |
| 75-50-3 | Trimethylamine | 59.11 | 1.10 | 2.0 | 11.6 | 48 | 25 | 0.14 | 0.74 | Dense | 4.8 | 0.12 |
| 689-97-4 | Vinyl acetylene | 52.08 | 1.13 | 2.2 | 31.7 | 47 | 24 | 0.13 | 0.69 | Dense | 5.4 | 0.086 |
| 75-01-4 | Vinyl chloride | 62.50 | 1.18 | 3.6 | 33.0 | 92 | 26 | 0.16 | 0.50 | Dense | 2.4 | 0.14 |
| 75-02-5 | Vinyl fluoride | 46.04 | 1.20 | 2.6 | 21.7 | 49 | 23 | 0.17 | 0.57 | Dense | 0.28 | 0.37 |
| 75-38-7 | Vinylidene fluoride | 64.04 | 1.16 | 5.5 | 21.3 | 140 | 27 | 0.22 | 0.42 | Dense | 1.8 | 0.50 |
| 107-25-5 | Vinyl methyl ether | 58.08 | 1.12 | 2.6 | 39.0 | 62 | 25 | 0.17 | 0.57 | Dense | 3.7 | 0.093 |

**Notes:**

NA: Data not available

[a] "Buoyant" in the Reference Table column refers to the tables for neutrally buoyant gases and vapors; "Dense" refers to the tables for dense gases and vapors. See Appendix D, Section D.4.4, for more information on the choice of reference tables.

[b] Gases that are lighter than air may behave as dense gases upon release if liquefied under pressure or cold; consider the conditions of release when choosing the appropriate table.

[c] Reported to be spontaneously combustible.

[d] Much lighter than air; table of distances for neutrally buoyant gases not appropriate.

[e] Pool formation unlikely.

[f] Calculated at 293 K (25•C) with the following exceptions:

Acetylene factor at 250 K as reported in TNO, *Methods for the Calculation of the Physical Effects of the Escape of Dangerous Material* (1980).
Ethylene factor calculated at critical temperature, 282 K.
Methane factor calculated at critical temperature, 191 K.
Silane factor calculated at critical temperature, 270 K.

[g] Use GF for gas leaks under choked (maximum) flow conditions.

## Exhibit C-3
## Data for Flammable Liquids

| CAS Number | Chemical Name | Molecular Weight | Flammability Limit (Vol%) | | LFL (mg/L) | Liquid Factors | | Density Factor | Liquid Leak Factor (LLF)[a] | Reference Table[b] | Pool Fire Factor (PFF) |
|---|---|---|---|---|---|---|---|---|---|---|---|
| | | | Lower (LFL) | Upper (UFL) | | Ambient (LFA) | Boiling (LFB) | | | | |
| 590-21-6 | 1-Chloropropylene | 76.53 | 4.5 | 16.0 | 140 | 0.11 | 0.15 | 0.52 | 45 | Dense | 3.2 |
| 60-29-7 | Ethyl ether | 74.12 | 1.9 | 48.0 | 57 | 0.11 | 0.15 | 0.69 | 34 | Dense | 4.3 |
| 75-08-1 | Ethyl mercaptan | 62.14 | 2.8 | 18.0 | 71 | 0.10 | 0.13 | 0.58 | 40 | Dense | 3.3 |
| 78-78-4 | Isopentane | 72.15 | 1.4 | 7.6 | 41 | 0.14 | 0.15 | 0.79 | 30 | Dense | 6.1 |
| 78-79-5 | Isoprene | 68.12 | 2.0 | 9.0 | 56 | 0.11 | 0.14 | 0.72 | 32 | Dense | 5.5 |
| 75-31-0 | Isopropylamine | 59.11 | 2.0 | 10.4 | 48 | 0.10 | 0.13 | 0.71 | 33 | Dense | 4.1 |
| 75-29-6 | Isopropyl chloride | 78.54 | 2.8 | 10.7 | 90 | 0.11 | 0.16 | 0.57 | 41 | Dense | 3.1 |
| 563-46-2 | 2-Methyl-1-butene | 70.13 | 1.4 | 9.6 | 40 | 0.12 | 0.15 | 0.75 | 31 | Dense | 5.8 |
| 107-31-3 | Methyl formate | 60.05 | 5.9 | 20.0 | 140 | 0.10 | 0.13 | 0.50 | 46 | Dense | 1.8 |
| 504-60-9 | 1,3-Pentadiene | 68.12 | 1.6 | 13.1 | 44 | 0.10 | 0.14 | 0.72 | 33 | Dense | 5.3 |
| 109-66-0 | Pentane | 72.15 | 1.3 | 8.0 | 38 | 0.077 | 0.15 | 0.78 | 30 | Dense | 5.8 |
| 109-67-1 | 1-Pentene | 70.13 | 1.5 | 8.7 | 43 | 0.13 | 0.15 | 0.77 | 31 | Dense | 5.8 |
| 646-04-8 | 2-Pentene, (E)- | 70.13 | 1.4 | 10.6 | 40 | 0.10 | 0.15 | 0.76 | 31 | Dense | 5.6 |
| 627-20-3 | 2-Pentene, (Z)- | 70.13 | 1.4 | 10.6 | 40 | 0.10 | 0.15 | 0.75 | 31 | Dense | 5.6 |
| 75-76-3 | Tetramethylsilane | 88.23 | 1.5 | NA | 54 | 0.17 | 0.17 | 0.59 | 40 | Dense | 6.3 |
| 10025-78-2 | Trichlorosilane | 135.45 | 1.2 | 90.5 | 66 | 0.18 | 0.23 | 0.37 | 64 | Dense | 0.68 |
| 109-92-2 | Vinyl ethyl ether | 72.11 | 1.7 | 28.0 | 50 | 0.10 | 0.15 | 0.65 | 36 | Dense | 4.2 |
| 75-35-4 | Vinylidene chloride | 96.94 | 7.3 | NA | 290 | 0.15 | 0.18 | 0.44 | 54 | Dense | 1.6 |

NA: Data not available

# CHAPTER 5: MANAGEMENT SYSTEM

## 5.1 GENERAL INFORMATION (§68.15)

If you have at least one Program 2 or Program 3 process (see Chapter 2 for guidance on determining the Program levels of your processes), the management system provision in § 68.15 requires you to:

Develop a management system to oversee the implementation of the risk management program elements;

Designate a qualified person or position with the overall responsibility for the development, implementation, and integration of the risk management program elements; and

Document the names of people or positions and define the lines of authority through an organizational chart or other similar document, if you assign responsibility for implementing individual requirements of the risk management program to people or positions other than the person or position with overall responsibility for the risk management program.

### ABOUT THE MANAGEMENT SYSTEM PROVISION

Management commitment to process safety is a critical element of your facility's risk management program. Management commitment should not end when the last word of the risk management plan is composed. For process safety to be a constant priority, your facility must remain committed to every element of the risk management program.

This rule takes an integrated approach to managing risks. Each element must be implemented on an ongoing, daily basis and become a part of the way you operate. Therefore, your commitment and oversight should be continuous.

By satisfying the requirements of this provision, you are ensuring that:

- The risk management program elements are integrated and implemented on an ongoing basis; and

- All groups within a source understand the lines of responsibility and communication.

## 5.2 HOW TO MEET THE MANAGEMENT SYSTEM REQUIREMENTS

We understand that the sources covered by this rule are diverse and that you are in the best position to decide how to appropriately implement and incorporate the risk management program elements at your facility; therefore, we sought to maximize your flexibility in complying with this program.

### WHAT DOES THIS MEAN FOR ME AS A SMALL FACILITY?

As a small facility that must comply with this provision, you most likely have one or two Program 2 or 3 processes. To begin, you may identify either the qualified person or position with overall responsibility for implementing the risk management program elements at your facility. As a small facility, it may make sense and be practical to identify the name of the qualified person, rather than the position. Recognize that the only element of your management system that you must report in the RMP is the name of the qualified person or position with overall responsibility. Further, changes to this data element in your RMP do not require that you update your RMP.

**Identification of a qualified individual or position with overall responsibility may be all you need to do** if the person or position named directly oversees the employees operating and maintaining the processes. You must define the lines of authority with an organizational chart or similar document only if you choose to assign responsibility for specific elements of the risk management program to persons or positions other than the person with overall responsibility. For a small facility, with few employees, it is likely that you will meet the requirements of this provision by identifying the one person or position with the overall responsibility of implementing the risk management program elements. If this is the case, you need not develop an organizational chart. For this reason, this chapter does not provide an example organizational chart for a small facility.

Even if you meet the requirements of this section by naming a single person or position, it is important to recognize that the person or position assigned the responsibility of overseeing implementation must have the ability and resources to ensure that your facility and employees carry out the risk management program, particularly the prevention elements, on an continuing basis. Key to the effectiveness of the rule is integrated management of the program elements.

### WHAT DOES THIS MEAN FOR ME AS A MEDIUM OR LARGE FACILITY?

As a medium or large facility you may have more managerial turnover than smaller sites. For this reason, it may make more sense at your facility to identify a position, rather than the name of the specific person, with overall responsibility for the risk management program elements. Remember that the only element of your management system that you must report in the RMP is the name of the qualified person or position with overall responsibility. Also note that changes to this data element in your RMP do not require you to update your RMP.

*Lines of Authority*

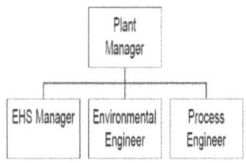

As a relatively large or complex facility, you will likely choose to identify several people or positions to supervise the implementation of the various elements of the program; therefore, you must define the lines of authority through an organizational chart or similar document. Further, we expect that most facilities your size already have an interest in formalizing internal communication and have likely developed and maintained some type of documentation defining positions and responsibilities. Any internal documents you currently have should be the starting point for defining the lines of authority at your facility. You may find that you can simply use or update current documents to satisfy this part of the management system provision. Exhibit 5-1 provides a sample of another type of documentation you may use in addition to or as a replacement for an organization chart.

Defining the lines of authority and roles and responsibilities of staff that oversee the risk management program elements will help to:

- • Ensure effective communication about process changes between divisions;

- • Clarify the roles and responsibilities related to process safety issues at your facility;

- • Avoid problems or conflicts among the various people responsible for implementing elements of the risk management program;

- • Avoid confusion and allow those responsible for implementation to work together as a team; and

- • Ensure that the program elements are integrated into an ongoing approach to identifying hazards and managing risks.

Remember that all of the positions you identify in your documentation will report their progress to the person with overall responsibility for the program. However, nothing in the risk management program rule prohibits you from satisfying the management provision by assigning process safety committees with management responsibility, provided that an organizational chart or similar document identifies the names or positions and lines of authority.

## EXHIBIT 5-1
## SAMPLE MANAGEMENT DOCUMENTATION

| Position | Primary Responsibility | Changes | Responsibility re: Changes |
|---|---|---|---|
| Operations Manager | Developing OPs<br>Oversight of operation<br>On-the-job training<br>On-the-job competency testing<br>Process Safety Information<br>Selecting participants for PHAs, incident investigations<br>Develop management of change and pre-startup procedures | New Equipment<br>New Process Chemistry<br>New Process Parameters<br>New Procedures<br>Change in Process Utilization | Inform head of training<br>Inform head of maintenance<br>Inform lead for PHAs<br>Inform hazmat team as needed<br>Inform contractors |
| Training Supervisor | Develop, track, oversee operator training program<br>Track competency testing<br>Set up and track operator refresher training<br>Set up training for maintenance<br>Work with contractors | New Equipment<br>New Process Chemistry<br>New Process Parameters<br>New Procedures<br>Change in Process Utilization<br>New regulatory requirements | Revise training and refresher training courses<br>Revise maintenance courses, as needed<br>Inform other leads of need for additional training |
| Maintenance Supervisor | Develop maintenance schedules<br>Oversee and document maintenance<br>Revise schedules as needed | New Equipment<br>New Process Chemistry<br>New Process Parameters<br>New Procedures<br>Change in Process Utilization | Inform operations manager of potential problem areas<br>Inform training supervisor of any training revisions<br>Inform contractors<br>Revise schedules |
| Hazmat Team Chief | Develop and exercise ER plan<br>Train responders<br>Test and maintain ER equipment<br>Coordinate with public responders<br>Select participants in accident investigations | New Equipment<br>New Process Chemistry<br>New Process Parameters<br>New Procedures<br>Change in Process Utilization<br>New regulatory requirements | Revise the ER plan as needed<br>Inform operations manager of problems created by changes<br>Work with training supervisor to revise training of team and others |

## EXHIBIT 5-1
## SAMPLE MANAGEMENT DOCUMENTATION

| Position | Primary Responsibility | Changes | Responsibility re: Changes |
|---|---|---|---|
| Health and Safety Officer | Oversee implementation of RMP<br>Develop accident investigation procedures<br>Oversee compliance audits<br>Develop employee participation plans<br>Conduct contractor evaluations<br>Track regulations | New Equipment<br>New Process Chemistry<br>New Process Parameters<br>New Procedures<br>Change in Process Utilization<br>New regulatory requirements | Inform all leads of new requirements and assign responsibilities<br>Ensure that everyone is informed of changes and that changes are incorporated in programs as needed |

# CHAPTER 6: PREVENTION PROGRAM (PROGRAM 2)

## 6.1    ABOUT THE PROGRAM 2 PREVENTION PROGRAM

If your warehouse is ineligible for Program 1 and the substances you have above the threshold are not covered by OSHA's PSM standard, you have a Program 2 process. For most facilities covered by EPA's rule, the prevention program will be slightly different for each covered process because the hazards and equipment will be different and, therefore, the training and procedures will differ. For warehouses that simply store materials (as opposed to repackaging them), however, the prevention program is likely to be essentially the same for all covered substances. Procedures for moving and stacking containers, operating forklifts and other equipment, and segregating substances will be common to all substances stored. If you start storing a new class of hazardous substances you may have to address segregation issues, but once you have, your procedures and safety information will mainly be the same.

Because of this common approach to prevention, you will probably want to treat your whole building as one process. You should address any differences in the hazard review and safety information, but you should develop a single prevention program that includes all covered substances within a building. If you have more than one building at your facility, you will need to develop separate prevention programs for each building with regulated substances above the threshold. Procedures that are common across buildings need not be duplicated.

The Chemical Manufacturers' Association (CMA), in coordination with the International Warehouse Logistics Association (IWLA) has developed a *Warehouse Assessment Protocol*. The Protocol includes some items that are not covered by the rule (e.g., package labeling), but many of the checklists will be useful in developing your prevention program. Reviewing and adapting applicable parts of the Protocol to your specific operations can save you time while helping you identify issues of concern.

There are seven elements in the Program 2 prevention program, which is Subpart C of part 68. Exhibit 6-1 sets out each of the seven elements and corresponding section numbers.

You must integrate these seven elements into a risk management program that you and your staff implement on a daily basis. Understanding and managing risks must become part of the way you operate. Doing so will provide benefits beyond accident prevention as well. Preventive maintenance and routine inspections will lessen the number of equipment failures and down time.

**EXHIBIT 6-1**
**SUMMARY OF PROGRAM 2 PREVENTION PROGRAM**

| Number | Section Title |
|--------|---------------|
| § 68.48 | Safety Information |
| § 68.50 | Hazard Review |
| § 68.52 | Operating Procedures |
| § 68.54 | Training |
| § 68.56 | Maintenance |
| § 68.58 | Compliance Audits |
| § 68.60 | Incident Investigation |

## 6.2    SAFETY INFORMATION (§ 68.48)

The purpose of this requirement is for you to understand the equipment and chemicals you have, know what limits they place on your operations, and adopt accepted standards and codes where they apply. Having up-to-date information about your process is the foundation of an effective prevention program. Many elements (especially the hazard review) depend on the accuracy and thoroughness of the information this element requires you to provide.

### WHAT DO I NEED TO DO?

You must compile and maintain safety information related to the regulated substances and process equipment for each Program 2 process. You probably have much of this information already, because you would have developed it to comply with OSHA or other rules. EPA has limited the information to what is likely to apply to the processes covered under the Program 2 program. Exhibit 6-2 gives a brief summary of the safety information requirements for Program 2.

### HOW DO I START?

**MSDSs.** You are required to maintain Material Safety Data Sheets under the OSHA Hazard Communication Standard (HCS) (29 CFR 1910.1200). If you are a public warehouse, you should obtain the MSDSs from your customers. If you do not have an MSDS for a regulated substance, you should contact your customer or the manufacturer for a copy. Because the rule states that you must have an MSDS that meets OSHA requirements you may want to review the MSDS to ensure that it is, in

**EXHIBIT 6-2**
**SAFETY INFORMATION REQUIREMENTS**

| You must compile and maintain this safety information: | You must ensure: | You must update the safety information if: |
|---|---|---|
| • Material Safety Data Sheets<br>• Maximum intended inventory<br>• Safe upper and lower parameters<br>• Equipment specifications<br>• Codes & standards used to design, build, and operate the process and building | • That the process is designed in compliance with recognized codes and standards | • There is a *major change* at your business that makes the safety information inaccurate |

fact, complete. Besides the chemical name, the MSDS must have physical and chemical characteristics (e.g., flash point, vapor pressure), the physical hazards (e.g., flammability, reactivity), the health hazards, the routes of entry, exposure limits (e.g., the OSHA permissible exposure level), precautions for safe handling, generally applicable control measures, and emergency and first aid procedures. (See 29 CFR 1910.1200(g) for the complete set of requirements for an MSDS.)

MSDSs also are available from a number of websites. The University of Vermont provides access to three university maintained MSDS collections through its website, http://www.hazard.com. The on-line databases usually have multiple copies of MSDSs for each substance and can help you find an MSDS that is well organized and easy to read. EPA has not verified the accuracy or completeness of MSDSs on any of these sites nor does it endorse any particular version of an MSDS.

**Maximum Inventory**. You must document the maximum intended inventory of any vessel that contains a regulated subject is part of a covered process. This requirement, when applied to warehouses, means that you must document the sizes of vessels that you store. Your customers can provide information on the capacity of the drums, barrels, cylinders, etc., that they store at your facility. You may also want to consider documenting the maximum storage capacity of areas where you store regulated substances. If you are doing predictive filing, as described in Chapter 1, section 1.8, you will want to keep a record of the sizes of vessels you may be storing.

You may want to check with the trade association or standards groups, such as NFPA, that develop standards for your industry to determine if there are any limitations on inventories. For example, fire codes may limit the size of individual flammable storage areas to less than 40,000 square feet. Codes or standards may set stack height limits. These standards will limit your maximum inventories.

**Storage and Process Limits**. You must document the safe upper and lower temperatures and pressures, process flows, and compositions (the last three items will generally not be applicable to warehouses).

Every substance you store or use will have limits on temperature, which will be determined by both the properties of the substance and the vessels. If you do not know these limits, you should contact your customer, the substance manufacturer (if different), or your trade association. They will be able to provide the data you need. It is important that you know these limits so you can avoid situations where these limits may be violated. Many people are aware of the dangers of exposing their vessels to high temperatures, but extreme low temperatures also may pose hazards you should know about.

**Equipment Specifications**. You must document any equipment you use to store, repackage, or move regulated substances. Equipment specifications will usually include information on the materials of construction, actual design, and tolerances. The vendor should be able to provide this information; you may have the specifications in your files from the time of purchase. You are not expected to develop engineering drawings of your equipment to meet this requirement. For warehouses, this requirement will apply mainly to forklifts and other equipment used to lift or move drums, barrels, pallets, etc., as well as storage racks. It is important that you understand the limitations on this equipment so that it can be operated and used properly.

The actual containers for the regulated substances should be designed to meet DOT performance oriented packaging rules. You need only ensure that containers you store meet DOT specifications; you do not need to maintain copies of the DOT specifications unless you package regulated substances at your warehouse.

**Codes and Standards**. You must document the codes and standards you used to design and build your facility and that you follow to operate. These codes will include the electrical and building codes that you must comply with under state laws. Besides the construction of the building, you should consider racks that you use for storage, sprinkler systems, heating and ventilation systems, and any other equipment or design features that affect the safety of your warehouse. Exhibit 6-3 lists some codes that may be relevant to your operation.

Note that the National Fire Protection Association (NFPA) codes may have been adopted as state or local codes. The American National Standards Institute (ANSI) is an umbrella, standards-setting organization, which imposes a specific process for gaining approval of standards and codes. ANSI codes may include codes and standards also issued by other organizations, which are incorporated by reference.

The CMA *Warehouse Assessment Protocol* has a section on loss prevention (section 4 of the protocol) that can help you identify areas of concern on design.

**EXHIBIT 6-3**
**CODES AND STANDARDS**

| Organization | Subject/Codes |
|---|---|
| American National Standards Institute (ANSI) | Piping, Electrical, Power wiring, Instrumentation, Lighting, Product storage and handling, Insulation and fireproofing, Painting and coating, Ventilation, Noise and Vibration, Fire protection equipment, Safety equipment, Pumps, Compressors, Motors, Refrigeration equipment, Pneumatic conveying |
| National Fire Protection Association (NFPA) | Fire pumps, Combustible liquid code, Flammable liquid code, Plant equipment and layout, Electrical system design, Shutdown systems, Venting requirements, Gas turbines and engines, Storage tanks, Gas code |
| American Society for Testing Materials (ASTM) | Inspection and testing, Noise and vibration, Materials of construction, Piping materials and systems, Instrumentation |

## HOW DO I DOCUMENT ALL THIS?

EPA does not expect you to develop piles of papers to document your safety information. Your MSDS(s) are usually three or four pages long. You only have to keep them on file, as you already do for OSHA. Equipment specifications are usually on a few sheets or a booklet provided by the vendor; you need only keep these on file. You can probably document the other information on a single sheet that simply lists each of the required items and any codes or standards that apply. See Exhibit 6-4 for a sample. Maintain that sheet in a file and update it whenever any item changes or new equipment is added.

The equipment specifications and list of standards and codes will probably meet the final requirement, that you ensure that your process is designed in compliance with recognized and generally good engineering practices. If you have any doubt that you are meeting this requirement, you should contact your trade association to determine if there are practices or standards that you are not aware of that may be useful in your operation.

**EXHIBIT 6-4**
**SAMPLE SAFETY INFORMATION SHEET**

| REQUIREMENT | CURRENT DATA/LIMITS |
|---|---|
| MSDSs on file<br>Nitric Acid<br>Hydrochloric Acid<br>Hydrofluoric Acid<br>Acrylonitrile<br>Flammable mixtures | Data of last update:<br>1994<br>1996<br>1995<br>1997<br>5 mixtures (1998, 1997, 1999) |
| Maximum Intended Inventory | Largest Vessel:  55-gallon drums<br>Maximum storage in any section 1,000 drums<br>Maximum area storage for  flammables 30,000 square feet.<br>Aerosol (flammable) storage less than 100 cubic feet |
| Temperature | Upper:<br>Lower: |
| Equipment Specifications<br> Fork lifts<br> Sprinkler system<br>  Wet system<br>  Foam system<br>  Rack system<br> Storage racks<br> Exhaust fans<br> Fire extinguishers<br> Alarm system | Specifications on file:<br>Last update, 1992<br><br>Construction drawings and specifications<br>Construction drawings and specifications<br>Construction drawings and specifications<br>Manufacturer's specifications (1985)<br>Manufacturer's specifications (1993)<br>Manufacturer's specifications (1995)<br>Manufacturer's specifications (1985) |
| CODES AND STANDARDS | |
| Building construction<br> Floor<br> Interior walls<br> Ceiling<br> Fire doors | State building and fire code met |
| Electrical | State electrical code met |
| Sprinkler system | State building code; NFPA met |
| Ventilation system | State building code met |
| Racks | |
| Stack heights, separations | |

After you have documented your safety information, you should double check it to be sure that the files you have reflect the equipment you are currently using. It is important to keep this information up to date. Whenever you replace equipment, be sure that you put the new equipment specifications in the file and consider whether any of your other prevention elements need to be reviewed to reflect the new equipment.

## 6.3    HAZARD REVIEW (§ 68.50)

The hazard review will help you determine whether you're meeting applicable codes and standards, identify and evaluate the types of potential failures, and focus your emergency response planning efforts.

### WHAT DO I NEED TO DO?

The hazard review is key to understanding your operation and continuing to operate safely. You must identify and review specific hazards and safeguards of your Program 2 processes. Exhibit 6-5 summarizes things you must do for a hazard review.

**EXHIBIT 6-5**
**HAZARD REVIEW REQUIREMENTS**

| Conduct a review & identify... | Use a guide for conducting the review. | Document results & resolve problems. | Update your hazard review. |
|---|---|---|---|
| • The hazards associated with the Program 2 process & regulated substances. <br> • Opportunities for equipment malfunction or human error that could cause a release. <br> • Safeguards that will control the hazards or prevent the malfunction or error. <br> • Steps to detect or monitor releases. | • You may use any checklist (such as you might in a model risk management program) to conduct the review. <br> • For a process designed to industry standards like NFPA-58 or Federal /state design rules, check the equipment to make sure that it's fabricated, installed, and operated properly. | • Your hazard review must be documented and you must show that you have addressed problems. | • You must update your review at least once every five years or whenever there is a major change in the process. <br> • You must resolve problems identified in the new review *before* you startup the changed process. |

### How Do I Start?

There are three possible approaches to conducting a hazard review; which you use will depend on your particular situation.

**Processes designed to industry-specific codes**. If all or part of your warehouse and its operation was designed and built to comply with a federal or state standard for that operation or an industry-specific design code, your hazard review will be relatively simple. The standard-setting organization has already conducted a hazard review, identified the hazards, and designed the equipment and operating requirements to minimize the risks. You can use the code or standard as a checklist. The purpose of your review is to ensure that your equipment still meets the code and is being operated in appropriate ways.

**Industry checklist.** CMA's *Warehouse Assessment Protocol,* particularly the Warehouse Assessment (as opposed to the Management Systems Assessment), can provide the basis for a hazard review checklist. CMA and IWLA have already identified what your general hazards are and what types of equipment and procedures you should be using. Your job is to use the checklist to decide if you meet the requirements and, if you do not, whether you should. In some cases, your individual circumstances may make a checklist item unnecessary. You should tailor this checklist to add chemical-specific concerns. For example, if you handle a wide range of chemicals across hazard classes, you will want to be sure that these materials are segregated properly. The segregation criteria you use should be documented. If you have an operating engineer on staff, he or she may be able to conduct the review. If you do not have any technical staff, your vendor or trade association may be able to help you. If you seek outside help, however, work with them so that you understand what they find.

**Develop your own checklist**. If you do not choose to use the CMA protocol or industry standards, you will have to conduct your own hazard review. As discussed in the requirements section, the review must identify:

- • The hazards of the substance and process;
- • Possible equipment failures or human errors that could lead to a release;
- • Safeguards used to prevent failures or errors; and
- • Steps needed to detect or monitor releases.

You will probably be able to define the hazards of the substances using the MSDSs, which list the hazardous properties of the substances. The hazards of the process (as opposed to the equipment) will be limited because you may not actually handle the substance outside of the container. However, if you mix or repackage chemicals, or if you fail to segregate hazard classes, you may have process hazards that you need to define. Your safety information should help.

The next step may be to conduct a simplified What If process, where your technical staff ask for each piece of equipment and procedure, "What if this fails?" and "What if the operator fails to do this?" Most industry standards and codes have already considered these questions and developed responses, in terms of design and operating practices. If you are doing this on your own, the important thing to remember is that you should not assume that something will not happen. Ask why something could not happen and whether the safeguards that you think protect the equipment or operator are really adequate. In many cases, they may be adequate, but it is useful to ask, to force yourself to examine your own assumptions.

From this exercise, you should develop a checklist of items that you need to check. For example, you may have decided that your racks can hold a certain weight. The checklist would then include an item to check procedures to be sure that they reflect this limit. You may have identified puncturing drums with a forklift as the most likely operator error. Your checklist might then include both a check of operating procedures that address proper practices, plus a check of the width of corridors separating racks or pallets to ensure that forklift operators have enough space to maneuver.

When you finish the checklist, it is useful to show it to your operators. They are familiar with the equipment and may be able to point out other areas of concern. A review with your vendors or trade association may also help; their wider knowledge of the industry may give them ideas about failures you may not have experienced or considered.

## CAUTION

Whichever approach you use, remember, you should consider external events as well as internal failures. If you are in an area subject to earthquakes, hurricanes, floods, or heavy snow you should examine whether your warehouse would survive these natural events without releasing the substance. You should consider the potential impacts of lightning strikes and power failures (e.g., if you lost heating in midwinter would that create dangerous situations?). These considerations may not be part of standard checklists. If you use these standards, you may have to modify them to address these site-specific concerns. Never use someone else's checklist blindly. You must be sure that it addresses all of your potential problems.

## DOCUMENTING THE REVIEW

You should maintain a copy of the checklist you used. The easiest way to document findings is to enter them into the checklist after each item. This approach will give you a simple, concise way of keeping track of findings and recommendations. Exhibit 6-6 provides a sample of part of a checklist (adapted from the CMA Warehouse Assessment Protocol, 1996). You may also want to create a separate document of recommendations that require implementation or other resolution. EPA does not require that you implement every recommendation. It is up to you to decide which recommendations are necessary and feasible. You may decide that other steps are as effective as the recommended actions or that the risk is too low to merit the expense. You must document your decision on each recommendation.

**EXHIBIT 6-6**
**SAMPLE CHECKLIST**

| Storage and Handling | Yes | No | Comments |
|---|---|---|---|
| Are chemicals segregated from foods/consumer goods? | | | |
| Are chemicals segregated by hazard class? | | | |
| Are damaged containers marked and segregated? | | | |
| Are product temperature specifications followed? | | | |
| Are there floor markings to indicate storage spaces, aisles, staging areas, and routes? | | | |
| Are products stacked properly to height specifications in accordance with fire regulations? | | | |
| Are there indications of exceeding height requirements, such as crushed boxes? | | | |
| Are aisle distances between stacking racks appropriate for safe access with mechanical handling equipment? | | | |
| Is aisle distance maintained for safe access for fire fighting? | | | |
| Is there at least one meter between the top of the stack and sprinkler heads? | | | |
| Are products stored outside of the pathway of forced air conditioning and heating units? | | | |
| Are products stored in areas other than on the floor? | | | |
| Is there a designated area for drums or intermediate bulk containers stored outside? | | | |
| Are empty pallets stored in accordance with fire regulations? | | | |
| Are container labels visible? | | | |

## UPDATES

You must update the review every five years or whenever a major change occurs. For most warehouses, major changes will be limited. If you start storing a new substance, particularly if it is in a hazard class you have not handled before, you would want to consider whether the new type of hazard requires any additional actions (e.g., different type of fire suppression system, new segregation patterns). In most cases, adding new regulated substances in a hazard class you already handle (flammable liquids, acids) will not be considered a major change. Even if the changes prove to be minor and do not require an update, you should examine the process carefully before starting. You will operate more safely if you take the time to evaluate the hazards before proceeding.

## 6.4   OPERATING PROCEDURES (§ 68.52)

Written operating procedures describe what tasks a process operator must perform, set safe process operating parameters that must be maintained, and set safety precautions for operations and maintenance activities. These procedures are the guide for telling your employees how to work safely everyday, giving everyone a quick source of information that can prevent or mitigate the effects of an accident, and providing workers and management with a standard against which to assess performance.

### WHAT DO I NEED TO DO?

You must prepare written operating procedures that give workers clear instruction for safely conducting activities involving a covered process. You may use standardized procedures developed by industry groups or provided in model risk management programs as the basis for your operating procedures, but be sure to check that these standard procedures are appropriate for your activities. If necessary, you must update your Program 2 operating procedures whenever there is a major change and before you startup the changed process. Exhibit 6-7 briefly summarizes what your operating procedures must address.

Your operating procedures must be:

- Appropriate for your equipment and operations;
- Complete, and
- Written in language that is easily understood by your operators.

The procedures do not have to be long. If you have simple equipment that requires a few basic steps, that is all you have to cover.

## EXHIBIT 6-7
## OPERATING PROCEDURES REQUIREMENTS

| Steps for each operating phase | Operating limits |
|---|---|
| • Initial startup<br>• Normal operations<br>• Temporary operations<br>• Emergency shutdown<br>• Emergency operations<br>• Normal shutdown<br>• Startup following a normal or emergency shutdown or a major change | • Consequences of deviating<br>• Steps to avoid, correct deviations |

### WHERE DO I START

If you already have written procedures, you may not have to do anything more. Review the procedures. If you are satisfied that they meet the criteria listed above, you are finished. You may want to check them against any recommended procedures provided by equipment manufacturers, trade associations, or standard setting organizations, but you are not required to do so. You are responsible for ensuring that the procedures explain how to operate your equipment and store chemicals safely.

If you do not have written procedures, you may want to check with equipment manufacturers, trade associations, or standard setting organizations. They may have recommended practices and procedures that you can adapt. Do not accept anyone else's procedures without checking to be sure that they are appropriate for your particular equipment and uses and are written in language that your operators will understand. You may also want to review any requirements imposed under state or federal rules. For example, if you are subject to federal, state, or local rules for loading and unloading of hazardous materials, those rules may dictate some procedures. Copies of these rules may be sufficient for those operations.

### WHAT DO THESE PROCEDURES MEAN?

The rule lists eight procedures. Not all of them will be applicable to you if you only store substances. The following is a brief description to help you decide whether you need to develop procedures for each item. If a particular element does not apply, do not spend any time on it. We do not expect you to create a document that is meaningless to you. You should spend your time on items that will be useful to you.

**Initial Startup.** This item will only apply to you if you repackage or mix chemicals. For most warehouses this item is not applicable. If you handle the chemicals outside of the containers, as opposed to simply storing and moving the containers, this item covers all the steps you need to take before you start a process for the first time. You should include all the steps needed to check out equipment as well as the steps needed to start the process itself.

**Normal Operations**. These procedures should cover your basic operations. These are your core procedures that you expect your operators to follow on a daily basis to run your warehouse safely. For a warehouse, these would include the following:

- • Segregation and storage procedures
- • Use of forklifts
- • Loading and unloading
- • Examination for damage and labeling
- • Stock controls
- • Site security
- • Bracing and stacking
- • Hot work
- • Handling damaged containers

You may also have to cover the HVAC system if failure of this system could lead to a release.

Some of these operations are covered by federal, state, or local rules (e.g., loading and unloading may be covered by US DOT; hot work is covered by OSHA). Your procedures should represent compliance with any applicable rules.

**Temporary Operations**. These operations are short-term; they will usually occur either when your regular process is down or when additional capacity is needed for a limited period. The procedures should cover the steps you need to take to ensure that these operations will function safely. The procedures will generally cover pre-startup checks and determinations (e.g., can the material be segregated properly?). The actual operating procedures for running the temporary process will be written as the operation is put into place.

You may need to consider procedures to ensure that if a new substance or product is brought into the warehouse for temporary storage, the necessary steps are taken before that storage to ensure that it is safe (e.g., barrels are not stacked too high, or located with incompatible substances).

**Normal Shutdown**. These procedures may not apply to warehouses unless you repackage. If you do not repackage, you may not need procedures for this step unless you use automatic equipment for moving containers

**Startup following a normal or emergency shutdown or a major change**. For most warehouses, these procedures are likely to be similar to those for initial startup.

Startup procedures following an emergency shutdown or a major change may include more equipment checks because you may need to check new or repaired equipment on a more frequent basis. You should include all the steps your workers should take to ensure that the process can operate safely. These procedures may not apply to warehouses in most instances.

**Consequences of Deviations**. Your operating procedures should tell the workers what will happen if something starts to go wrong. For example, if a rack appears to be sagging inward, the operator must know (1) whether this poses a problem that must be addressed and (2) what steps to take to correct the problem or otherwise respond to it. You should include this information in each of the other procedures (startup, normal operations, shutdowns), rather than as separate documents.

If you have substances with a distinctive odor, color, or other characteristic that operators will be able to sense, you should include in your procedures information about what to do if they notice leaks. Frequently, people are the most sensitive leak detectors. Take advantage of their abilities to catch leaks before they become serious.

**Equipment Inspections**. You should include steps for routine inspection of equipment by operators as part of your other procedures. These inspections cover the items that operators should look for on a daily basis to be sure that the equipment is running safely (e.g., vibration checks, leakage, overheating equipment). These inspections are not the same as those detailed checks that maintenance workers will perform, but rather are the "eyeball," "sound," and "feel" tests that experienced operators do often without realizing it. Most likely, your warehouse is already doing OSHA pre-checks and checks after work shifts. If you need further assistance, your operators, your vendors, and your trade association can help you define the things that should trigger concern: How much vibration is normal? What does a smoothly running motor sound like?

## CMA PROTOCOL

The CMA *Warehouse Assessment Protocol* provides a checklist of operational practices in its Management Systems Assessment. You may want to review this list; some of the items on the list are not specifically covered by the rule (e.g., traffic office procedures), but may be important to efficient running of your warehouse. For warehouses, more than for many other businesses covered by this rule, the total operation of the business is relevant to safety. Although many of the substances you handle will not be subject to this rule, you are likely to use the same procedures that you use for covered substances for the other chemicals you store.

## UPDATING PROCEDURES

You must update your procedures whenever you change your process in a way that alters the steps needed to operate safely. A change in the process, for a warehouse, is likely to involve either the introduction of new equipment or introduction of a new class of chemicals. If you add new equipment, you will need to expand your procedures or develop a separate set to cover the new items. If you store a class of chemicals you have not handled before, you will need to inform your workers of the hazards and make sure that these substances are segregated properly. Storing containers of a chemical you have not handled before, but which is part of the same class (e.g., flammables) that you already handle, would not be considered a change unless the chemical had some other hazard of concern that you have not handled before.

## WHAT KIND OF DOCUMENTS DO I HAVE TO KEEP?

You must maintain your current set of operating procedures. You are not required to keep old versions; in fact, you should avoid doing so because keeping copies of outdated procedures may cause confusion. You should date all procedures so you will know when they were last updated.

## 6.5   TRAINING (§ 68.54)

Training programs often provide immediate benefits because trained workers have fewer accidents, damage less equipment, and improve operational efficiency. Training gives workers the information they need to understand how to operate safely and why safe operations are necessary. A training program, including refresher training, is the key to ensuring that the rest of your prevention program is effective. You already have some type of training program because you must conduct training to comply with OSHA's Hazard Communication standard (29 CFR 1910.1200) and DOT training requirements.

## WHAT DO I NEED TO DO?

You must train all new workers in your operating procedures developed under the previous element; if any of your more experienced workers need training on these procedures, you should also train them. Any time the procedures are revised, you must train everyone using the new procedures. At least once every three years, you should provide refresher training on the operating procedures even if they have not changed. The training must cover all parts of the operating procedures, including information on the consequences of deviations and steps needed to address deviations. New hires should be trained before being allowed to operate equipment or handle regulated substances.

For current workers, you may certify in writing that the employees have the "required knowledge, skills, and abilities to safely carry out the duties and responsibilities as provided in the operating procedures." This "grandfather clause" means that you do not need to conduct additional training for employees who are employed prior to June 21, 1999, and who have the appropriate knowledge and skills to operate covered processes safely, in accordance with the operating procedures. This certification should be kept in your files; you do not need to submit it to EPA.

You are not required to provide a specific amount of training or type of training. You should develop a training approach that works for you. If you are a small facility, one-on-one training and on-the-job training may work best. Larger facilities may want to provide classroom training or video courses developed by vendors or trade associations before moving staff on to supervised work. You may have senior operators present the training or use trainers provided by vendors or other outside sources. The form and the length of the training will depend on your resources and your processes. If you can teach someone the basics in two hours and move them on to supervised work, that is all right. The important thing is that your workers understand how to operate safely and can carry out their tasks properly. We are interested in the results of the training, not the details of how you achieve them. Find a system that works for you. Exhibit 6-8 lists things that you may find useful in developing your training program.

## How Does This Training Fit with Other Required Training?

You are required by OSHA to provide training under the hazard communication standard; this training covers the hazards of the chemicals and steps to take to prevent exposures. DOT has required training for loading and unloading of hazardous materials. Some of that training will cover items in your operating procedures. You do not need to repeat that training to meet EPA's requirements. You may want to integrate the training programs, but you do not have to do so.

## What Kind of Documentation Do I Need to Keep?

You are not required to maintain documentation of your training program. You may, however, want to keep an attendance log for any formal training courses and refresher training to ensure that everyone who needs to be trained is trained. Such logs will help you when you do a compliance audit; without such logs you will have to rely on your memory and the memory of your operators. Again, you are not required to keep them for this rule.

**EXHIBIT 6-8
TRAINING CHART**

| | |
|---|---|
| • **Who needs training?** | Clearly identify the employees who need to be trained and the subjects to be covered. |
| • **What are the objectives?** | Specify learning objectives, and write them in clear, measurable terms before training begins. Remember that training must address the process operating procedures. |
| • **How will you meet the training objectives?** | Tailor the specific training modules or segments to the training objectives. Enhance learning by including hands-on training like using simulators whenever appropriate. Make the training environment as much like the working environment as you can, consistent with safety. Allow your employees to practice their skills and demonstrate what they know. |
| • **Is your training program working?** | Evaluate your training program periodically to see if your employees have the skills and know the routines required under your operating procedures. Make sure that language or presentation are not barriers to learning. Decide how you will measure your employees's competence. |
| • **How will your program work for new hires and refresher training?** | Make sure all workers – including maintenance and contract employees – receive initial and refresher training. If you make changes to process chemicals, equipment, or technology, make sure that involved workers understand the changes and the effects on their jobs. |

## 6.6   MAINTENANCE (§ 68.56)

You have several elements you must satisfy: you must develop maintenance procedures, train your workers in these procedures, and carry out inspections and tests on your equipment; if you use a contractor for maintenance, you must ensure that the contractors are able to follow your procedures. Maintenance procedures should cover routine maintenance, inspection, and testing. For warehouses, maintenance will apply primarily to equipment used to move storage containers (lifts, conveyors, ladders, dock equipment). If you repackage regulated substances, equipment use to repackage will be covered.

### WHAT DO I NEED TO DO?

You must prepare and implement procedures for maintaining the mechanical integrity of process equipment, and train your workers in the maintenance procedures. You may use procedures or instructions from equipment vendors, in Federal or state regulations, or in industry codes as the basis of your maintenance program. You should develop a schedule for inspecting and testing your equipment based on manufacturers' recommendations or your own experience. Exhibit 6-9 briefly summarized the elements of a maintenance program that would satisfy EPA's rule.

## How Do I Start?

Your first steps will probably be to determine whether you already meet all these requirements. If you review your existing written procedures and determine that they are appropriate, you do not need to revise or rewrite them. If your workers are already trained in the procedures and carry them out, you may not need to do anything else.

If you do not have written procedures, you will need to develop them. Your equipment vendors may be able to provide procedures and maintenance schedules. Using these as the basis of your program is acceptable. Your trade association may also be able to help you with industry-specific checklists. If there are existing standards, your trade association can provide you with the references. Copies of these may form the basis for your maintenance program. If there are federal or state regulations that require certain maintenance, you should use these as well.

### EXHIBIT 6-9
### MAINTENANCE GUIDELINES

| Written procedures | Training | Inspection & testing |
|---|---|---|
| • You may use someone else's procedures as the basis for your program. If you choose to develop your own, you must write them down. | • Train process maintenance employees in process hazard and how to avoid or correct an unsafe condition.<br>• Make sure this training covers the procedures applicable to safe job performance. | • Inspect & test process equipment.<br>• Use recognized and generally accepted good engineering practices.<br>• Follow a schedule that matches the manufacturer's recommendations or that prior operating experience indicates is necessary. |

You need to determine if procedures provided by vendors, manufacturers, trade associations, or others are appropriate for your operation. If you are operating in a standard way (e.g., using your equipment in the way it was designed for), you may assume that these other procedures will work for you. If you are using equipment for purposes other than those for which it was designed, your best option is obtain appropriate equipment. In the interim, you should consult with the equipment manufacturer or vendor to decide whether your use is safe and whether you need to upgrade maintenance practices or increase the frequency and rigor of inspection and testing.

## TRAINING

Once you have written procedures, you must ensure that your maintenance workers are trained in the procedures and in the hazards of the process. As with the training discussed in the previous section, how you provide this training is up to you. We believe that you are in the best position to decide how to train your workers. Vendors may provide the training or videos; you may already provide training on hazards and how to avoid or correct them as part of Hazard Communication Standard training. You do not need to repeat this training to comply with this rule. New hires or temporary workers must be trained before they perform maintenance on covered equipment.

If you hire contractors to do your maintenance, you must ensure that they are trained to carry out the procedures. You can do this by providing training or by developing agreements with the contractor that gives you the assurance that only trained workers will sent to your site. In some cases, you may be able to rely on licenses (e.g., electricians).

## INSPECTION AND TESTING

You must establish a schedule for inspection and testing equipment associated with covered processes. You may obtain recommendations from manufacturers, vendors, or trade associations. You should, however, use your own experience as a basis for examining any schedules you obtain from others. Many things may affect whether a schedule is appropriate. The manufacturer may assume a certain rate of use. If your use (e.g., the hours per day a forklift is operated) varies considerably, the variations may affect the wear on the equipment. Extreme weather conditions may also impact wear on equipment.

Talk with your operators as you prepare or adopt these procedures and schedules. If their experience indicates that equipment fails more frequently than the manufacturer expects, you should adjust the inspection schedule to reflect that experience. Your trade association may also be able to provide advice on these issues.

## WHAT KIND OF DOCUMENTATION MUST I KEEP?

You must keep your written procedures and schedules as well as any agreements you have with contractors. You are not required to keep training logs or maintenance logs to comply with this rule. You may, however, want to maintain such logs for your own use. Without some record, you will have to rely on workers' memories about when something was last checked. As workers leave or change jobs at your company, it can be difficult to keep track of when inspections and tests were done.

Maintaining a record of when something was last done or is scheduled to be done next can help keep your program working smoothly.

## 6.7    COMPLIANCE AUDITS (§ 68.58)

Any risk management program should be reviewed periodically to ensure that employees and contractors are implementing it properly. A compliance audit is a way for you to evaluate and measure the effectiveness of your risk management program. An audit reviews each of the prevention program elements to ensure that they are up-to-date and are being implemented and will help you identify problem areas and take corrective actions. As a result, you'll be running a safer operation.

### WHAT DO I NEED TO DO?

At least every three years, you must certify that you have evaluated compliance with EPA's requirements for the prevention program for each covered process. At least one person on your audit team must be knowledgeable about the process. You must develop a report of your findings, determine and document an appropriate response to each finding, and document that you have corrected any deficiency.

The purpose of the compliance audit is to ensure that you are continuing to implement the risk management program as required. Remember, the risk management program is an on-going process; it is not a set of documents that you develop and put on a shelf in case the government inspects your site. To be in compliance (and gain the benefits) procedures must be followed on a daily basis; documents must be kept up to date. The audit will check these items and provide you with items that need to be improved.

You must check each of the items in the prevention program. Because you have simple procedures, the audit should not take a long time. You may want to use the CMA protocol as the basis of your audit or tailor it to fit your operation.

Once you have the checklist, you, your chief operator, or some other person who is knowledgeable about your process, singly or as a team, should walk through the facility and check on each of the items, writing down comments and recommendations. You may want to talk with employees to determine if they have been trained and are familiar with the procedures.

You must respond to each of the findings and document what actions, if any, you take to address problems. You should take steps to correct any deficiencies you find.

You may choose to have the audit conducted by a qualified outside party. For example, you may have someone from another part of your company do the audit or hire an expert in warehousing. If you do either of these, you should have someone work with the person, both to understand the findings and answer questions.

Remember, this is an audit of compliance with the prevention program of this rule. You may choose to expand the scope of the audit to cover your compliance with other parts of the rule and the overall safety of your operation, but you are not required to do so.

### WHAT KIND OF DOCUMENTATION MUST I KEEP?

You must keep a written record of the findings and actions for five years. You may also want to keep a record of who conducted the audit, but you are not required to do this. Exhibit 6-10 provides a sample format for documenting the audit and subsequent actions.

---

**Q and A**
**AUDITS**

**Q.** Do the compliance audits cover all of the Part 68 requirements or just the prevention program requirements?

**A.** The compliance audit applies only to the requirements of the prevention programs under Subpart C. If you have a Program 2 process you must certify that you have evaluated compliance with the Program 2 prevention program provisions at least every three years to verify that the procedures and practices developed under the rule are adequate and are being followed. You may want to expand your audit to check other part 68 elements but you are not required to do so.

---

## 6.8    INCIDENT INVESTIGATION (§ 68.60)

Incidents can provide valuable information about site hazards and the steps you need to take to prevent accidental releases. Often, the immediate cause of an incident is the result of a series of other problems that need to be addressed to prevent recurrences. For example, an operator's mistake may be the result of poor training. Equipment failure may result from improper maintenance or misuse. Without a thorough investigation, you may miss the opportunity to identify and solve these problems.

**EXHIBIT 6-10**
**SAMPLE AUDIT CHECKLIST**
**FOR SAFETY INFORMATION AND HAZARD REVIEWS**

| Element | Yes/No/NA | Action/Completion Data |
|---|---|---|
| **Safety Information** | | |
| MSDSs updated? | | |
| Maximum intended inventory determined? | | |
| Determined<br>Safe upper and lower temperature?<br>Segregation of incompatible substances | | |
| Equipment specifications<br>Forklifts<br>Fire suppression systems<br>Ventilation system | | |
| **Hazard Review** | | |
| Are incompatible materials appropriately segregated? | | |
| Is the fire suppression system appropriate for materials stored? | | |
| Are stack heights in accordance with industry standards and codes? | | |
| Has equipment been inspected to determine if it is operated according to industry standards and codes? | | |
| Are the results of the inspections documented? | | |
| Have inspections been conducted after every major change? | | |

## WHAT DO I NEED TO DO?

You must investigate each incident that resulted in, or could have resulted in a "catastrophic release of a regulated substance." A catastrophic release is one that presents an imminent and substantial endangerment to public health and the environment. The easiest way to understand imminent and substantial endangerment is to consider whether the release could have exposed the public to levels that exceed the toxic or flammable endpoints. If a release had that potential, even if no such exposure occurred (because of favorable weather conditions or because the adjoining facilities were unoccupied at the time), you should investigate. Most warehouse accidents will not meet this criterion; minor spills of toxic substances that are contained within the warehouse building are unlikely to represent a potential catastrophic release. Minor fires, however, may represent potential catastrophic releases if the fire had the potential to spread and release toxic substances. Spills of toxic regulated substances outside may pose a threat to the public and should be investigated. Exhibit 6-11 briefly summarizes the steps you must take for investigating incidents.

**EXHIBIT 6-11**
**INCIDENT INVESTIGATION REQUIREMENTS**

| | |
|---|---|
| • **Initiate an investigation promptly**. | Begin investigating no later than 48 hours following the incident. |
| • **Summarize the investigation in a report**. | Among other things, this report will include the factors contributing to the incident. Remember that identifying the root cause may be more important than identifying the initiating event. Remember, also, that the purpose of the report is to help management take corrective action. |
| • **Address the team's findings and recommendations.** | Establish a system to address the incident report findings and recommendations and document resolutions and corrective actions. |
| • **Review the report with your staff and contractors.** | You must share the report - its findings and recommendations - with affected workers whose job tasks are relevant to the incident. |
| • **Retain the report.** | Keep incident investigation summaries for five years. |

## How Do I Start?

You should start with a simple set of procedures that you will use to begin an investigation. You may want to assign someone to be responsible for compiling the initial incident data and putting together the investigation team. If you have a small facility, your "team" may be one person who works with the local responders, if they were involved.

The purpose of the investigation is to find out what went wrong and why, so you can prevent it from happening again. Do not stop at the obvious failure or "initiating event" (e.g., the hose was clogged, the operator forgot to check the connection); try to determine why the failure occurred. If you write off the accident as operator error alone you miss the chance to take the steps needed to prevent such errors the next time. Similarly, if equipment fails, you should try to decide whether it had been used or maintained improperly.

Remember, your goals are to prevent accidents, not to blame someone, and correct any problems in your prevention program. In this way, you can prevent recurrences.

In many cases, an investigation will not take long. If you have a complex facility, if equipment has been severely damaged, or the workers seriously hurt, an investigation may take several days. You should talk with the operators who were in the area at the time and check records on maintenance (another reason for keeping logs). If equipment has failed in an unusual way, you may need to talk to the manufacturer and your trade association to determine if similar equipment has suffered similar failures.

You must develop a summary of the accident and its causes and make recommendations to prevent recurrences. You must address each recommendation and document the resolution and any actions taken. Finally, you must review the findings with operators affected by the findings.

## What Kind of Documentation Must I Keep?

You must maintain the summary of the accident, recommendations, and actions. A sample format is shown in Exhibit 6-12 that combines all of these in a single form. Note that the form also includes accident data that you will need for the five-year accident history. These data are not necessarily part of the incident investigation report, but including them will create a record you can use later to create the accident history.

## 6.9   CONCLUSION

Many of you will need to do little that's new to comply with the Program 2 prevention program, because you already are complying with many program elements through other Federal rules, state requirements, and industry-specific codes and standards. And if you've voluntarily implemented OSHA's PSM standard for your Program 2 process, you'll meet the lesser Program 2 prevention program requirements. No matter what choices you make in complying with the Program 2 prevention program, keep these things in mind:

Integrate the elements of your prevention program. For Program 2 owners and operators, a major change in any single element of your program should lead to a review of other elements to identify any effect caused by the change.

Make accident prevention an institution at your site. Like the entire risk management program, a prevention program is more than a collection of written documents. It is a way to make safe operations and accident prevention the way you do business everyday.

Check your operations on a continuing basis and ask if you can improve them to make them safer as well as more efficient.

### EXHIBIT 6-12
### SAMPLE INCIDENT INVESTIGATION REPORT

| Hydrofluoric Acid Release | | |
|---|---|---|
| Date: May 15, 1998; 3 pm | Substance: Hydrofluoric acid (70%) | Quantity: 1800 pounds |
| Duration: 2 hours | Weather: 82· F, 8 mph winds, WSW | |
| Description: | A forklift punctured two 55-gallon drums of HF and severely damaged two other drums on the pallet, which then split open as they fell off the loading dock. Five workers and two local responders were treated for exposure. Neighboring facilities were notified to shelter in place. | |
| **Findings** | **Recommendations** | **Actions** |
| The forklift controls stuck. | Institute more frequent inspections and tests of the forklifts. | Changed inspection and testing intervals; revised procedures; conducted training on new procedures |

| Hydrofluoric Acid Release | | |
|---|---|---|
| Operator and other workers left the scene to protect themselves. It took 15 minutes for the hazmat staff to suit up and begin responding. | Conduct exercises quarterly for hazmat staff. Conduct refresher training for other staff on evacuation and notification procedures. | Exercise schedule established. Refresher training provided; safety meetings added and held on a monthly basis to review safety issues |
| Inadequate quantities of neutralizer were available. Supply had not been replenished after several minor spills. | Check and replenish supply monthly or after each use. | Routine checks added to work order schedule. |

# CHAPTER 7: PREVENTION PROGRAM (PROGRAM 3)

Warehouses are subject to Program 3 only if they are subject to the OSHA PSM standard. Many of you will need to do little that is new to comply with the Program 3 prevention program, because you already have the OSHA PSM program in place. Whether you're building on the PSM standard or creating a new program, keep these things in mind.

- EPA and OSHA have different legal authority — EPA for offsite consequences, OSHA for on-site consequences. If you are already complying with the PSM standard, your process hazard analysis (PHA) team may have to assess new hazards that could affect the public or the environment offsite. Protection measures that are suitable for workers (e.g., venting releases to the outdoors) may be the very kind of thing that imperils the public.

- Integrate the elements of your prevention program. You must ensure that a change in any single element of your program leads to a review of other elements to identify any effect caused by the change.

- Most importantly, make accident prevention an institution at your site. Like the entire risk management program, a prevention program is more than a collection of written documents. It is a way to make safe operations and accident prevention the way you do business everyday.

## 7.1 PROGRAM 3 PREVENTION PROGRAM AND OSHA PSM

The Program 3 prevention program includes the requirements of the OSHA PSM standard. Whenever we could, EPA used OSHA's language verbatim. However, there were a few terms that EPA had to change to reflect the differences between its authority and OSHA's. For example, OSHA regulates to protect workers; EPA's responsibility is to protect public health and safety and the environment. Therefore, an "owner or operator" subject to EPA's rule must investigate catastrophic releases "that present(s) (an) imminent and substantial endangerment to public health and the environment," but an OSHA "employer" would focus its concerns on the workplace. To clarify these distinctions, we deleted specific references to workplace impacts and "safety and health" contained in OSHA's PSM standards. We also used different schedule dates and references where appropriate. Exhibit 7-1 compares terms in EPA's rule with their counterparts in the OSHA PSM standard.

**EXHIBIT 7-1**
**COMPARABLE EPA AND OSHA TERMS**

| OSHA TERM | EPA TERM |
|---|---|
| Highly hazardous substance | Regulated substance |
| Employer | Owner or operator |
| Facility | Stationary source |
| Standard | Rule or part |

There are twelve elements in the Program 3 prevention program. Each element corresponds with a section of subpart D of part 68. Exhibit 7-2 sets out each of the twelve elements, the corresponding section numbers, and OSHA references. Two OSHA elements are not included. Emergency response is dealt with separately in part 68; the OSHA trade secrets requirement (provision of trade secret information to employees) is beyond EPA's statutory authority.

**EXHIBIT 7-2**
**SUMMARY OF PROGRAM 3 PREVENTION PROGRAM**
**(40 CFR PART 68, SUBPART D)**

| SECTION | TITLE | OSHA PSM REFERENCE |
|---|---|---|
| § 68.65 | Process Safety Information | PSM standard § 1910.119(d). |
| § 68.67 | Process Hazard Analysis (PHA) | PSM standard § 1910.119(e). |
| § 68.69 | Operating Procedures | PSM standard § 1910.119(f). |
| § 68.71 | Training | PSM standard § 1910.119(g). |
| § 68.73 | Mechanical Integrity | PSM standard § 1910.119(j). |
| § 68.75 | Management of Change | PSM standard § 1910.119(l). |
| § 68.77 | Pre-Startup Review | PSM standard § 1910.119(I). |
| § 68.79 | Compliance Audits | PSM standard § 1910.119(o). |
| § 68.81 | Incident Investigation | PSM standard § 1910.119(m) |
| § 68.83 | Employee Participation | PSM standard § 1910.119(c). |
| § 68.85 | Hot Work Permit | PSM standard § 1910.119(k). |
| § 68.87 | Contractors | PSM standard § 1910.119(h). |

OSHA provided guidance on PSM in non-mandatory Appendix C to the standard. OSHA has reprinted this appendix as PSM Guidelines for Compliance (OSHA 3133). The OSHA guidance is reproduced, reordered to track part 68, in Appendix

D. The remainder of this chapter briefly outlines the major requirements and provides a discussion of any differences between EPA and OSHA. In some cases, further guidance is provided on the meaning of specific terms. For more detailed guidance, you should refer to the OSHA guidance in Appendix D.

---

**Q &A**
**PROCESS**

**Q.** I have a tank with more than 10,000 pounds of propane. I use the propane to heat the offices. The propane is not subject to PSM or the risk management program rule. The tank, however, is close to equipment that has chlorine above the applicable threshold and is subject to OSHA PSM and Program 3. Is the tank considered part of the chlorine process?

**A.** If a fire or explosion in the propane tank could cause a release of chlorine or other regulated substances or interfere with mitigation of such a release, the tank is considered part of the process. When you do your PHA for the process, you must evaluate how the propane tank could cause a release of chlorine and determine what steps may be needed to prevent such releases.

---

### PSM AND WAREHOUSES

As discussed in Chapter 2, you will be subject to OSHA PSM if you have more than a threshold quantity of most of the regulated toxic substances (OSHA does not cover the acids in solution, toluene diisocyanate, chloroform, arsenous trichloride, tetranitromethane, and titanium tetrachloride). If you have regulated flammables liquids stored in containers at atmospheric conditions, these are not subject to OSHA PSM under an exemption. Flammables stored under pressure or refrigeration would be covered by OSHA.

## 7.2    PROCESS SAFETY INFORMATION (§ 68.65)

Exhibit 7-3 briefly summarizes the process safety information requirements.

### PROCESS SAFETY INFORMATION AND WAREHOUSES

If you only store containers of regulated substances, many of the process safety information requirements will not be applicable to you. Process chemistry, block flow diagrams, P&IDs, relief systems, and safety systems are unlikely to be applicable to most warehouses. You will still need to define maximum inventories (see Chapter 6), safe process limits, and consequences of deviations. Materials of construction, electrical classification, ventilation system design, and design codes and standards will also apply.

## EXHIBIT 7-3
## PROCESS SAFETY INFORMATION REQUIREMENTS

| For chemicals, you must complete information on: | For process technology, you must provide: | For equipment in the process, you must include information on: |
|---|---|---|
| • Toxicity<br>• Permissible exposure limits<br>• Physical data<br>• Reactivity<br>• Corrosivity<br>• Thermal & chemical stability<br>• Hazardous effects of inadvertent mixing of materials that could foreseeably occur | • A block flow diagram or simplified process flow diagram<br>• Information on process chemistry<br>• Maximum intended inventory of the EPA-regulated chemical<br>• Safe upper & lower limits for such items as temperature, pressure, flows, or composition<br>• An evaluation of the consequences of deviation | • Materials of construction<br>• Piping & instrument diagrams (P&IDs)<br>• Electrical classification<br>• Relief system design & design basis<br>• Ventilation system design<br>• Design codes & standards employed<br>• Safety systems<br>• Material and energy balances for processes built after June 21, 1999 |

### WHERE TO GO FOR MORE INFORMATION

**Diagrams.** You may find it useful to consult Appendix B of OSHA's PSM final rule, computer software programs that do P&IDs, or other diagrams.

**Guidance and Reports.** Various engineering societies issue technical reports relating to process design. Other sources you may find useful include:

• • *Guidelines for Process Safety Documentation*, Center for Chemical Process Safety of the American Institute of Chemical Engineers 1995.

• • *Emergency Relief System Design Using DIERS Technology,* American Institute of Chemical Engineers, 1992.

• • *Emergency Relief Systems for Runaway Chemical Reactions and Storage Vessels: A Summary of Multiphase Flow Methods*, American Institute of Chemical Engineers, 1986.

• • *Guidelines for Pressure Relief and Emergency Handling Systems,* Center for Chemical Process Safety of the American Institute of Chemical Engineers, 1998.

• • *Loss Prevention in the Process Industries*, Volumes I, II, and III, Frank P. Lees, Butterworths: London 1996.

---

**Qs & As**
**PROCESS SAFETY INFORMATION**

**Q.** What does "materials of construction" apply to and how do I find this information?

**A.** You must document the materials of construction for all process equipment in a covered process. For example, you need to know the materials of construction for process vessels, storage vessels, piping, hoses, valves, and flanges. Equipment specifications should provide this information.

**Q.** What does "electrical classification" mean?

**A.** Equipment and wiring for locations where fire and explosion hazards may exist must meet requirements based on the hazards. Each room, section, or area must be considered separately. Equipment should be marked to show Class, Group, and operating temperature or temperature range. You must determine the appropriate classification for each area and ensure that the equipment used is suitable for that classification. The equipment covered includes transformers, capacitors, motors, instruments, relays, wiring, switches, fuses, generators, lighting, alarms, remote controls, communication, and grounding. Electrical classification will be included in equipment specifications.

**Q.** What do I have to do for material and energy balances?

**A.** For new processes, you must document both material and energy inputs and outputs of a process. For example, you would document the quantity of a regulated substance added to the process, the quantity consumed during the process, and the quantity that remains in the output. This requirement will not generally apply to storage processes.

---

## 7.3    PROCESS HAZARD ANALYSIS (§ 68.67)

### PHA AND WAREHOUSES

PHA checklists will generally work well for warehouses. See Chapter 6 for guidance on identifying or developing a checklist.

Exhibit 7-4 provides a summary of the requirements for process hazard analyses (PHAs).

### EPA/OSHA DIFFERENCES

You can use a PHA conducted under the OSHA PSM standard as your initial process hazard analysis. All OSHA PHAs must have been completed by May 1997. Therefore, the only "new" PHAs will be for non-OSHA Program 3 processes. If the process is subject to OSHA PSM, you can update and revalidate your PHA on OSHA's schedule.

## EXHIBIT 7-4
## PROCESS HAZARD ANALYSIS REQUIREMENTS

| The PHA must cover:: | Techniques must be one or more of: | Other requirements: |
|---|---|---|
| • Hazards of the process<br>• Identification of previous, potentially catastrophic incidents<br>• Engineering and administrative controls applicable to the hazards<br>• Consequence of failure of controls<br>• Siting<br>• Human factors<br>• Qualitative evaluation of health and safety impacts of control failure | • What If<br>• Checklist<br>• What If/Checklist<br>• Hazard and Operability Study (HAZOP)<br>• Failure Mode and Effects Analysis (FMEA)<br>• Fault Tree Analysis<br>• Appropriate equivalent methodology | • Analysis must be done by a team, one member of which has experience in the process, one member of which is knowledgeable in the PHA technique<br>• A system must be developed for addressing the team's recommendations and documenting resolution and corrective actions taken<br>• The PHA must be updated at least once every five years<br>• PHAs and documentation of actions must be kept for the life of the process |

**Offsite impacts.** You should consider offsite impacts when you conduct a PHA under EPA's rule (except for an initial PHA where are using the PHA conducted for OSHA PSM). If you are in the Program 3 prevention program because you must comply with the PSM standard, you may not have fully considered offsite consequence because the focus of PSM is worker protection. Practically speaking, however, there should be few instances where the scenarios considered for OSHA fail to address offsite impacts. A well-done PHA should identify all failure scenarios that could lead to significant exposure of workers, the public, or the environment. The only issue that may require further consideration for part 68 processes is whether any protection measures that were adequate for worker safety are inadequate for public and environmental safety.

Consider two circumstances — one where OSHA's PSM standard and EPA's risk management program rule lead to the same result, and another where protecting workers could mean endangering the public and the environment. For flammables, any scenario that could affect the public almost certainly would have the potential to affect workers; measures taken to protect your employees likely will protect the public and the environment. For toxics under PSM, however, you may plan to address a loss of containment by venting toxic vapors to the outside air. In each circumstance, a PHA should define how the loss of containment could occur. However, for EPA, the PHA team should reassess venting as an appropriate mitigation measure.

**Updating and revalidating your PHA.** For EPA, you must complete the initial PHA for each Program 3 process not later than June 21, 1999, and update it at least once every five years. You may complete an initial PHA before that date. You may use an OSHA PHA as your initial PHA, and update and revalidate it every five years on the OSHA schedule. A PHA completed after August 19, 1996 (the effective date of part 68) should consider offsite impacts.

## REJECTING TEAM RECOMMENDATIONS

You may not always agree with your PHA team's recommendations and may wish to reject a recommendation. OSHA's compliance directive CPL 2-2.45A-revised states that you may decline a team recommendation if you can document one of the following: (1) the analysis upon which the recommendation is based contains factual errors; (2) the recommendation is not necessary to protect the health of employees or contractors; (3) an alternative measure would provide a sufficient level of protection; or (4) the recommendation is infeasible. For part 68, you should also consider whether recommendations are not necessary to protect public health and the environment.

## UPDATING YOUR PHA

You should update or revalidate your PHA whenever there is a new hazard or risk created by changes to your process. Such changes might include introducing a new process, process equipment, or regulated substance; altering process chemistry that results in any change to safe operating limits; or other alteration that introduces a new hazard. You might, for example, introduce a new hazard if you installed a gas pipeline next to a storage tank containing a regulated substance. Other candidates could be making changes in process constituents that increase the possibility of runaway reactions or polymerization. EPA recommends that you consider revalidating your PHA whenever adjoining processes create a hazard. Remember that you have a general duty to prevent accidents and ensure safety at your source, which may require you to take steps beyond those specified in the risk management program rule.

## WHERE TO GO FOR MORE INFORMATION

Appendix 7-A of this chapter provides a summary of each of the techniques, a description of the types of processes for which they may be appropriate, and estimates about the time and staff required for each.

Part 68 and OSHA PSM require that whichever technique or techniques you use, you must have at least one person on the PHA team who is trained in the use of the technique. Training on such techniques is available from a number of professional organizations as well as private companies. You may have staff members who are capable of providing this training as well. Many trade associations publish detailed

---

## Qs & As
### OFFSITE CONSEQUENCES

**Q.** What does EPA mean by "consider offsite consequences"?  Do we have to do an environmental impact assessment (EIA)?

**A.** EPA does not expect you to do an EIA.  Potential consequences to the public and the environment are already analyzed in the offsite consequence analysis.  In the PHA, EPA only expects you to identify any failure scenarios that could lead to public exposures and to examine whether your strategies are adequate to reduce the risk of such exposures.

**Q.** If I need to revise a PHA to consider offsite consequences, when do I have to do that?

**A.** In general, for a PHA completed to meet the requirements of OSHA PSM, you should revise the PHA to consider offsite consequences when you update that PHA.  Any PHA for a covered process completed or updated for OSHA PSM after August 19, 1996, when part 68 was effective, should examine offsite consequences.  For example, if you completed an initial PHA for OSHA PSM in May 1993, OSHA requires that you update that PHA by May 1998.  In that update, you should consider offsite consequences.  If you complete your initial PHA for OSHA in May 1995, you must update it by May 2000; PHAs conducted for part 68 must include consideration of offsite consequences at that time.

---

guidance on methods for conducting a process hazard analysis.  You might find the following documents useful.

- *Guidelines for Hazard Evaluation Procedures, 2nd Ed. with Worked examples,* Center for Chemical Process Safety of the American Institute of Chemical Engineers 1992.

- *Evaluating Process Safety in the Chemical Industry*, Chemical Manufacturers Association.

- *Loss Prevention in the Process Industries, Volumes I, II, and III*, Frank P. Lees, Butterworths: London 1996.

- *Management of Process Hazards (RP 750)*, American Petroleum Institute.

- *Risk-Based Decision Making (Publication 16288)*, American Petroleum Institute.

## 7.4   OPERATING PROCEDURES (§ 68.69)

### OPERATING PROCEDURES AND WAREHOUSES

Chapter 6 of this document provides descriptions of what each operating phase and when these phases may not apply to warehouse operations.

Exhibit 7-5 summarizes what your operating procedures must address. Operating procedures must be readily accessible to workers who operate or maintain the process. You must review operating procedures as often as necessary to assure that they reflect current practices and any changes to the process or facility. You must certify annually that the operating procedures are current and accurate.

### EXHIBIT 7-5
### OPERATING PROCEDURES REQUIREMENTS

| Steps for each operating phase | Operating limits | Safety & health considerations | Safety systems & their functions |
|---|---|---|---|
| • Initial startup<br>• Normal operations<br>• Temporary operations<br>• Emergency shutdown<br>• Emergency operations<br>• Normal shutdown<br>• Startup following a turnaround or emergency shutdown<br>• Lockout/tagout<br>• Confined space entry<br>• Opening process equipment or piping<br>• Entrance into the facility | • Consequences of deviations<br>• Steps to avoid, correct deviations | • Chemical properties & hazards<br>• Precautions for preventing chemical exposure<br>• Control measures for exposure<br>• QC for raw materials and chemical inventory<br>• Special or unique hazards | • Address whatever is applicable |

### WHERE TO GO FOR MORE INFORMATION

Chapter 7 of this document provides descriptions of each operating phase and when these phases may not apply to certain operations.

- • *Guidelines for Process Safety Fundamentals for General Plant Operations,* Center for Chemical Process Safety of the American Institute of Chemical Engineers 1995.

- • *Guidelines for Safe Process Operations and Maintenance,* Center for Chemical Process Safety of the American Institute of Chemical Engineers 1995.

- *Guidelines for Writing Effective Operating and Maintenance Procedures*, Center for Chemical Process Safety of the American Institute of Chemical Engineers 1996.

## 7.5    TRAINING (§ 68.71)

You are required to train new operators on the operating procedures and cover health and safety hazards, emergency operations, and safe work practices applicable to the employee's tasks. For workers involved in operating the process before June 21, 1999, you may certify in writing that they are competent to operate the process safely, in accordance with the operating procedures. At least every three years you must provide refresher training (you must consult with employees involved in operating the process to determine the appropriate frequency). Finally, you are required to determine that each operator has received and understood the training and keep a record for each employee with the date of the training and the method used to verify that the employee understood the training.

### WHERE TO GO FOR MORE INFORMATION

- *Guidelines for Process Safety Fundamentals for General Plant Operations*, Center for Chemical Process Safety of the American Institute of Chemical Engineers 1995.

- *Guidelines for Technical Planning for On-Site Emergencies*, Center for Chemical Process Safety of the American Institute of Chemical Engineers 1995.

- *Federally Mandated Training and Information (Publication 12000)*, American Petroleum Institute.

## 7.6    MECHANICAL INTEGRITY (§ 68.73)

You must have a mechanical integrity program for pressure vessels and storage tanks, piping systems, relief and vent systems and devices, emergency shutdown systems, controls, and pumps. Exhibit 7-6 briefly summarizes the other requirements for your mechanical integrity program.

### WHERE TO GO FOR MORE INFORMATION

**Guidance and Reports.** Other sources of guidance and reports you may find useful include:

- *Guidelines for Process Equipment Reliability Data with Data Tables*, Center for Chemical Process Safety of the American Institute of Chemical Engineers 1989.

- *Guidelines for Process Safety Documentation*, Center for Chemical Process Safety of the American Institute of Chemical Engineers 1995.

## EXHIBIT 7-6
## MECHANICAL INTEGRITY CHART

| Written procedures | Training | Inspection & testing | Equipment deficiencies | Quality assurance |
|---|---|---|---|---|
| • Establish & implement written procedures to maintain the integrity of process equipment. | • Train process maintenance employees in an overview of the process and its hazards.<br>• Make sure this training covers the procedures applicable to safe job performance. | • Inspect & test process equipment.<br>• Use recognized and generally accepted good engineering practices.<br>• Follow a schedule that matches the manufacturer's recommendations or more frequently if prior operating experience indicates is necessary.<br>• Document each inspection & test with: Date, inspector name, equipment identifier, test or inspection performed, results. | • Correct equipment deficiencies before further use of process equipment or whenever necessary to ensure safety. | • Establish a QA program for new construction & equipment, newly installed equipment, maintenance materials, and spare parts & equipment. |

- *Pressure Vessel Inspection Code: Maintenance Inspection, Rating, Repair, and Alteration (API 510)*, American Petroleum Institute.

- *Tank Inspection, Repair, Alteration, and Reconstruction (Std 653)*, American Petroleum Institute.

## 7.7 MANAGEMENT OF CHANGE (§ 68.75)

Exhibits 7-7 briefly summarizes EPA's MOC requirements

**EXHIBIT 7-7**
**MANAGEMENT OF CHANGE REQUIREMENTS**

| MOC procedures must address: | Employees affected by the change must: | Update process safety information if: | Update operating procedures if: |
|---|---|---|---|
| • Technical basis for the change<br><br>• Impact on safety and health<br><br>• Modifications to operating procedures<br><br>• Necessary time period for the change<br><br>• Authorization requirements for proposed change | • Be informed of the change before startup<br><br>• Trained in the change before startup | • A change covered by MOC procedures results in a change in any PSI required under EPA's rule (see § 67.65) | • A change covered by MOC procedures results in a change in any operating procedure required under EPA's rule (see § 67.69) |

**MOC AND WAREHOUSES**

If you only store substances at your warehouse, management of change will apply primarily to the use of new equipment and the handling of a new class of substances. If your procedures and practices do not change with the introduction of a new substance or class of substances, you may not need to take any steps under management of change. These changes will not trigger an RMP update if you have included the new regulated substance in your existing RMP under predictive filing (see Chapter 1).

**WHERE TO GO FOR MORE INFORMATION**

• • *Management of Change in Chemical Plants: Learning from Case Histories*, Center for Chemical Process Safety of the American Institute of Chemical Engineers 1993.

• • *Plant Guidelines for Technical Management of Chemical Process Safety*, Center for Chemical Process Safety of the American Institute of Chemical Engineers 1992.

• • *Management of Process Hazards (RP 750)*, American Petroleum Institute.

## 7.8 PRE-STARTUP REVIEW (§ 68.77)

You must conduct your pre-startup safety review for new stationary sources or modified stationary sources when the modification is significant enough to require a change in safety information under the management of change element. You must conduct your pre-startup review before you introduce a regulated substance to a process, and you must address the items listed in Exhibit 7-8.

**EXHIBIT 7-8**
**PRE-STARTUP REVIEW REQUIREMENTS**

| Design Specifications | Adequate Procedures | PHA/MOC | Training |
|---|---|---|---|
| • Confirm that new or modified construction and equipment meet design specifications. | • Ensure that procedures for safety, operating, maintenance, and emergencies are adequate and in place. | Perform a PHA and resolve or implement any recommendations for new process. Meet management of change requirements for modified process. | • Confirm that each employee involved in the process has been trained completely. |

## 7.9 COMPLIANCE AUDITS (§ 68.79)

You must conduct an audit of the process to evaluate compliance with the prevention program requirements at least once every three years. At least one person involved in the audit must be knowledgeable in the process. You must develop a report of the findings and document appropriate responses to each finding and document that deficiencies have been addressed. The two most recent audit reports must be kept on-site.

### WHERE TO GO FOR MORE INFORMATION

• *Guidelines for Auditing Process Safety Management Systems*, Center for Chemical Process Safety of the American Institute of Chemical Engineers 1993.

## 7.10 INCIDENT INVESTIGATION (§ 68.81)

Exhibit 7-9 briefly summarizes the steps you must take for investigating incidents.

You must investigate each incident which resulted in, or could have resulted in, a "catastrophic release of a regulated substance." A catastrophic release is one that "presents an imminent and substantial endangerment to public health and the environment." Although the rule requires you to investigate only those incidents which resulted in, or could reasonably have resulted in a catastrophic release, EPA encourages you to investigate all accidental releases. Investigating minor accidents or near misses can help you identify problems that could result in major releases if left unaddressed.

### WHERE TO GO FOR MORE INFORMATION

•  • *Guidelines for Investigating Chemical Process Incidents*, Center for Chemical Process Safety of the American Institute of Chemical Engineers 1992.

•  • *Guide for Fire and Explosion Investigations (NFPA 921)*, National Fire Protection Association.

### EXHIBIT 7-9
### INCIDENT INVESTIGATION REQUIREMENTS

| | |
|---|---|
| • Initiate an investigation promptly. | Begin investigating no later than 48 hours following the incident. |
| • Establish a knowledgeable investigation team. | Establish an investigation team to gather the facts, analyze the event, and develop the how and why of what went wrong.  At least one team member must have knowledge of the process involved.  Consider adding other workers in the process area where the incident occurred. Their knowledge will  be significant and should give you the fullest insight into the incident. |
| • Summarize the investigation in a report. | Among other things, the report must identify the factors contributing to the incident.  Remember that identifying the root cause may be more important than identifying the initiating event.  The report must also include any recommendations for corrective actions. Remember that the purpose of the report is to help management take corrective action. |
| • Address the team's findings and recommendations. | Establish a system to address promptly and resolve the incident report findings and recommendations; document resolutions and corrective actions. |
| • Review the report with your staff and contractors. | You must share the report - its findings and recommendations - with affected workers whose job tasks are relevant to the incident. |
| • Retain the report. | Keep incident investigation reports for five years. |

## 7.11   EMPLOYEE PARTICIPATION (§ 68.83)

Exhibit 7-10 briefly summarizes what you must do.

**EXHIBIT 7-10**
**EMPLOYEE PARTICIPATION REQUIREMENTS**

| | |
|---|---|
| • Write a plan. | Develop a written plan of action regarding how you will implement employee participation. |
| • Consult with employees. | Consult your employees and their representatives regarding conducting and developing PHAs and other elements of process safety management in the risk management program rule. |
| • Provide access to information. | Ensure that your employees and their representatives have access to PHAs and all other information required to be developed under the rule. |

## 7.12   HOT WORK PERMITS (§ 68.85)

Exhibit 7-11 briefly summarizes how to meet the hot work permit requirement.

**EXHIBIT 7-11**
**HOT WORK PERMITS REQUIREMENTS**

| | |
|---|---|
| • Issue a hot work permit. | You must issue this permit for hot work conducted on or near a covered process. |
| • Implement fire prevention and protection. | You must ensure that the fire prevention and protection requirements in 29 CFR 1910.252(a) are implemented before the hot work begins. The permit must document this. |
| • Indicate the appropriate dates. | The permit should indicate the dates authorized for hot work. |
| • Identify the work. | The permit must identify the object on which hot work is to be performed. |
| • Maintain the permit on file. | You must keep the permit on file until workers have completed the hot work operations. |

### WHERE TO GO FOR MORE INFORMATION

- *Standard for Fire Prevention in Use of Cutting and Welding Processes (NFPA 518)*, National Fire Protection Association.

- *Standard for Welding*, Cutting and Brazing, 29 CFR 1910 Subpart Q.

## 7.13   CONTRACTORS (§ 68.87)

Exhibit 7-12 summarizes both yours and the contractors' responsibilities where contractors perform maintenance or repair, turnaround, major renovation, or specialty work on or adjacent to a covered process.

## EXHIBIT 7-12
## CONTRACTORS CHART

| You must... | Your contractor must... |
|---|---|
| • **Check safety performance.** When selecting a contractor, you must obtain and evaluate information regarding the safety performance of the contractor. | • **Ensure training for its employees.** The contractor must train its employees to ensure that they perform their jobs safely and in accordance with your source's safety procedures. |
| • **Provide safety and hazards information.** You must inform the contractor of potential fire, explosion, or toxic release hazards; and of your emergency response activities as they relate to the contractor's work and the process. | • **Ensure its employees know process hazards and applicable emergency actions.** The contractor must assure that contract employees are aware of hazards and emergency procedures relating to the employees' work. |
| • **Ensure safe practices.** You must ensure that you have safe work practices to control the entrance, presence, and exit of contract employees in covered process areas. | • **Document training.** The contractor must prepare a record documenting and verifying adequate employee training. |
| • **Verify that the contractor acts responsibly.** You must verify that the contractor is fulfilling its responsibilities. | • **Ensure its employees are following your safety procedures.** |
| | • **Inform you of hazards.** The contractor must tell you of any unique hazards presented by its work or of any hazards it finds during performance. |

### EPA/OSHA DIFFERENCES

EPA has no authority to require that you maintain an occupational injury and illness log for contract employees. Be aware, however, that OSHA does have this authority, and that the PSM standard does set this requirement. (See 29 CFR 1910.119(h)(2)(vi)).

### WHERE TO GO FOR MORE INFORMATION

- *Contractor and Client Relations to Assure Process Safety*, Center for Chemical Process Safety of the American Institute of Chemical Engineers 1996.

- *API/CMA Managers Guide to Implementing a Contractor Safety Program (RP 2221)*, American Petroleum Institute.

- *Improving Owner and Contractor Safety Performance (RP 2220)*, American Petroleum Institute.

# APPENDIX 7-A
# PHA TECHNIQUES

This appendix provides descriptions of each of the PHA techniques listed in the OSHA PSM standard and § 68.67. These descriptions include information on what each technique is, which types of processes they may be appropriate for, what their limitations are, and what level of effort is typically associated with each. This information is based on *Guidelines for Hazard Evaluation Procedures*, 2nd Ed., published by AIChE/CCPS. If you are interested in more detailed discussion and worked examples, you should refer to the AIChE/CCPS volume.

Neither the information below nor the full AIChE/CCPS volume will provide you with enough information to conduct a PHA. The rule requires that your PHA team include at least one person trained in the technique you use. Training in PHA techniques is available from a number of organizations. If you must conduct multiple PHAs, you are likely to need to update your PHAs frequently, or you have a complex process that will take several weeks to analyze, you may want to consider training one or more of your employees. If you have a single process that is unlikely to change more than once every five years, you may find it more cost-effective to hire a trained PHA leader.

## DESCRIPTIONS OF TECHNIQUES

## CHECKLISTS

Checklists are primarily used for processes that are covered by standards, codes, and industry practices — for example, storage tanks designed to ASME standards, ammonia handling covered by OSHA (29 CFR 1910.111). Checklists are easy to use and can help familiarize new staff with the process equipment. AIChE/CCPS states that checklists are a highly cost-effective way to identify customarily recognized hazards. Checklists are dependent on the experience of the people who develop them; if the checklist is not complete, the analysis may not identify hazardous situations.

Checklists are created by taking the applicable standards and practices and using them to generate a list of questions that seek to identify any differences or deficiencies. If a checklist for a process does not exist, an experienced person must develop one based on standards, practices, and facility or equipment experience. A completed checklist usually provides "yes," "no," "not applicable," and "need more information" answers to each item. A checklist analysis involves touring the process area and comparing equipment to the list.

AIChE/CCPS estimates that for a small or simple system a checklist will take 2 to 4 hours to prepare, 4 to 8 hours to evaluate the process, and 4 to 8 hours to document the results. For larger or more complex processes, a checklist will take 1 to 3 days to prepare, 3 to 5 days to evaluate, and 2 to 4 days to document.

## WHAT-IF

A What-If is a brainstorming approach in which a group of people familiar with the process ask questions about possible deviations or failures. These questions may be framed as What-If, as in "What if the pump fails?" or may be expressions of more general concern, as in "I worry about contamination during unloading." A scribe or recorder takes down all of the questions on flip charts or a computer. The questions are then divided into specific areas of investigation, usually related to consequences of interest. Each area is then addressed by one or more team members.

What-If analyses are intended to identify hazards, hazardous situations, or accident scenarios. The team of experienced people identifies accident scenarios, consequences, and existing safeguards, then suggest possible risk reduction alternatives. The method can be used to examine deviations from design, construction, modification, or operating intent. It requires a basic understanding of the process and an ability to combine possible deviations from design intent with outcomes. AIChE describes this as a powerful procedure if the staff are experienced; "otherwise, the results are likely to be incomplete."

A What-If usually reviews the entire process, from the introduction of the chemicals to the end. The analysis may focus on particular consequences of concern. AIChE provides the following example of a What-If question: "What if the raw material is the wrong concentration?" The team would then try to determine how the process would respond: "If the concentration of acid were doubled, the reaction could not be controlled and a rapid exotherm would result." The team might then recommend steps to prevent feeding wrong concentrations or to stop the feed if the reaction could not be controlled.

A What-If of simple systems can be done by one or two people; a more complex process requires a larger team and longer meetings. AIChE/CCPS estimates that for a small or simple system a What-If analysis will take 4 to 8 hours to prepare, 1 to 3 days to evaluate the process, and 1 to 2 days to document the results. For larger or more complex processes, a What-If will take 1 to 3 days to prepare, 4 to 7 days to evaluate, and 4 to 7 days to document.

## WHAT-IF/CHECKLIST

A What-If/Checklist combines the creative, brainstorming aspects of the What-If with the systematic approach of the Checklist. The combination of techniques can compensate for the weaknesses of each. The What-If part of the process can help the team identify hazards and accident scenarios that are beyond the experience of the team members. The checklist provides a more detailed systematic approach that can fill in gaps in the brainstorming process. The technique is generally used to identify the most common hazards that exist in a process. AIChE states that it is often the first PHA conducted on a process, with subsequent analyses using more detailed approaches.

The purpose of a What-If/Checklist is to identify hazards and the general types of accidents that could occur, evaluate qualitatively the effects of the effects, and determine whether safeguards are adequate. Usually the What-If brainstorming precedes the use of the checklist, although the order can be reversed.

The technique usually is performed by a team experienced in the design, operation, and maintenance of the process. The number of people required depends on the complexity of the process. AIChE/CCPS estimates that for a small or simple system a What-If/Checklist analysis will take 6 to 12 hours to prepare, 6 to 12 hours to evaluate the process, and 4 to 8 hours to document the results.  For larger or more complex processes, a What-If/Checklist will take 1 to 3 days to prepare, 4 to 7 days to evaluate, and 1 to 3 weeks to document.

## HAZOP

The Hazard and Operability Analysis (HAZOP) was originally developed to identify both hazards and operability problems at chemical process plants, particularly for processes using technologies with which the plant was not familiar.  The technique has been found to be useful for existing processes as well.  A HAZOP requires an interdisciplinary team and an experienced team leader.

The purpose of a HAZOP is to review a process or operation systematically to identify whether process deviations could lead to undesirable consequences.  AIChE states that the technique can be used for continuous or batch processes and can be adapted to evaluate written procedures.  It can be used at any stage in the life of a process.

HAZOPs usually require a series of meetings in which, using process drawings, the team systematically evaluates the impact of deviations.  The team leader uses a fixed set of guide words and applies them to process parameters at each point in the process.  Guide words include "No," "More," "Less," "Part of," "As well as," Reverse," and "Other than."  Process parameters considered include flow, pressure, temperature, level, composition, pH, frequency, and voltage.  As the team applies the guide words to each process step, they record the deviation, with its causes, consequences, safeguards, and actions needed, or the need for more information to evaluate the deviation.

HAZOPs require more resources than simpler techniques.  AIChE states that a simple process or a review with a narrow scope may be done by as few as three or four people, if they have the technical skills and experience.  A large or complex process usually requires a team of five to seven people.  AIChE/CCPS estimates that for a small or simple system a HAZOP analysis will take 8 to 12 hours to prepare, 1 to 3 days to evaluate the process, and 2 to 6 days to document the results.  For larger or more complex processes, a HAZOP will take 2 to 4 days to prepare, 1 to 3 weeks to evaluate, and 2 to 6 weeks to document.

## FAILURE MODE AND EFFECTS ANALYSIS (FMEA)

A Failure Mode and Effects Analysis (FMEA) evaluates the ways in which equipment fails and the system's response to the failure.  The focus of the FMEA is on single equipment failures and system failures.  An FMEA usually generates recommendations for increasing equipment reliability.  FMEA does not examine human errors directly, but will consider the impact on equipment of human error.  AIChE states that FMEA is "not efficient for identifying an exhaustive list of combinations of equipment failures that lead to accidents."

An FMEA produces a qualitative, systematic list of equipment, failure modes, and effects.  The analysis can easily be updated for design or systems changes.  The FMEA usually produces a table that, for each item of equipment, includes a description, a list of failure modes, the effects of each failure, safeguards that exist, and actions recommended to address the failure.  For example, for pump operating normal, the failure modes would include fails to stop when required, stops when required to run, seal leaks or ruptures, and pump case leaks or ruptures.  The effects would detail both the immediate effect and the impact on other equipment. Generally, when analyzing impacts, analysts assume that existing safeguards do not work, AIChE states that "more optimistic assumptions may be satisfactory as long as all equipment failure modes are analyzed on the same basis."

An FMEA requires an equipment list or P&ID, knowledge of the equipment, knowledge of the system, and responses to equipment failure. AIChE states that on average, an hour is sufficient to analyze two to four pieces of equipment. AIChE/CCPS estimates that for a small or simple system an FMEA will take 2 to 6 hours to prepare, 1 to 3 days to evaluate the process, and 1 to 3 days to document the results. For larger or more complex processes, an FMEA will take 1 to 3 days to prepare, 1 to 3 weeks to evaluate, and 2 to 4 weeks to document.

## FAULT TREE ANALYSIS (FTA)

A Fault Tree Analysis (FTA) is a deductive technique that focuses on a particular accident or main system failure and provides a method for determining causes of the event. The fault tree is a graphic that displays the combinations of equipment failures and human errors that can result in the accident. The FTA starts with the accident and identifies the immediate causes. Each immediate cause is examined to determine its causes until the basic causes of each are identified. AIChE states that the strength of FTA is its ability to identify combinations of basic equipment and human failures that can lead to an accident, allowing the analyst to focus preventive measures on significant basic causes.

AIChE states that FTA is well suited for analyses of highly redundant systems. For systems vulnerable to single failures that can lead to accidents, FMEA or HAZOP are better techniques to use. FTA is often used when another technique has identified an accident that requires more detailed analysis. The FTA looks at component failures (malfunctions that require that the component be repaired) and faults (malfunctions that will remedy themselves once the conditions change). Failures and faults are divided into three groups: primary failures and faults occur when the equipment is operating in the environment for which it was intended; secondary failures and faults occur when the system is operating outside of intended environment; and command faults and failures are malfunctions where the equipment performed as designed but the system that commanded it malfunctioned.

An FTA requires a detailed knowledge of how the plant or system works, detailed process drawings and procedures, and knowledge of component failure modes and effects. AIChE states that FTAs need well trained and experienced analysts. Although a single analyst can develop a fault tree, input and review from others is needed

AIChE/CCPS estimates that for a small or simple system an FTA will take 1 to 3 days to prepare, 3 to 6 days for model construction, 2 to 4 days to evaluate the process, and 3 to 5 days to document the results. For larger or more complex processes, an FTA will take 4 to 6 days to prepare, 2 to 3 weeks for model constructions, 1 to 4 weeks to evaluate, and 3 to 5 weeks to document.

## OTHER TECHNIQUES

The rule allows you to use other techniques if they are functionally equivalent. The AIChE Guidelines includes descriptions of a number of other techniques including Preliminary Hazard Review, Cause-Consequence Analysis, Event Tree Analysis, and Human Reliability Analysis. You may also develop a hybrid technique that combines features of several techniques or apply more than one technique.

## SELECTING A TECHNIQUE

Exhibit 7A-1 is adapted from the AIChE Guidelines and indicates which techniques are appropriate for particular phases in a process's design and operation.

### EXHIBIT 7A-1
### APPLICABILITY OF PHA TECHNIQUES

|  | Checklist | What-If | What-If-Checklist | HAZOP | FMEA | FTA |
|---|---|---|---|---|---|---|
| R&D |  | • • |  |  |  |  |
| Design | • • | • • | • • |  |  |  |
| Pilot Plant Operation | • • | • • | • • | • • | • • | • • |
| Detailed Engineering | • • | • • | • • | • • | • • | • • |
| Construction/Start-Up | • • | • • | • • |  |  |  |
| Routine Operation | • • | • • | • • | • • | • • | • • |
| Modification | • • | • • | • • | • • | • • | • • |
| Incident Investigation |  | • • |  | • • | • • | • • |
| Decommissioning | • • | • • | • • |  |  |  |

### FACTORS IN SELECTING A TECHNIQUE

Type of process will affect your selection of a technique. AIChE states that most of the techniques can be used for any process, but some are better suited for certain processes than others. FMEA efficiently analyzes the hazards associated with computer and electronic systems; HAZOPs do not work as well with these. Processes or storage units designed to industry or government standards can be handled with checklists.

AIChE lists What-If, What-If/Checklist, and HAZOP as better able to handle batch processes than FTA or FMEA because the latter do not easily deal with the need to evaluate the time-dependent nature of batch operations.

Analysis of multiple failure situations is best handled by FTA. Single-failure techniques, such as HAZOP and FMEA, are not normally used to handle these although they can be extended to evaluate a few simple accident situations involving more than one event.

AIChE states that when a process has operated relatively free of accidents for a long time, the potential for high consequence events is low, and there have been few changes to invalidate the experience base, the less exhaustive techniques, such as a Checklist, can be used. When the opposite is true, the more rigorous techniques are more appropriate.

A final factor in selecting a technique is time required for various techniques. Exhibit 7A-2 summarizes AIChE's estimates of the time required for various steps. The full team is usually involved in the evaluation step; for some techniques, only the team leader and scribe are involved in the preparation and documentation steps.

## EXHIBIT 7A-2
## TIME AND STAFFING FOR PHA TECHNIQUES

|  | Checklist | What-If | What-If Checklist | HAZOP | FMEA | FTA |
|---|---|---|---|---|---|---|
| **Simple/Small System** | | | | | | |
| # Staff | 1-2 | 2-3 | 2-3 | 3-4 | 1-2 | 2-3 |
| Preparation | 2-4 h | 4-8 h | 6-12 h | 8-12 h | 2-6 h | 1-3 d |
| Modeling | | | | | | 3-6 d |
| Evaluation | 4-8 h | 1-3 d | 6-12 h | 1-3 d | 1-3 d | 2-4 d |
| Documentation | 4-8 h | 1-2 d | 4-8 h | 2-6 d | 1-3 d | 3-5 d |
| **Large/Complex Process** | | | | | | |
| # Staff | 1-2 | 3-5 | 3-5 | 5-7 | 2-4 | 2-5 |
| Preparation | 1-3 d | 1-3 d | 1-3 d | 2-4 d | 1-3 d | 4-6 d |
| Modeling | | | | | | 2-3 w |
| Evaluation | 3-5 d | 4-7 d | 4-7 d | 1-3 w | 1-3 w | 1-4 w |
| Documentation | 2-4 d | 4-7 d | 1-3 w | 2-6 w | 2-4 w | 3-5 w |

h = hours        d = days (8 hours)        w = weeks (40 hours)

# CHAPTER 8: EMERGENCY RESPONSE PROGRAM

If you have at least one Program 2 or Program 3 process at your facility, then part 68 may require you to implement an emergency response program, consisting of an emergency response plan, emergency response equipment procedures, employee training, and procedures to ensure the program is up-to-date. This requirement applies if your employees will respond to some releases involving regulated substances. (See the box on the next page for more information on What is Response?)

EPA recognizes that, in some cases (particularly for retailers and other small operations with few employees), it may not be appropriate for employees to conduct response operations for releases of regulated substances. For example, it would be inappropriate, and probably unsafe, for an ammonia retailer with only one full-time employee to expect that a tank fire could be handled without the help of the local fire department or other emergency responders. EPA does not intend to force such facilities to develop emergency response capabilities. At the same time, you are responsible for ensuring effective emergency response to any releases at your facility. If your local public responders are not capable of providing such response, you must take steps to ensure that effective response is available (e.g., by hiring response contractors).

## 8.1 NON-RESPONDING FACILITIES (§ 68.90(b))

EPA has adopted a policy for non-responding facilities similar to that adopted by OSHA in its Hazardous Waste Operations and Emergency Response (HAZWOPER) Standard (29 CFR 1910.120), which allows certain facilities to develop an emergency action plan to ensure employee safety, rather than a full-fledged emergency response plan. If your employees will not respond to accidental releases of regulated substances, then you need not comply with the emergency response plan and program requirements. Instead, you are simply required to coordinate with local response agencies to ensure that they will be prepared to respond to an emergency at your facility. (You may want to briefly review the program design issues discussed in 8.2 prior to making this decision.) This will help to ensure that your community has a strategy for responding to and mitigating the threat posed by a release of a regulated substance from your facility. To do so, you must ensure that you have set up a way to notify emergency responders when there is need for a response. Coordination with local responders also entails the following steps:

- If you have a covered process with a regulated toxic, work with the local emergency planning entity to ensure that the facility is included in the community emergency response plan prepared under EPCRA regarding a response to a potential release.

- If you have a covered process with a regulated flammable, work with the local fire department regarding a response to a potential release.

---

**What is "Response"?**

EPA interprets "response" to be consistent with the definition of response specified under OSHA's HAZWOPER Standard. OSHA defines emergency response as "a response effort by employees from outside the immediate release area or by other designated responders ... to an occurrence which results, or is likely to result, in an uncontrolled release of a hazardous substance." The key factor here is that responders are designated for such tasks by their employer. This definition *excludes* "responses to incidental releases of hazardous substances where the substance can be absorbed, neutralized, or otherwise controlled at the time of release by employees in the immediate release area, or by maintenance personnel" as well as "responses to releases of hazardous substances where there is no potential safety or health hazard (i.e., fire, explosion, or chemical exposure)." Thus, if you expect your employees to take action to end a small leak (e.g., shutting a valve) or clean up a spill that does not pose an immediate safety or health hazard, this action could be considered an incidental response and you would not need to develop an emergency response program if your employees are limited to such activities.

However, due to the nature of the regulated substances subject to EPA's rule, only the most minor incidents would be included in this exception. In general, most activities will qualify as a response due to the immediacy of the dispersion of a toxic plume or spread of a fire, the volatilization of a spill, and the threat to people on and off site. As a result, if you will have your employees involved in any substantial way in responding to releases, you will need to develop an emergency response program. Your emergency response procedures need only apply to "response" actions; other activities will be described in your maintenance and operating procedures.

---

Although you do not need to describe these activities in your risk management plan, to document your efforts you should keep a record of:

- The emergency contact (i.e., name or organization and number) that you will call for a toxic or flammable release, and

- The organization that you worked with on response procedures.

The remainder of this chapter is applicable only to those facilities which will conduct a more extensive level of response operations. As noted above, you may want to review the next section before making a decision on whether the facility will take responsibility for conducting any response activities.

## 8.2    ELEMENTS OF AN EMERGENCY RESPONSE PROGRAM (§ 68.95)

If you will respond to releases of regulated substances with your own employees, your emergency response program must consist of the following elements:

- An emergency response plan (maintained at the facility) that includes:

- • Procedures for informing the public and emergency response agencies about releases,
- • Documentation of proper first aid and emergency medical treatment necessary to treat human exposures, and
- • Procedures and measures for emergency response.

- Procedures for using, inspecting, testing, and maintaining your emergency response equipment;

- Training for all employees in relevant procedures; and

- Procedures to review and update, as appropriate, the emergency response plan to reflect changes at the facility and ensure that employees are informed of changes.

Finally, your plan must be coordinated with the community plan developed under the Emergency Planning and Community Right-to-Know Act (EPCRA, also known as SARA Title III). In addition, at the request of local emergency planning or response officials, you must provide any information necessary for developing and implementing the community plan.

In keeping with the approach outlined in Chapter 6, EPA is not requiring facilities to document training and maintenance activities. However, as noted above, facilities must maintain an on-site emergency response plan as well as emergency response equipment maintenance and program evaluation procedures.

Although EPA's required elements are essential to any emergency response program, they are not comprehensive guidelines for creating an adequate response capability. Rather than establish another set of federal requirements for an emergency response

---

### What is a Local Emergency Planning Committee?

Local emergency planning committees (LEPCs) were formed under the Emergency Planning and Community Right-to-Know Act (EPCRA) of 1986. The committees are designed to serve as a community forum for issues relating to preparedness for emergencies involving releases of hazardous substances in their jurisdictions. They consist of representatives from local government (including law enforcement and firefighting), local industry, transportation groups, health and medical organizations, community groups, and the media. LEPCs:

- • Collect information from facilities on hazardous substances that pose a risk to the community;
- • Develop a contingency plan for the community based on this information; and
- • Make information on hazardous substances available to the general public.

Contact the mayor's office or the county emergency management office for more information on your LEPC.

program, EPA has limited the provisions of its rule to those the CAA mandates. If you have a regulated substance on site, you are already subject to at least one emergency response rule: OSHA's emergency action plan requirements (29 CFR 1910.38). Under OSHA HAZWOPER, any facility that handles "hazardous substances" (a broad term that includes all of the CAA regulated substances and thus applies to all facilities with covered processes) must comply with either 29 CFR 1910.38(a) or 1910.119(q). If you have a hazmat team, you are subject to the 29 CFR 1910.119(q) requirements. If you determine that the emergency response programs you have developed to comply with these other rules satisfy the elements listed at the beginning of this section, you will not have to do anything additional to comply with these elements. Additional guidance on making this decision is provided in section 8.5 of this chapter.

In addition, be careful not to confuse writing a set of emergency response procedures in a plan with developing an emergency response program. An emergency response plan is only one element of the integrated effort that makes an emergency response program. Although the plan outlines the actions and equipment necessary to respond effectively, training, program evaluation, equipment maintenance, and coordination with local agencies must occur regularly if your plan is to be useful in an emergency: The goal of the program is to enable you to respond quickly and effectively to any emergency. The documents listed in Exhibit 8-1 may be helpful in developing specific elements of your emergency response program.

Finally, remember that under the General Duty Clause of CAA section 112(r)(1) you are responsible for ensuring that any release from your processes can be handled effectively. If you plan to rely on local responders for some or all of the response, you must determine that those responders have both the equipment and training needed to do so. If they do not, you must take steps to meet any needs, either by developing your own response capabilities, developing mutual aid agreements with other facilities, hiring response contractors, or providing support to local responders so they can acquire equipment or training.

### RELATIONSHIP TO HAZWOPER

If you choose to establish and maintain onsite emergency response capabilities, then you will be subject to the detailed provisions of the OSHA or EPA HAZWOPER Standard. HAZWOPER covers preparing an emergency response plan, employee training, medical monitoring of employees, recordkeeping, and other issues. Call your state or federal district OSHA office (see Appendix C) for more information on complying with the HAZWOPER Standard.

## Exhibit 8-1
## Federal Guidance on Emergency Planning and Response

*Hazardous Materials Emergency Planning Guide* (NRT-1), National Response Team, March 1987. Although designed to assist communities in planning for hazmat incidents, this guide provides useful information on developing a response plan, including planning teams, plan review, and ongoing planning efforts.

*Criteria for Review of Hazardous Materials Emergency Plans* (NRT-1A), National Response Team, May 1988. This guide provides criteria for evaluating response plans.

*Integrated Contingency Plan*, National Response Team, (61 FR 28642, June 5, 1996). This provides guidance on how to consolidate multiple plans developed to comply with various federal regulations into a single, functional emergency response plan..

*Emergency Response Guidebook*, U.S. Department of Transportation, 2000. This guidebook lists over 1,000 hazardous materials and provides information on their general hazards and recommended isolation distances.

*Response Information Data Sheets* (RIDS), US EPA and National Oceanic and Atmospheric Administration. Developed for use with the Computer-Aided Management of Emergency Operations (CAMEO) software, these documents outline the properties, hazards, and basic safety and response practices for thousands of hazardous chemicals.

## Qs & As
## Emergency Response and Warehouses

**Q.** Does the emergency response program apply to specific covered processes?

**A.** The requirements for the emergency response program are intended to apply across all covered processes at a facility. Although certain elements of the program (e.g., how to use specific items of response equipment) may differ from one process to another, EPA does not intend or expect you to develop a separate emergency response program for each covered process. With this in mind, you should realize that your emergency response program will probably apply to your entire facility, although technically it need only apply to covered processes.

**Q.** My customers control emergency response for their stored materials. How do I reflect that in my response program?

**A.** Your emergency response plan should outline who is responsible for handling responses and provide contact numbers for customers who will handle responses at your warehouse. You should be sure that your local responders understand how you and your customers coordinate responses.

## 8.3   DEVELOPING AN EMERGENCY RESPONSE PROGRAM

The development of an emergency response program should be approached systematically. As described in section 8.2, all facilities complying with these emergency response program provisions will already be subject to OSHA HAZWOPER. As a result, you are likely to fall into one of two groups:

- You have already met several federal requirements for emergency planning and are interested in developing an integrated program to minimize duplication (section 8.4).

- You have a pre-existing emergency response program (perhaps based on an internal policy decision) and need to determine what additional activities you will need to conduct (section 8.5).

### STEPS FOR GETTING STARTED

The following steps outline a systematic approach that can serve as the framework for the program development process in each of these cases. Following these initial steps will allow you to conduct the rest of the process more efficiently.

**Form an emergency response program team.** The team should consist of employees with varying degrees of emergency response responsibilities, as well as personnel with expertise from each functional area of your facility. You should consider including persons from the following departments or areas:

- Maintenance;
- Operations or line personnel;
- Upper and line management;
- Legal;
- Fire and hazmat response;
- Environmental, health, and safety affairs;
- Training;
- Security;
- EPCRA section 302 emergency coordinator (if one exists);
- Public relations; and
- Personnel.

Of course, the membership of the team will need to be more or less extensive depending on the scope of the emergency response program. A three-member team may be appropriate for a small facility with a couple of process operators cross-trained as fire responders, while a facility with its own hazmat team and environmental affairs department may need a dozen representatives.

**Collect relevant facility documents.** Members of the development team should collect and review all of the following:

- Existing emergency response plans and procedures;

- • Submissions to the LEPC under EPCRA sections 302 and 303;
- • Hazard evaluation and release modeling information;
- • Hazard communication and emergency response training;
- • Emergency drill and exercise programs;
- • After-action reports and response critiques; and
- • Mutual aid agreements.

**Identify existing programs to coordinate efforts.** The team should identify any related programs from the following sources:

- Corporate- and industry-sponsored safety, training, and planning efforts; and

- Federal, state, and local government safety, training, and planning efforts (see Exhibit 8-2).

---

### Exhibit 8-2
### Federal Emergency Planning Regulations

The following is a list of some of the federal emergency planning regulations:

- EPA's Oil Pollution Prevention Regulation (SPCC and Facility Response Plan Requirements) - 40 CFR part 112.7(d) and 112.20-.21;
- • EPA's Risk Management Programs Regulation - 40 CFR part 68;
- • OSHA's Emergency Action Plan Regulation - 29 CFR 1910.38(a);
- • OSHA's Process Safety Standard - 29 CFR 1910.119;
- • OSHA's HAZWOPER Regulation - 29 CFR 1910.120;
- • OSHA's Fire Brigade Regulation - 29 CFR 1910.156;
- • EPA's Emergency Planning and Community Right-to-Know Act Requirements - 40 CFR part 355. (These planning requirements apply to communities, rather than facilities, but will be relevant when facilities are coordinating with local planning and response entities).
- • EPA's Storm Water Regulations - 40 CFR 122.26.

Facilities may also be subject to state and local planning requirements.

---

**Determine the status of each required program element.** Using the information collected, you should assess whether each required program element (see section 8.2) is:

- • In place and sufficient to meet the requirements of part 68;

- • In place, but not sufficient to meet the requirements of Part 68; or

- • Not in place.

This examination will shape the nature of your efforts to complete the emergency response program required under the risk management program. For example, if you are already in compliance with OSHA's HAZWOPER Standard, you have probably satisfied most, if not all, of the requirements for an emergency response program. Section 8.6 explains the intent of each of EPA's requirements to help you determine whether you are already in compliance.

**Take additional actions as necessary.**

### TAILORING YOUR PROGRAM TO YOUR HAZARDS

If your processes and chemicals pose a variety of hazards, it may be necessary to tailor some elements of your emergency response program to these specific hazards. Unless each part of your program element is appropriate to the release scenarios that may occur, your emergency response program cannot be fully effective. Your program should include core elements that are appropriate to most of the scenarios, supplemented with more specific response information for individual scenarios. This distinction should be reflected in your emergency response plan, which should explain when to access the general and specific response information. To do this, you will need to consider the following four steps:

- • Identify and characterize the hazards for each covered process. The process hazards analysis (see Chapter 7) or hazard review (see Chapter 6), and offsite consequence analysis (see Chapter 4) should provide this information.

- • For each program element, compare the activities involved in responding to each type of accident scenario and decide if they are different enough to require separate approaches. For example, response equipment and training will likely be different for releases of toxic versus flammable gases.

- • For those program elements that may be chemical- or process-specific, identify what and how systems and procedures need to be modified. For example, if existing mitigation systems are inadequate for responding to certain types of releases, you will need to consider what additional types of equipment are needed.

- • Consider possible causes of emergencies in developing your emergency response program. You should consider both the hazards at your facility and in the surrounding environment. In making this determination, you should consider your susceptibility to:

  - • Fires, spills, and vapor releases;
  - • Floods, temperature extremes, tornadoes, earthquakes, and hurricanes;
  - • Loss of utilities, including power failures; and
  - • Train derailments, bomb threats, and other man-made disasters.

## 8.4   INTEGRATION OF EXISTING PROGRAMS

A number of other federal statutes and regulations require emergency response planning (see Exhibit 8-2). On June 5, 1996, the National Response Team (NRT), a multi-agency group chaired by EPA, published the Integrated Contingency Plan Guidance in the *Federal Register* (61 FR 28642). This guidance is intended to be used by facilities to prepare emergency response plans for responding to releases of oil and hazardous substances. The guidance provides a mechanism for consolidating multiple plans that you prepared to comply with various regulations into a single, functional emergency response plan or integrated contingency plan (ICP).

The ICP guidance does not change existing regulatory requirements; rather, it provides a format for organizing and presenting material currently required by regulations. Individual regulations are often more detailed than the ICP guidance. To ensure full compliance, you will still need to read and comply with all of the federal regulations that apply. The guidance contains a series of matrices designed to assist you in consolidating various plans while documenting compliance with these federal requirements.

The NRT and the agencies responsible for reviewing and approving plans to which the ICP option applies have agreed that integrated response plans prepared according to the guidance will be acceptable and the federally preferred method of response planning. The NRT anticipates that future development of all federal regulations addressing emergency response planning will incorporate use of the ICP guidance.

## 8.5   HAVE I MET PART 68 REQUIREMENTS?

EPA believes that the creation of multiple response plans to meet slightly different federal or state standards is counterproductive, diverting resources that could be used to develop better response capabilities. Therefore, as part of the overall effort to reduce the imposition of potentially duplicative or redundant federal requirements, EPA has limited its requirements for the emergency response program to the general provisions mandated by Congress, as described in Section 8.2.

As a result, EPA believes that facilities subject to other federal emergency planning requirements may have already met the requirements of these regulations. For example, plans developed to comply with other EPA contingency planning requirements and the OSHA HAZWOPER rule (29 CFR 1910.120) will likely meet the requirements for the emergency response plan (and most of the requirements for the emergency response program). The following discussion presents some general guidance on what actions you need to take for each of the required elements.

### EMERGENCY RESPONSE PLAN

If you already have a written plan to comply with another planning regulation, you do not need to write another plan, but only add to it as necessary to cover the elements listed on the next page.

*Keep in mind:* At a minimum, your plan must describe:

•  •     Your procedures for informing the public and offsite emergency response agencies of a release. This must include the groups and individuals that will be contacted and why, the means by which they will be contacted, the time frame for notification, and the information that will be provided.

•  •     The proper first aid and emergency medical treatment for employees, first responders, and members of the public who may have been exposed to a release of a regulated substance. This must include standard safety precautions for victims (e.g., apply water to exposed skin immediately) as well as more detailed information for medical professionals. You must also indicate who is likely to be responsible for providing the appropriate treatment: an employee, an employee with specialized training, or a medical professional.

•  •     Your procedures for emergency response in the event of a release of a regulated substance. This must include descriptions of the actions to be taken by employees and other individuals on-site over the entire course of the release event:

   •  •     Activation of alarm systems and interpretation of signals;
   •  •     Safe evacuation, assembly, and return;
   •  •     Selection of response strategies and incident command structure;
   •  •     Use of response equipment and other release mitigation activities; and
   •  •     Post-release equipment and personnel cleanup and decontamination.

## PLANNING COORDINATION

One of the most important issues in an emergency response program is deciding which response actions will be assigned to employees and which will be handled by offsite personnel. As a result, talking to public response organizations will be critical when you develop your emergency response procedures. Although EPA is not requiring you to be able to respond to a release alone, you should not simply assume that local responders will be able to manage an emergency. You must work with them to determine what they can do, and then expand your own abilities or establish mutual aid agreements or contracts to handle those situations for which you lack the appropriate training or equipment.

If you have already coordinated with local response agencies on how to respond to potential releases of regulated substances and you have ensured an effective response, you do not need to take any further action.

*Keep in mind:* Your coordination must involve planning for releases of regulated substances from all covered processes and must cover:

• • What offsite response assistance you will require for potential release scenarios, including fire-fighting, security, and notification of the public;

• • How you will request offsite response assistance; and

• • Who will be in charge of the response operation and how will authority be delegated down the internal and offsite chain of command.

Coordination equivalent to that required for planning for extremely hazardous substances under EPCRA sections 302-303 will be considered sufficient to meet this requirement. A more detailed discussion of this element is provided in 8.6.

## EMERGENCY EQUIPMENT

If you already have written procedures for using and maintaining your emergency response equipment, you do not need to write new procedures.

*Keep in mind:* Your procedures must apply to any emergency equipment relevant to a response involving a covered process, including all detection and monitoring equipment, alarms and communications systems, and personal protective equipment not used as part of normal operations (and thus not subject to the prevention program requirements related to operating procedures and maintenance). The procedures must describe:

• How and when to use the equipment properly;

• How and when the equipment should receive routine maintenance; and

• How and when the equipment should be inspected and tested for readiness.

Written procedures comparable to those necessary for process-related equipment under the OSHA PSM Standard and EPA's Program 2 and 3 Prevention Programs will be considered sufficient to meet this requirement.

## EMPLOYEE TRAINING

If you already train your employees in how to respond to (or evacuate from) releases of regulated substances, then you do not need a new training program.

*Keep in mind:* Your training must address the actions to take in response to releases of regulated substances from all covered processes. The training should be based directly on the procedures that you have included in your emergency response plan and must be given to all employees and contractors on site. Individuals should receive training appropriate to their responsibilities:

• If they will only need to evacuate, then their training should cover when and how to evacuate their location.

- If they may need to activate an alarm system in response to a release event, then their training should cover the location of the alarms and when and how to use the alarm system.

- If they will serve on an emergency response team, then their training should cover the location of all emergency equipment, how to use it, and how the incident command system works.

Emergency response training conducted in compliance with the OSHA HAZWOPER Standard and 29 CFR 1910.38 will be considered sufficient to meet this requirement.

### RESPONSE PLAN EVALUATION

If you already have a formal practice for regular review and updates of your plan based on changes at the facility, you do not need to develop additional procedures.

*Keep in mind:* You must also identify the types of changes to the facility that would cause the plan to be updated (e.g., a new covered process) and include a method of communicating any changes to the plan to your employees (e.g., through training). You may want to set up a regular schedule on which you review your entire emergency response plan and identify any special conditions (e.g., a drill or exercise) that could result in an interim review.

## 8.6  COORDINATION WITH LOCAL EMERGENCY PLANNING ENTITIES (§ 68.95(c))

Once you determine that you have at least one covered process, you should open communications with local emergency planning and response officials, including your local emergency planning committee if one exists. Because your LEPC consists of representatives from many local emergency planning and response agencies, it is likely to be the best source of information on the critical emergency response issues in your community. However, in some cases, there may not be an active LEPC in your community. If so, or if your state has not designated your community as an emergency planning district under EPCRA, you will likely need to contact local agencies individually to determine which entities (e.g., fire department, emergency management agency, police department, civil defense office, public health agency) have jurisdiction for your facility.

### KEY COORDINATION ISSUES

If you have any of the toxic regulated substances above the threshold quantity, you should have already designated an emergency coordinator to work with the LEPC on chemical emergency preparedness issues (a requirement for certain facilities regulated under EPCRA). If you have not (or if your facility has only regulated flammable substances), you may want to do so at this time. The emergency coordinator should be the individual most familiar with your emergency response program (e.g., the person designated as having overall responsibility for this program in your management system — see Chapter 5).

Involvement in the activities of your LEPC can have a dramatically positive effect on your emergency response program, as well as on your relationship with the surrounding community. Your LEPC can provide technical assistance and guidance on a number of topics, such as conducting response training and exercises, developing mutual aid agreements, and evaluating public alert systems. The coordination process will help both the community and the facility prepare for an emergency, reducing expenditures of time and money, as well as helping eliminate redundant efforts.

You should consider providing the LEPC with draft versions of any emergency response program elements related to local emergency planning efforts. This submission can initiate a dialogue with the community on potential program improvements and lead to coordinated training and exercise efforts. In return, your LEPC can support your emergency response program by providing information from its own emergency planning efforts, including:

• • Data on wind direction and weather conditions, or access to local meteorological data, to help you make decisions related to the evacuation of employees and public alert notification;

• • Lists of emergency response training programs available in the area for training police, medical, and fire department personnel, to help you identify what training is already available;

• • Schedules of emergency exercises designed to test the community response plan to spur coordinated community-facility exercises;

• • Lists of emergency response resources available from both public and private sources to help you determine whether and how a mutual aid agreement could support your program; and

• • Details on incident command structure, emergency points of contact, availability of emergency medical services, and public alert and notification systems.

Upon completion of your emergency response plan, you should coordinate with the LEPC, local response organizations, local hospitals, and other response organizations (e.g., state hazmat team) and offer them a copy of the plan. In some instances, only a portion of the plan may be of use to individuals or organizations; in such cases, you should consider making only that portion of the plan available. For instance, it may be appropriate to send a hospital only the sections of your plan that address emergency medical procedures and decontamination.

You may also want to provide your LEPC and local response entities with a description of your emergency response program elements, as well as any important subsequent amendments or updates, to ensure that the community is aware of the scope of your facility response efforts prior to an emergency. Although the summary of your emergency response program will be publicly available as part of your RMP, this information may not be as up-to-date or as comprehensive. Remember, the LEPC has been given the authority under EPCRA and Clean Air Act regulations to request any information necessary for preparing the community response plan.

---

### Planning for Flammable Substances

In the case of regulated flammable substances, the fire department with jurisdiction over your facility may already be conducting fire prevention inspections and pre-planning activities under its own authority. Your participation in these efforts (as requested) will allow local responders to gather the information they need and prepare for an emergency. If there is no local fire department, or if there is only a volunteer fire department in your area, you may need to contact other local response or planning officials (e.g., police) to determine how you can work with the community.

---

# CHAPTER 9: RISK MANAGEMENT PLAN (PART 68, SUBPART G)

You must submit one risk management plan (RMP) to EPA for all of your covered processes (§ 68.150). EPA has developed a system to submit your RMP via the Internet (RMP*eSubmit) for your use. Your RMP is due no later than the latest of the following dates:

- The date on which your facility first has a regulated substance above the threshold quantity in a process; or

- Three years after the date on which a regulated substance is first listed by EPA, if your facility has that regulated substance above the threshold quantity.

You must fully update and resubmit all nine sections of your RMP at least every five years. Under certain circumstances, RMPs must be updated and resubmitted before their five-year anniversary (see 40 CFR §68.190(b)). Your five-year anniversary date is listed in the notification letter or e-mail that was sent to you after you submitted your last RMP.

The Agency has an Internet-based system, called RMP*eSubmit, for submitting a new RMP or making changes or corrections to an existing RMP. RMP*eSubmit is available on EPA's Central Data Exchange (CDX) – the Agency's electronic reporting site. If you are new to submitting RMPs, you should go to EPA's public Web site at www.epa.gov/emergencies/rmp to find instructions on how to obtain a CDX account and register as an RMP submitter or certifier in CDX.

Facilities that submit RMPs and later change their process(es) in ways that make them no longer subject to part 68 (e.g., switching to unregulated substances or reducing inventories of regulated substances to less than threshold quantities) must de-register their RMP within six months of making the change.

Finally, facilities submitting changes to their RMPs must identify the type of change being submitted and the reason for the change (e.g., submission of an updated RMP as a result of a process change at the plant). This information will help implementing agencies understand the nature of the RMP submission being made.

## 9.1 ELEMENTS OF THE RMP

The length and content of your RMP will vary depending on the number and program levels of the covered processes at your facility. See Chapter 2 for detailed guidance on how to determine the program levels of each of the covered processes at your facility.

Any facility with one or more covered processes must include in its RMP:

- An executive summary (§ 68.155);

- The registration for the facility (§ 68.160);

- The certification statement (§ 68.185);

- A worst-case scenario analysis for each Program 1 process; at least one worst-case scenario analysis to cover all Program 2 and 3 processes involving regulated toxic substances; at least one worst-case scenario analysis to cover all Program 2 and 3 processes involving regulated flammables (§ 68.165(a));

- The five-year accident history for each process (§ 68.168); and

- Information concerning emergency response at the facility (§ 68.180).

Any facility with at least one covered process in Program 2 or 3 must also include:

- At least one alternative release scenario analysis for each regulated toxic substance in Program 2 or 3 processes and at least one alternative release scenario analysis to cover all regulated flammables in Program 2 or 3 processes (§ 68.165(b));

- A summary of the prevention program for each Program 2 process (§ 68.170); and

- A summary of the prevention program for each Program 3 process (§ 68.175).

Subpart G of part 68 (see Appendix A) describes the data required for each of the elements. RMP*eSubmit and its accompanying user's manual contain more detailed instructions. RMP*eSubmit is designed to limit the number of text entries. For example, the rule requires you to report on the major hazards identified during a PHA or hazard review and on public receptors potentially affected by worst-case and alternative release scenarios. RMP*eSubmit provides a list of options for you to check for these elements. Except for the executive summary, the RMP consists primarily of yes/no answers, numerical information (e.g., dates, quantities, distances), and a few text answers (e.g., names, addresses, chemical identity). Where possible, RMP*eSubmit provides "pick lists" to help you complete the form. For example, RMP*eSubmit provides a list of regulated substances and automatically fills in the CAS numbers when you select a substance. The RMP*eSubmit User's Manual and on-line help screens explain each data element and provide guidance on acceptable data entry. Access to both RMP*eSubmit and the RMP*eSubmit User's Manual is available free of charge – for further instructions visit EPA's Web site at www.epa.gov/emergencies/rmp.

## 9.2    RMP SUBMISSION

### ELECTRONIC SUBMISSION

EPA has made RMP*eSubmit available to complete and file your RMP. RMP*eSubmit does the following:

- Provides a user-friendly, Web-based RMP Submission System;

- Performs data quality checks, accepts limited graphics, and provides on-line help including defining data elements and providing instructions;

- Online reporting simplifies the process. It saves you time, and improves data quality and security;

- EPA uses industry-standard technology, including encryption used by most commercial banks, as well as stringent user ID and password protocols to protect your information; and

- You will be able to access your RMP online at any time.

For a facility to submit their RMP, the certifier will be required to have a CDX account. Some facilities may use CDX to report other data to the Agency, and if your facility's certifying official already has a CDX account, he/she can use it to obtain access to RMP*eSubmit. If you are new to CDX, visit EPA's public Web site at www.epa.gov/emergencies/rmp, to learn how to obtain a CDX account and gain access to RMP*eSubmit.

## HARD COPY SUBMISSION

If you are unable to submit electronically you may fill out the Paper Submission form available in the RMP*eSubmit User's Manual and send it in with your RMP. See the RMP*eSubmit User's Manual for more information on the Paper Submission form. If you submit on paper, you will need to use the official form. If you do not use the official form, your RMP cannot be processed. You can download the RMP*eSubmit User's Manual free of charge at www.epa.gov/emergencies/rmp.

## IMPORTANT REMINDERS

If you submit on paper, do not forget your certification letter. A certification letter is required for all RMP submissions. See the RMP*eSubmit User's Manual for more information on the certification letter.

If you use the RMP*eSubmit system, you will certify your RMP electronically – however, you will be required to submit a one-time Electronic Signature Agreement form prior to the first time you submit an RMP using RMP*eSubmit. The CDX system will guide you through the ESA process when you first register for CDX access as an RMP certifying official.

EPA's old PC-based electronic reporting software – RMP*Submit – will no longer be available for download from the EPA Web site beginning in March 2009.

## WHERE DO I SEND MY RMP?

If you use RMP*eSubmit, your RMP will be submitted over the Internet. The process will require you to complete and mail a one-time Electronic Signature Agreement to EPA prior to your first RMP submission using RMP*eSubmit.

If you submit a hard copy RMP, you must also include a certification letter (see the RMP*eSubmit User's Manual for more information on the certification letter). After completing the certification letter and RMP forms, you should send them together via certified mail to:

RMP Reporting Center
P.O. Box 10162
Fairfax, VA 22038

For courier and FEDEX packages, the address is:

RMP Reporting Center
c/o CGI Federal, Inc.
12601 Fair Lakes Circle
Fairfax, VA 22033

## 9.3    ISSUES PERTAINING TO SUBMISSIONS OF AND ACCESS TO CLASSIFIED, CONFIDENTIAL BUSINESS INFORMATION (CBI), AND TRADE SECRETS

### WHAT SHOULD I DO ABOUT CLASSIFIED INFORMATION?

Only Federal agencies and their contractors at Federal facilities may make claims of classified information. If you have such a claim, EPA urges you not to submit the information you claim as classified as part of your RMP. If any classified information is critical to the clarity and completeness of any part of the RMP, you should submit that information separately, on paper, in an annex to the RMP. Any annex marked as classified will be reviewed only by Federal and state representatives who have received security clearances and are thereby authorized to review such information.

### WHAT SHOULD I DO ABOUT CONFIDENTIAL BUSINESS INFORMATION (CBI)?

Under CAA section 114(c), 40 CFR part 2 and part 68, you may claim some information included in your RMP as CBI if you are able to show that the information meets the substantive criteria set forth in 40 CFR 2.301. These criteria generally require that the data be commercial or financial in nature, that they not be available to the public through other means, that you take appropriate steps to prevent disclosure, and that disclosure of the data would be likely to cause substantial harm to your competitive position. Review of any CBI claims will be handled as provided for in 40 CFR part 2. However, part 68 provides that certain RMP data elements may not be claimed as CBI because they do not convey any business-sensitive information. EPA has developed specific procedures for submission of CBI claims for RMPs. See §§ 68.151 and 68.152 for details on what data may be claimed as CBI and how to assert such claims. It is worth noting that few CBI claims have been asserted since RMPs were first submitted in 1999.

In general the part 68 procedures provide that:

- Owners or operators must substantiate CBI claims at the time they make the claim by providing documentation demonstrating that the claim meets the criteria set forth in 40 CFR 2.301.

- Substantiating information may be claimed confidential by marking it as CBI. Information that is not so marked will be treated as public and may be disclosed without notice to the submitter. If substantiating information is claimed confidential, the owner or operator must provide a sanitized and unsanitized version of the substantiating information.

- The owner, operator, or senior official with management responsibility of the stationary source must sign a certification that the signer has personally examined the information submitted and that, based on inquiry of the persons who compiled the information, the information is true, accurate, and complete, and that those portions of the substantiation claimed as confidential would, if disclosed, reveal trade secrets or other confidential business information.

If your RMP will include CBI, contact the RMP Reporting Center by phone at 703-227-7650 or by e-mail at RMPRC@epacdx.net to obtain instructions on how to complete your submission.

## 9.4    RMP UPDATES, CORRECTIONS AND DE-REGISTRATIONS (§ 68.190)

Whether and when you are required to fully update and resubmit, correct, or de-register your RMP is based on what changes occur at your facility. Please refer to the Exhibit 9-1 and note that you are required to take action with regard to your RMP on the earliest of the dates that apply to your facility. In some cases, changes at the facility may require only a partial revision of the RMP or a simple correction of administrative or emergency contact information. Exhibit 9-1 also covers these situations.

Corrections to submissions made in hardcopy form or via the Agency's previous electronic reporting software (RMP*Submit) will need to be made in RMP*eSubmit. Regardless of how you submitted your last RMP, all data for your current RMP will be available via RMP*eSubmit, and can be corrected on-line as necessary.

### Can I File Predictively?

Predictive filing is an option that allows you to submit an RMP that includes regulated substances that may not be held at the facility at the time of submission. This option is intended to assist facilities such as chemical warehouses, chemical distributors, and batch processors whose operations involve highly variable types and quantities of regulated substances, but who are able to forecast their inventory with some degree of accuracy. Under § 68.190, you are required to update and resubmit your RMP no later than the date on which a new regulated substance is first present in a covered process above a threshold quantity. By using predictive filing, you will not be required to update and re-submit your RMP every time you receive a new regulated substance if that substance was included in your latest RMP submission (as long as you receive it in a quantity that does not trigger a revised offsite consequence analysis as provided in § 68.36). RMP*eSubmit contains an option to indicate that your RMP is a predictive filing.

If you use predictive filing, you must implement your Risk Management Program and prepare your RMP exactly as you would if you actually held all of the substances included in the RMP. This means that you must meet all rule requirements for each regulated substance for which you file, whether or not that substance is actually held on site at the time you submit your RMP. Depending on the substances for which you file, this may require you to perform additional worst-case and alternative-case scenarios and to implement additional prevention program elements. If you use this option, you must still update and resubmit your RMP if you receive a regulated substance that was not included in your latest RMP. You must also continue to comply with the other update requirements stated in § 68.190.

## EXHIBIT 9-1
### RMP UPDATES, CORRECTIONS AND DE-REGISTRATIONS

| Change That Occurs | Date by Which You Must Update, Correct or De-register your RMP |
|---|---|
| No changes occur | At least once every 5 years from its initial submission or most recent update |
| A newly regulated substance is first listed by EPA | Within 3 years of the date EPA listed the newly regulated substance if your facility has more than a threshold quantity of that substance in a process |
| A regulated substance first becomes present above its threshold quantity in:<br>–     a process already covered; or<br>–     a new process | On or before the date the quantity of the regulated substance exceeds the threshold in the process |
| A change occurs at your facility that requires a revised PHA or hazard review | Within 6 months of the change |
| A change occurs at or near your facility that requires a revised offsite consequence analysis (e.g., you increase your inventory of a regulated substance such that it increases the distance to the endpoint by a factor of 2 or more, or a new public receptor is constructed near your facility) | Within 6 months of the change |
| A change occurs that alters the Program level that previously applied to any covered process | Within 6 months of the change |
| An accidental release meeting the reporting criteria of § 68.42 occurs at your facility | Add to and correct accident history information and incident investigation data elements within 6 months of the date of the accident (revising other RMP sections is not required unless facility changes resulting from an accident trigger a full update) |
| Facility emergency contact information changes | Correct the emergency contact information within one month of the change (revising other RMP elements is not required). |
| Minor administrative change (i.e., correct a clerical error or supply additional information) | Correct the information as soon as practicable (revising other RMP elements is not required). |
| A change occurs that makes the facility no longer subject to the requirement to submit an RMP | Submit a de-registration letter to EPA within 6 months of the change, indicating that the RMP is no longer required |

## HOW DO I DE-REGISTER?

If your facility is no longer covered by this rule, you must submit a letter to the RMP Reporting Center within six months indicating that your stationary source is no longer covered.

## RECURRING ACCIDENT PREVENTION PROGRAM REQUIREMENTS

Don't forget that in addition to updating your RMP submission, the Risk Management Program regulation contains various recurring implementation requirements for covered facilities' accident prevention and emergency response programs. When you update your RMP, you should ensure that you are up to date with implementation of these requirements and that your updated RMP reflects the most recent information for your prevention and emergency response programs. The following is a list of some key elements in your RMP that you should review, as well as recurring prevention and emergency response program implementation requirements that you should make sure are completed:

- Review and update your offsite consequence analyses (OCA) at least once every 5 years (40 CFR 68.36).

  For your worst-case and alternate-case scenarios, you should review your documentation to determine that the parameters and assumptions used in the analyses are still appropriate. Such assumptions include the use of any administrative controls or passive mitigation, the estimated quantity released, the release rate, and the duration of release. The results of this review should be documented and maintained as part of your RMP records. Any changes to the scenarios resulting from this review, including changes in the distance-to-endpoint, should be reported in your updated submission.

  You should review the data used to identify and estimate population and environmental receptors to be sure that it is current. For example, new construction in your area may have resulted in public receptors closer to your facility than was the case when you first conducted the OCA for your facility. Also, you may have used Census data to estimate the residential population within the distance to endpoint. If so, you should update this estimate based on the latest Census data. Census data can be found in publications of the US Census Bureau. These publications, including the County and City Data Book, are available on the Census Bureau website (www.census.gov) and in public libraries.

- For Program 2, review and update your hazard review at least once every 5 years (40 CFR 68.50).

  The review and any updates of the hazard review, as well as resolution of any problems identified, must be documented. You should report the date of your most recent hazard review update, and the completion date for any changes resulting from the hazard review, in your RMP update.

- For Program 3, update and revalidate your process hazard analysis (PHA) at least once every 5 years (40 CFR 68.67).

  This revalidation must be conducted by a team with expertise in engineering and process operations. The team must include at least one employee who has experience and knowledge specific to the process being evaluated. Also, one member of the team must be knowledgeable in the specific PHA methodology being used. The revalidation is intended to assure that the PHA is consistent with the current process.

  To revalidate your PHA, you should evaluate your current process hazard analysis for accuracy and completeness. This evaluation should include checking that all modifications to your process are reflected in the PHA; evaluating the process safety information to ensure that it is complete, current, and accurate; verifying that operating procedures are adequate, up-to-date, and implemented; documenting that PHA recommendations have been incorporated into equipment design, process conditions, mechanical integrity, operating procedures, training, and emergency response; verifying that recommendations have been implemented; reviewing incident investigation reports. Updated and revalidated PHAs completed to comply with OSHA's Process Safety Management Standard (29 CFR 1910.119(e)) are acceptable to meet this requirement as long as they also considered hazards that could result in off-site consequences.

  The revalidation and any updates of the process hazard analyses, as well as resolution of any recommendations, must be documented. This documentation must be retained as part of your RMP records for the life of the process. You should report the date of your most recent process hazard update, and the completion date for any changes resulting from the process hazard update, in your RMP update.

- The Risk Management Program requires several aspects of your prevention program to be periodically implemented or reviewed. The most recent dates for these activities should be reported in your RMP update:

  Training in operating procedures (40 CFR 68.54 and 68.71): For both Program 2 and 3, you are required to provide refresher training at least every three years, and more often if necessary.

  Compliance audits (40 CFR 68.58 and 68.79): For both Program 2 and 3, you are required to audit your procedures and practices for compliance with the Risk Management Program regulations at least every three years to verify their adequacy and implementation.

  Maintenance/Mechanical Integrity (40 CFR 68.56 and 68.73): For both Program 2 and 3, you are required to inspect and test your process equipment according to the schedule that you have established based on good engineering practices.

  Operating procedures (40 CFR 68.89): For Program 3 only, you are required to certify annually that your operating procedures are current and accurate.

  Management of change (40 CFR 68.75): For Program 3 only, you are required to update your process safety information and your procedures and practices for a covered process in the event of any change to the process chemicals, technology, equipment, or procedures.

- Correction of your five-year accident history.

  You must submit a correction that revises your five-year accident history within six months of an accidental release of a regulated substance from a covered process that resulted in deaths, injuries, or significant property damage on site, or known off-site deaths, injuries, evacuations, sheltering in place, property damage, or environmental damage.

- Verify your process information.

  Although you have an ongoing responsibility to monitor whether changes to your process or to the quantities stored or handled alter your program level eligibility, your five-year update is an opportunity to verify that each covered process still meets the eligibility criteria for its program level.

- Review your emergency response program or coordination with local officials.

  If your employees will take part in responding to accidental releases, you are required to periodically review and update, as appropriate, your emergency response program and to notify your employees of any changes to your emergency response plan (40CFR 68.95). You must include the date of your most recent review of your emergency response program and most recent training in your re-submission. You should contact your Local Emergency Planning Committee (LEPC) to verify whether your facility is currently included in the community emergency response plan. You should also review and update your procedures for notifying emergency responders in an emergency. These last two steps are particularly important if your employees will not respond to accidental releases.

# CHAPTER 10: IMPLEMENTATION

## 10.1   IMPLEMENTING AGENCY

The implementing agency is the federal, state, or local agency that is taking the lead for implementation and enforcement of part 68. The implementing agency will review RMPs, select some RMPs for audits, and conduct on-site inspections. The implementing agency should be your primary contact for information and assistance.

### WHO IS MY IMPLEMENTING AGENCY?

Under the CAA, EPA will serve as the implementing agency until a state or local agency seeks and is granted delegation under CAA section 112(l) and 40 CFR part 63, subpart E. You should check with the EPA Regional Office to determine if your state has been granted delegation or is in the process of seeking delegation. The Regional Office will be able to provide contact names at the state or local level. See http://www.epa.gov/swercepp/pubs/112r-sts/112r-sts.html for addresses and contact information for EPA Regions and state implementing agencies.

### IF THE PROGRAM IS DELEGATED, WHAT DOES THAT MEAN?

To gain delegation, a state or local agency must demonstrate that it has the authority and resources to implement and enforce part 68 for all covered processes in the state or local area. Some states may, however, elect to seek delegation to implement and enforce the rule for only sources covered by an operating permit program under Title V of the CAA. When EPA determines that a state or local agency has the required authority and resources, EPA may delegate the program. If the state's rules differ from part 68 (a state's rules are allowed to differ in certain specified respects, as discussed below), EPA will adopt, through rulemaking, the state program as a substitute for part 68 in the state, making the state program federally enforceable. In most cases, the state will take the lead in implementation and enforcement, but EPA maintains the ability to enforce part 68 in states in which EPA has delegated part 68. Should EPA decide that it is necessary to take an enforcement action in the state, the action would be based on the state rule that EPA has adopted as a substitute for part 68. Similarly, citizen actions under the CAA would be based on the state rules that EPA has adopted.

Under 40 CFR 63.90, EPA will not delegate the authority to add or delete substances from § 68.130. EPA has proposed, in revisions to part 63, that the authority to revise Subpart G (relating to RMPs) will not be delegated. Even if your state or local authority is the implementing agency, you must file your RMP with EPA (see Chapter 9). You should check with your state to determine whether you need to file additional data for state use or submit amended copies of the RMP with the state to cover state elements or substances.

*If your state has been granted delegation, it is important that you contact them to determine if the state has requirements in addition to those in part 68. State rules may be more stringent than part 68. This document does not cover state requirements.*

---

**Qs & As**
**DELEGATION**

**Q.** In what ways may state rules be more stringent? Does this document provide guidance on state differences?

**A.** States may impose more detailed requirements, such as requiring more documentation or more frequent reporting, specifying hours of training or maintenance schedules, imposing equipment requirements or call for additional analyses. Some states are likely to cover at least some additional chemicals and may use lower thresholds. This document does not cover state differences.

**Q.** Will the general duty clause be delegated?

**A.** The general duty clause (CAA section 112(r)(1)) is not included in part 68 and, therefore, will not be delegated. States, however, may adopt their own general duty clause under state law.

---

## 10.2 REVIEWS/AUDITS/INSPECTIONS (§ 68.220)

The implementing agency is required under part 68 to review and conduct audits of RMPs. Reviews are relatively quick checks of the RMPs to determine whether they are complete and whether they contain any information that is clearly problematic. For example, if an RMP for a process containing flammables fails to list fire and explosion as a hazard in the prevention program, the implementing agency may flag that as a problem. The RMP data system will perform some of the reviews automatically by flagging RMPs submitted without necessary data elements completed.

Facilities may be selected for audits based on any of the following criteria, set out in §68.220:

- • Accident history of the facility
- • Accident history of other facilities in the same industry
- • Quantity of regulated substances handled at the site
- • Location of the facility and its proximity to public and environmental receptors
- • The presence of specific regulated substances
- • The hazards identified in the RMP
- • A plan providing for random, neutral oversight

### WHAT ARE AUDITS AND HOW MANY WILL BE CONDUCTED?

Under the CAA and part 68, audits are conducted on the RMP. Audits will generally be reviews of the RMP to review its adequacy and require revisions when necessary to ensure compliance with part 68. Audits will help identify whether the underlying risk management program is being implemented properly. The implementing agency will look for any inconsistencies in the dates reported for compliance with prevention program elements. For example, if you report that the date of your last revision of operating procedures was in June 1998 but your training program was last reviewed or revised in December 1994, the implementing agency will ask why the training program was not reviewed to reflect new operating procedures.

The agency will also look at other items that may indicate problems with implementation. For example, if you are reporting on a distillation column at a refinery, but used a checklist as your PHA technique, or you fail to list an appropriate set of process hazards for the process chemicals, the agency may seek further explanations as to why you reported in the way you did. The implementing agency may compare your data with that of other facilities in the same industrial sector using the same chemicals to identify differences that may indicate compliance problems.

If audits indicate potential problems, they may lead to requests for more information or to on-site inspections. If the implementing agency determines that problems exist, it will issue a preliminary determination listing the necessary revisions to the RMP, an explanation of the reasons for the revisions, and a timetable. Section 68.220 provides details of the administrative procedures for responding to a preliminary determination.

The number of audits conducted will vary from state to state and from year to year. Neither the CAA nor part 68 sets a number or percentage of facilities that must be audited during a year. Implementing agencies will set their own goals, based on their resources and particular concerns.

## WHAT ARE INSPECTIONS?

Inspections are site visits to check on the accuracy of the RMP data and on the implementation of all part 68 elements. During inspections, the implementing agency will probably review the documentation for rule elements, such as the PHA reports, operating procedures, maintenance schedules, process safety information, and training. Unlike audits, which focus on the RMP but may lead to determinations concerning needed improvements to the risk management program, inspections will focus on the underlying risk management program itself.

Implementing agencies will determine how many inspections they need to conduct. Audits may lead to inspections or inspections may be done separately. Depending on the focus of the inspection (all covered processes, a single process, or particular part of the risk management program) and the size of the facility, inspections may take several hours to several weeks.

## 10.3   RELATIONSHIP WITH TITLE V PERMIT PROGRAMS

Part 68 is an applicable requirement under the CAA Title V permit program and must be listed in a Title V air permit. You do not need a Title V air permit solely because you are subject to part 68. If you are required to apply for a Title V permit because you are subject to requirements under some other part of the CAA, you must:

- List part 68 as an applicable requirement in your permit

- Include conditions that require you to either submit a compliance schedule for meeting the requirements of part 68 by the applicable deadlines or include compliance with part 68 as part of your certification statement.

You must also provide the permitting agency with any other relevant information it requests.

The RMP and supporting documentation are not part of the permit and should not be submitted to the permitting authority. The permitting authority is only required to ensure that you have submitted the RMP and that it is complete. The permitting authority may delegate this review of the RMP to other agencies.

If you have a Title V permit and it does not address the part 68 requirement, you should contact your permitting authority and determine whether your permit needs to be amended to reflect part 68.

## 10.4   PENALTIES FOR NON-COMPLIANCE

Penalties for violating the requirements or prohibitions of part 68 are set forth in CAA section 113. This section provides for both civil and criminal penalties. EPA may assess civil penalties of not more than $27,500 per day per violation. Any one convicted of knowingly violating part 68 may also be punished by a fine pursuant to Title 18 of the U.S. Code or by imprisonment for no more than five years, or both; anyone convicted of knowingly filing false information may be punished by a fine pursuant to Title 18 or by imprisonment for no more than two years.

**Qs & As**
**AUDITS**

**Q.** If we are a Voluntary Protection Program (VPP) facility under OSHA's VPP program, are we exempt from audits?

**A.** You are exempt from audits based on accident history of your industry sector or on random, neutral oversight. An implementing agency that is basing its auditing strategy on other factors may include your facility although EPA expects that VPP facilities will generally not be a high priority for audits unless they have a serious accident.

**Q.** If we have been audited by a qualified third party, for ISO 14001 certification or for other programs, are we exempt from audits?

**A.** No, but you may want to inform your implementing agency that you have gained such certification and indicate whether the third party reviewed part 68 compliance as part of its audit. The implementing agency has the discretion to determine whether you should be audited.

**Q.** Will we be audited if a member of the public requests an audit of our facility?

**A.** The implementing agency will have to decide whether to respond to such public requests. EPA's intention is that part 68 implementation reflect that hazards are primarily a local concern.

# CHAPTER 11: COMMUNICATION WITH THE PUBLIC

Once you have prepared and submitted your RMP, EPA will make it available to the public. Public availability of the RMP is a requirement under section 114(c) of the Clean Air Act (the Act provides for protection of trade secrets, and EPA will accordingly protect any portion of the RMP that contains Confidential Business Information). Therefore, you can expect that your community will discuss the hazards and risks associated with your facility as indicated in your RMP. You will necessarily be part of such discussions. The public and the press are likely to ask you questions because only you can provide specific answers about your facility and your accident prevention program. This dialogue is a most important step in preventing chemical accidents and should be encouraged. You should respond to these questions honestly and candidly. Refusing to answer, reacting defensively, or attacking the regulation as unnecessary are likely to make people suspicious and willing to assume the worst. A basic fact of risk communication is that trust, once lost, is very hard to regain. As a result, you should prepare as early as possible to begin talking about these issues with the community, Local Emergency Planning Committees (LEPCs), State Emergency Response Commissions (SERCs), other local and state officials, and other interested parties.

Communication with the public can be an opportunity to develop your relationship with the community and build a level of trust among you, your neighbors, and the community at large. By complying with the RMP rule, you are taking a number of steps to prevent accidents and protect the community. These steps are the individual elements of your risk management program. A well-designed and properly implemented risk management program will set the stage for informative and productive dialogue between you and your community. The purpose of this chapter is to suggest how this dialogue may occur. In addition, note that some industries have developed guidance and other materials to assist in this process; contact your trade association for more information.

## 11.1 BASIC RULES OF RISK COMMUNICATION

Risk communication means establishing and maintaining a dialogue with the public about the hazards at your operation and discussing the steps that have been or can be taken to reduce the risk posed by these hazards. Of particular concern under this rule are the hazards related to the chemicals you use and what would happen if you had an accidental release.

Many companies, government agencies, and other entities have confronted the same issue you may face: how to discuss with the public the risks the community is subject to. Exhibit 11-1 outlines seven "rules" of risk communication that have been developed based on many experiences of dealing with the public about risks.

A key message of these "rules" is the importance and legitimacy of public concerns. People generally are less tolerant of risks they cannot control than those they can. For example, most people are willing to accept the risks of driving because they have some control over what happens to them. However, they are generally more uncomfortable accepting the risks of living near a facility that handles hazardous chemicals if they feel that they have no control over whether the facility has an accident. The Clean Air Act's provision for public availability of RMPs gives public an opportunity to take part in reducing the risk of chemical accidents that might occur in their community.

---

**Exhibit 11-1:  Seven Cardinal Rules of Risk Communication**

1.  Accept and involve the public as a legitimate partner

2.  Plan carefully and evaluate your efforts

3.  Listen to the public's specific concerns

4.  Be honest, frank, and open

5.  Coordinate and collaborate with other credible sources

6.  Meet the needs of the media

7.  Speak clearly and with compassion

---

## HAZARDS VERSUS RISKS

Dialogue in the community will be concerned with both hazards and risks; it is useful to be clear about the difference between them.

Hazards are inherent properties that cannot be changed.  Chlorine is toxic when inhaled or ingested; propane is flammable.  There is little that you can do with these chemicals to change their toxicity or flammability.  If you are in an earthquake zone or an area affected by hurricanes, earthquakes and hurricanes are hazards.  When you conduct your hazard review or process hazards analysis, you will be identifying your hazards and determining whether the potential exposure to the hazard can be reduced in any way (e.g., by limiting the quantity of chlorine stored on-site).

Risk is usually evaluated based on several variables, including the likelihood of a release occurring, the inherent hazards of the chemicals combined with the quantity released, and the potential impact of the release on the public and the environment.  For example, if a release during loading occurs frequently, but the quantity of chemical released is typically small and does not generally migrate off site, the overall risk to the public is low.  If the likelihood of a catastrophic release occurring is extremely low, but the number of people who could be affected if it occurred is large, the overall risk may still be low because of the low probability that a release will occur.  On the other hand, if a release occurs relatively frequently *and* a large number of people could be affected, the overall risk to the public is high.

The rule does not require you to assess risk in a quantitative way because, in most cases, the data you would need to estimate risk levels (e.g., one in 100 years) are not available. Even in cases where data such as equipment failure rates are available, there are large uncertainties in using that data to determine a numerical risk level for your facility, because your facility is probably not the same as other facilities, and your situation may be dynamic. Therefore, you may want to assign qualitative values (high, medium, low) to the risks that you have identified at your facility, but you should be prepared to explain the terms if you do. For example, if you believe that the worst-case release is very unlikely to occur, you must give good reasons; you must be able to provide specific examples of measures that you have taken to prevent such a release, such as installation of new equipment, careful training of your workers, rigorous preventive maintenance, etc. You should also be able to show documentation to support your claim.

## WHO WILL ASK QUESTIONS?

Your Local Emergency Planning Committee (LEPC) and other facilities can help you identify individuals in the following groups who may be reviewing RMP data and asking questions. Interested parties may include:

(1)     Persons living near the facility and elsewhere in the community or working at a neighboring facility

(2)     Local officials from zoning and planning boards, fire and police departments, health and building code officials, elected officials, and various county and state officials

(3)     Your employees

(4)     Special interest groups including environmental organizations, chambers of commerce, unions, and various civic organizations

(5)     Journalists, reporters, and other media representatives

(6)     Medical professionals, educators, consultants, neighboring companies and others with special expertise or interests

In general, people will be concerned about accident risks at your facility, how you manage the risks, and potential impacts of an accident on health, safety, property, natural resources, community infrastructure, community image, property values, and other matters. Those individuals in the public and private sector who are responsible for dealing with these impacts and the associated risks also will have an interest in working with you to address these risks.

## WHAT INFORMATION ABOUT YOUR FACILITY IS AVAILABLE TO THE PUBLIC?

Even though the non-confidential information you provide in your RMP is available to the public, it is likely that people will want additional information. Interested parties will know that you retain additional information at your facility (e.g., documentation of the results of the offsite consequence analysis reported in your RMP) and are required to make it available to EPA or its implementing agency during inspections or compliance audits. Therefore, they may request such information. EPA encourages you to provide public access to this information. If EPA or its implementing agency were to request this information, it would be available to the public under section 114(c) of the CAA.

The public may also be interested in other information relevant to risk management at your facility, such as:

- Submissions under sections 302, 304, 311-312, and 313 of the Emergency Planning and Community Right to Know Act (EPCRA) reporting on chemical storage and releases, as well as the community emergency response plan prepared under EPCRA section 303.

- Other reports on hazardous materials made, used, generated, stored, spilled, released and transported, that you submitted to federal, state, and local agencies.

- Reports on workplace safety and accidents developed under the Occupational Safety and Health Act that you provide to employees, who may choose to make the information publicly available, such as medical and exposure records, chemical data sheets, and training materials.

- Any other information you have provided to public agencies that can be accessed by members of the public under the federal Freedom of Information Act and similar state laws (and that may have been made widely available over the Internet).

- Any published materials on facility safety (either industry- or site-specific), such as agency reports on facility accidents, safety engineering manuals and textbooks, and professional journal articles on facility risk management.

## 11.2 SAMPLE QUESTIONS FOR COMMUNICATING WITH THE PUBLIC

Smaller businesses may not have the resources or time to develop the types of outreach programs, described later in this chapter, that many larger chemical companies have used to handle public questions and community relations. For many small businesses, communication with the public will usually occur when you are asked questions about information in your RMP. It is important that you respond to these questions constructively. Go beyond just answering questions; discuss what you have done to prevent accidents and work with the community to reduce risks. The people in your community will be looking to you to provide answers.

To help you establish a productive dialogue with the community, the rest of this section presents questions you are likely to be asked and a framework for answering them. These are elements of the public dialogue that you may anticipate. The person from your facility designated as responsible for communicating with the public should review the following and talk to other community organizations to determine which questions are most likely to be raised and identify other foreseeable issues. Remember that others in the community, notably LEPCs and other emergency management organizations are also likely to be asked these and other similar questions. You should consider the unique features of your facility, your RMP, and your historical relationship with the community (e.g., prior accidents, breakdowns in the coordination of emergency response efforts, and management-labor disputes), and work together with these other organizations to answer these questions for your situation and to resolve the issues associated with them.

### What does your worst-case release distance mean?

The distance is intended to provide an estimate of the maximum possible area that might be affected under catastrophic conditions. It is intended to ensure that no potential risks to public health are overlooked, but the distance to an endpoint estimated under worst-case conditions should not be considered a "public danger zone."

In most cases, the mathematical models used to analyze the worst-case release scenario as defined in the rule may overestimate the area that would be impacted by a release. In other cases, the models may underestimate the area. For distances greater than approximately six miles, the results of toxic gas dispersion models are especially uncertain, and you should be prepared to discuss such possibilities in an open, honest manner.

Reasons that modeling may underestimate the distance generally relate to the inability of some models to account for site-specific factors that might tend to increase the actual endpoint distance. For example, assume a facility is located in a river valley and handles dense toxic gases such as chlorine. If a release were to occur, the river valley could channel the toxic cloud much farther than it might travel if it were to disperse in a location with generally flat terrain. In such cases, the actual endpoint distance might be longer than that predicted using generic lookup tables.

Reasons that the area may be overestimated include:

••      For toxics, the weather conditions (very low wind speed, calm conditions) assumed for a worst-case release scenario are uncommon and probably would not last as long as the time the release would take to travel the distance estimated. If weather conditions are different, the distance would be much shorter.

••      For flammables, although explosions can occur, a release of a flammable is more likely to disperse harmlessly or burn. If an explosion does occur, however, this area could be affected by the blast; debris from the blast could affect an even broader area.

••      In general, some models cannot take into account other site-specific factors that might tend to disperse the chemicals more quickly and limit the distance.

Note: When estimating worst case release distances, the rule does not allow facilities to take into account active mitigation systems and practices that could limit the scope of a release. Specific systems (e.g., monitoring, detection, control, pressure relief, alarms, mitigation) may limit a release or prevent the failure from occurring. Also, if you are required to analyze alternative release scenarios (i.e., if your facility is in Program 2 or Program 3), these scenarios are generally more realistic than the worst case, and you can offer to provide additional information on those scenarios.

## *What does it mean that we could be exposed if we live/work/shop/go to school X miles away?*

*(For an accident involving a flammable substance):*

The distance means that people who are in that area around the facility could be hurt if the contents of a tank or other vessel exploded. The blast of the explosion could shatter windows and damage buildings. Injuries would be the result of the force of the explosion and of flying glass or falling debris.

*(For an accident involving a toxic substance):*

The distance is based on a concentration of the chemical that you could be exposed to for an hour without suffering irreversible health effects or other symptoms that would make it difficult for you to escape. If you are within that distance, you could be exposed to a greater concentration of the chemical. If you were exposed to higher levels for an extended period of time (10 minutes, 30 minutes, or longer), you could be seriously hurt. However, that does not mean that you would be. Remember, for worst case scenarios, the rule requires you to make certain conservative assumptions with respect to, for example, wind speed and atmospheric stability. If the wind speed is higher than that used in the modeling, or if the atmosphere is more unstable, a chemical release would be dispersed more quickly, and the distances would be much smaller and the exposure times would be shorter. If the question pertains to an alternative release scenario, you probably assumed typical weather conditions in the modeling. Therefore, the actual impact distance could be shorter or longer, and you should be prepared to acknowledge this and clearly explain how you chose the conditions for your release scenario.

In general, the possibility of harm depends on the concentration of the chemical you are exposed to and the length of time you are exposed.

### *IF THERE IS AN ACCIDENT, WILL EVERYONE WITHIN THAT DISTANCE BE HURT? WHAT ABOUT PROPERTY DAMAGE?*

In general, no. For an explosion, everyone within the circle would certainly feel the blast wave since it would move in all directions at once. However, while some people within the circle could be hurt, it is unlikely that everyone would be since some people would probably be in less vulnerable locations. Most injuries would probably be due to the effects of flying glass, falling debris, or impact with nearby objects.

Two types of chemicals may be modeled - toxics and flammables. Releases of flammables do not usually lead to explosions; released flammables are more likely to disperse without igniting. If the released flammable does ignite, a fire is more likely than an explosion, and fires are usually concentrated at the facility.

For toxic chemicals, whether someone is hurt by a release depends on many factors. First, the released chemicals would usually move in the direction of the wind (except for some dense gases, which may be constrained by terrain features to flow in a different direction). Generally, only people downwind from the facility would be at risk of exposure if a release occurred, and this is normally only a part of the population inside the circle. If the wind speed is moderate, the chemicals would disperse quickly, and people would be exposed to lower levels of the chemical. If the release is stopped quickly, they might be exposed for a very short period time, which is less likely to cause injury. However, if the wind speed is low or the release continues for a long time, exposure levels will be higher and more dangerous. The population at risk would be a larger proportion of the total population inside the circle. You should be prepared to discuss both possibilities.

Generally, it is the people who are closest to the facility — within a half mile or less — who would face the greatest danger if an accident occurred.

Damage to property and the environment will depend on the type of chemical released. In an explosion, environmental impacts and property damage may extend beyond the distance at which injuries could occur. For a vapor release, environmental effects and property damage may occur as a result of the reactivity or corrosivity of the chemical or toxic contamination.

### HOW SURE ARE YOU OF YOUR DISTANCES?

Perhaps the largest single difficulty associated with hazard assessment is that different models and modeling assumptions will yield somewhat different results. There is no one model or set of assumptions that will yield "certain" results. Models represent scientists' best efforts to account for all the variables involved in an accidental release. While all models are generally based on the same physical principles, dispersion modeling is not an exact science due to the limited opportunity for real-world validation of results. No model is perfect, and every model represents a somewhat different analytical approach. As a result, for a given scenario, people can use different consequence models and obtain predictions of the distance to the toxic endpoint that in some situations might vary by a factor of ten. Even using the same model, different input assumptions can cause wide variations in the predictions. It follows that, when you present a single predicted value as your best estimate of the predicted distance, others will be able to claim that the answer ought to be different, perhaps greater, perhaps smaller, depending on the assumptions used in modeling and the choice of model itself.

You therefore need to recognize that your predicted distance lies within a considerable band of uncertainty, and to communicate this fact to those who have an interest in your results. A neighboring facility handling the same covered substances as you do may have come up with a different result for the same scenario for these reasons.

If you use EPA's *RMP Offsite Consequence Analysis Guidance* or one of the industry-specific guidance documents that EPA has developed, you will be able to address the issue of uncertainty by stating that the results you have generated are conservative (that is they are likely to overestimate distances). However, if you use other models, you will have to provide your own assessment of where your specific prediction lies within the plausible range of uncertainties.

### WHY DO YOU NEED TO STORE SO MUCH ON-SITE?

If you have not previously considered the feasibility of reducing the quantity, you should do so when you develop your risk management program. Many companies have cited public safety concerns as a reason for reducing the quantities of hazardous chemicals stored on-site or for switching to non-hazardous substitutes. If you have evaluated your process and determined that you need a certain volume to maintain your operations, you should explain this fact to the public in a forthright manner. As appropriate, you should also discuss any alternatives, such as reducing storage quantities and scheduling more frequent deliveries. Perhaps these options are feasible - if so, you should consider implementing them; if not, explain why you consider these alternatives to be unacceptable. For example, in some situations, more frequent deliveries would mean more trucks carrying the substance through the community on a regular basis and a greater opportunity for smaller-scale releases because of more frequent loading and unloading.

### WHAT ARE YOU DOING TO PREVENT RELEASES?

If you have rigorously implemented your risk management program, this question will be your chance, if you have not already done so, to tell the community about your prevention activities, the safe design features of your operations, the specific activities that you are performing such as training, operating procedures, maintenance, etc., and any industry codes or standards you use to operate safely. If you have installed new equipment or safety systems, upgraded training, or had outside experts review your site for safety (e.g., insurance inspectors), you could offer to share the results. You may also want to mention state or federal rules you comply with.

### WHAT ARE YOU DOING TO PREPARE FOR RELEASES?

For such questions, you will need to talk about any coordination that you have done with the local fire department, LEPC, or mutual aid groups. Such coordination may include activities such as defining an incident command structure, developing notification protocols, conducting response training and exercises, developing mutual aid agreements, and evaluating public alert systems. This description is particularly important if your employees are not designated or trained to respond to releases of regulated substances.

If your employees will be involved in a response, you should describe your emergency response plan and the emergency response resources available at the facility (e.g., equipment, personnel), as well as through response contractors, if appropriate. You also may want to indicate the types of events for which such resources are applicable. Finally, indicate your schedule for internal emergency response training and drills and exercises and discuss the results of the latest relevant drill or exercise, including problems found and actions taken to address them.

### DO YOU NEED TO USE THIS CHEMICAL?

Again, if you have not yet considered the feasibility of switching to a non-hazardous substitute, you should do so when you develop your risk management program. Assuming that there is no substitute, you should describe why the chemical is critical to what you produce and explain what you do to handle it safely. If there are substitutes available, you should describe how you have evaluated such options.

*WHY ARE YOUR DISTANCES DIFFERENT FROM THE DISTANCES IN THE EPA LOOKUP TABLES?*

If you did your own modeling, this question may come up. You should be ready to explain in a general way how your model works and why it produces different results. EPA allows using other models (as long as certain parameters and conditions specified by the rule are met) because it realizes that EPA lookup table results will not necessarily reflect all site-specific conditions.

In addition, although all models are generally based on the same physical principles, dispersion modeling is not an exact science due to the limited opportunity for real-world validation of the results. Thus, the method by which different models combine the basic factors such as wind speed and atmospheric stability can result in distances that readily vary by a factor of two (e.g., five miles versus ten miles). The introduction of site-specific factors can produce additional differences.

EPA recognizes that different models will produce differing predictions of the distance to an endpoint, especially for releases of toxic substances. The Agency has provided a discussion of the uncertainties associated with the model it has adopted for the OCA Guidance. You need to understand that the distances produced by another model lie within a band of uncertainty and be able to demonstrate and communicate this fact to those who are reviewing your results.

*HOW LIKELY ARE THE WORST-CASE AND ALTERNATIVE RELEASE SCENARIOS?*

It is generally not possible to provide accurate numerical estimates of how likely these scenarios are. EPA has stated that providing such numbers for accident scenarios rarely is feasible because the data needed (e.g., on rates for equipment failure and human error) are not usually available. Even when data are available, there are large uncertainties in applying the data because each facility's situation is unique.

In general, the risk of the worst-case scenario is low. Although catastrophic vessel failures have occurred, they are rare events. Combining them with worst-case weather conditions makes the overall scenario even less likely. This does not mean that such events cannot or will not happen, however.

For the alternative scenario, the likelihood of the release is greater and will depend, in part, on the scenario you chose. If you selected a scenario based on your accident history or industry accident history, you should explain this to the public. You should also discuss any steps you are taking to prevent such an accident from recurring.

***IS THE WORST-CASE RELEASE YOU REPORTED REALLY THE WORST ACCIDENT YOU CAN HAVE?***

The answer to this question will depend on the type of facility you have and how you handle chemicals. EPA defined a specific scenario (failure of the single largest vessel) to provide a common basis of comparison among facilities nationwide. So, if you have only one vessel, EPA's worst case is likely to be the worst event you could have.

On the other hand, if you have a process which involves multiple co-located or interconnected vessels, it is possible that you could have an accident more severe than EPA's worst case scenario. If credible scenarios exist that could be more serious (in terms of quantities released or consequences) than the EPA worst case scenario, you should be ready to discuss them. For example, if you store chemicals in small containers such as 55-gallon drums, the EPA-defined worst-case release scenario may involve a limited quantity, but a fire or explosion at the facility could release larger quantities if multiple containers are involved. In this case, you should be ready to frankly discuss such a scenario with the public. If you take precautions to prevent such scenarios from occurring, you should explain these precautions also. If an accidental release is more likely to involve multiple drums than a single drum as a result, for example, of the drums being stored closely together, then you must select such a scenario as your alternative release scenario so that information on this scenario is available in your RMP.

Chemical manufacturers may want to talk about releases that could result from runaway reactions that could continue for several hours. This type of event could result in longer exposure times.

***WHAT ABOUT THE ACCIDENT AT THE [NAME OF SIMILAR FACILITY] THAT HAPPENED LAST MONTH?***

This question highlights an important point: you need to be aware of events in your industry (e.g., accidents, new safety measures) for two reasons. First, your performance likely will be compared to that of your competitors. Second, learning about the circumstances and causes of accidents at other facilities like yours can help you prevent such accidents from occurring at your facility.

You should be familiar with accidents that happen at facilities similar to yours, and you should have evaluated whether your facility is at risk for similar accidents. You should take the appropriate measures to prevent the accident from occurring and be prepared to describe these actions. If your facility has experienced a similar release in the past, this information may be documented in your accident history or other publicly available records, depending on the date and nature of the incident, the quantity released, and other factors. If you have already taken steps specifically designed to address this type of accident, whether as a result of this accident, a prior accident at your facility, or other internal decision-making, you should describe these efforts. If, based on your evaluation, you determine that the accident could not occur at your facility, you should discuss the pertinent differences between the two facilities and explain why you believe those differences should prevent the accident from occurring at your facility.

---

**WHAT ACTIONS HAVE YOU TAKEN TO INVOLVE THE COMMUNITY IN YOUR ACCIDENT PREVENTION AND EMERGENCY PLANNING EFFORTS?**

If you have not actively involved the community in accident prevention and emergency planning in the past, you should acknowledge this as an area where you could improve and start doing so as you develop your risk management program. First, you may want to begin participating in the LEPC and regional mutual aid organizations if you aren't doing so already. Other opportunities for community involvement are fire safety coordination activities with the local fire department, joint training and exercises with local public and private sector response personnel, the establishment of green fields between the facility and the community, and similar efforts.

When discussing accident prevention and emergency planning with the community, you should indicate any national programs in which you participate, such as the Chemical Manufacturers Association's Responsible Care program or Community Awareness and Emergency Response program or OSHA's Voluntary Protection Program. If fully implemented, these programs can help improve the safety of the facility and the community. You may have future plans to participate in areas described previously or have new initiatives associated with the risk management program. Be sure you ask what else the community would like you to do and explain how you will do it.

---

**CAN WE SEE THE DOCUMENTATION YOU KEEP ON SITE?**

If the requested information is not confidential business information, EPA encourages you to make it available to the public. Although you are not required to provide this information to the public, refusing to provide it simply because you are not compelled to is not the best approach. If you decide not to provide any or most of this material, you should have good reasons for not doing so and be prepared to explain these reasons to the public. Simply taking a defensive position or referring to the extent of your legal obligations is likely to threaten the effectiveness of your interaction with the community. Offer as much information as possible to the public; if particular documents would reveal proprietary information, try to provide a redacted copy, summary, or some other form that answers the community's concerns. You may want to work with your LEPC on this issue. You should also be aware that information that EPA or the implementing agency obtains as part of an inspection or investigation conducted under section 114 of the Clean Air Act would be available to the public under section 114(c) of the Act to the extent it does not reveal confidential business information.

---

## 11.3    COMMUNICATION ACTIVITIES AND TECHNIQUES

Although this section is most applicable to larger companies, small businesses may want to review it and use some of the ideas to expand their communications with the public. To prepare for effective communication with the community, you should:

(1)    Adopt an organizational policy that includes basic risk communication principles (see exhibit 11-1).

(2)      Assign responsibilities and resources to implement the policy.

(3)      Plan to use "best communication practices".

## ADOPT AN ORGANIZATIONAL COMMUNICATIONS POLICY

An organizational policy will support communication with the public on your RMP and make it an integral part of management practices. Otherwise, breakdowns are likely to occur, which could cause mistrust, hostility and conflicts.

A policy helps to establish communication as a normal organizational function and to present it as an opportunity rather than a burden or threat. The policy can be incorporated in an organization's policies, an approach taken by many companies who belong to the Responsible Care program of the Chemical Manufacturers Association (CMA). These companies have adopted CMA's Codes of Management Practices, which contain risk communication principles and practices.

Remember that what you communicate is more important than the type of communication policy or program you use, and what you actually *do* to maintain a safe facility is more important than anything you say. Your company's safety and prevention steps in your risk management program should serve as the core elements of any risk communication program.

## ASSIGN RESPONSIBILITIES AND RESOURCES

A policy is only a paper promise until it is regularly and effectively implemented. Thus, you should follow up your communication policy by (1) having top management participate at the outset and at key points throughout the communication process, and (2) assigning communication responsibilities within your organization and providing the necessary resources.

Experience has demonstrated that assigning responsibility to knowledgeable managers, plant engineers, and staff and encouraging participation by employees, (most of whom are likely to be community residents) is a good communications practice. Delegating communication functions to outside technical consultants, attorneys, and public relations specialists has repeatedly failed to impress the community and even tends to incur mistrust. (However, if you hired a firm with acknowledged expertise in dispersion modeling, you may want them on hand to help respond to technical questions.)

Communications staff will need work time and resources to prepare presentation materials, hold meetings with interested persons in the community, and do other work necessary to respond to questions and concerns and maintain ongoing dialogue. A training program in communication skills and incentives for good performance also may be advisable.

Organizations have a legitimate interest in preventing disclosure of confidential business information or statements that inadvertently and unfairly harm the organization or its employees. Thus, you should assure that your risk communication staff is instructed on how to deal with situations that pose these problems. This may mean that you have an internal procedure enabling your staff to bring such situations to top management and legal counsel for quick resolution, keeping in mind that unduly defensive or legalistic responses that result in restricting the amount of information that is provided can damage or destroy the risk communication process.

Your communication staff may find the following steps helpful in addressing the priority issues in the communication process:

Prior to RMP Submittal

- • Enlist employee support for, and involvement in, the communication process.

- • Build on work you have done with your LEPC, fire department, and local officials, and gain their insights.

- • Incorporate technical expertise, management commitment, and employee involvement in the risk communication process.

- • Use your RMP's executive summary to begin the dialogue with the community; be sure you have taken all of the steps you present.

- • Taking a community perspective, identify which data elements need to be clarified, interpreted, or amplified, and which are most likely to raise community concerns; then compile the information needed to respond and determine the most understandable methods (e.g., use of graphics) for presenting the information.

At Submittal

- • Review the RMP to assure that you are familiar with its data elements and how they were developed. In particular, review the hazard assessment, prevention, and response program features, as well as documentation of the methods, data, and assumptions used, especially if an outside consultant performed the analyses and developed these materials. You have certified their accuracy and your spokesperson should know them intimately, as they reflect your plan.

- • Review your performance in implementing the prevention and response programs and prepare to discuss problems identified and actions taken.

- • Review your performance in investigating accidents and prepare to discuss any corrective actions that followed.

Other Steps

- Identify the most likely concerns about risks identified in the RMP but not fully addressed, consult with management and safety engineering, and determine additional measures the organization will take to resolve these concerns.

- Avoid misrepresentations and minimize the roles of public relations specialists.

- Identify "best communication practices" (as described in the next section) and plan how to use them.

## USE "BEST COMMUNICATION PRACTICES"

Many facilities already have gained considerable experience in communicating with the public. Lessons from their experiences are described below. However, the value of these best practices and your credibility will depend on your facility's possession and ongoing demonstration of certain essential qualities:

- Top management commitment (e.g., owner and facility manager) to improving safety

- Honesty, openness, and concern for the community

- Respect for public concerns and perceptions

- Commitment to maintaining a dialogue with all sectors of the community, to learning from this dialogue, and to being prepared to change your practices to make your facility more safe

- Commitment to continuous improvement through internal procedures for evaluating incidents and promoting organizational learning

- Knowledge of safety issues and safety management methods

- Good working relationships with the LEPC, fire department, and other local officials

- Active support for the LEPC and related activities

- Employee support and commitment

- Continuation of commitment despite potential public hostility or mistrust

Another note: Because each facility and community involves a unique combination of factors, the practices used to achieve good risk communication in one case do not necessarily ensure the same quality result when used in another case. Therefore, while it is advisable for you to review such experience to identify "best communication practices," you should carefully evaluate such practices to determine if they can be adapted to fit your unique circumstances. For example, if your facility is in the middle of an urban area, you probably will use different approaches than you would use if it were located in an industrial area far from any residential populations. These practices are complementary approaches to delivering your risk management message and responding to the concerns of the community.

With these cautions in mind, a number of "best" practices are outlined below for consideration. First, you will want to establish formal channels for information-sharing and communication with stakeholders. The most basic approaches include the following:

- • Convene public meetings for discussion and dialogue regarding your risk management program and RMP and take steps to have the facility owner or manager and all sectors of the community participate, including minorities and low-income residents.

- • Arrange meetings with local media representatives to facilitate their understanding of your risk management program and the program summary presented in your RMP.

- • Establish a repository of information on safety matters for the LEPC and the public and, if electronic, provide software for public use. Some organizations also have provided computer terminals for public use in the community library or fire department.

Other, more resource-intensive activities of this type to consider include the following:

- Create and convene focus groups (small working groups) to facilitate dialogue and action on specific concerns, including technical matters, and take steps to assure that membership in each group reflects a cross section of the community and includes technically trained persons (e.g., engineers, medical professionals).

- Hold seminars on hypothetical release scenarios, prevention and response programs, applicable standards and industry practices, analytic methods and models (e.g., on dispersion of airborne releases, health effects of airborne concentrations), and other matters of special concern or complexity.

- Convene special meetings to foster dialogue and collaborations with the LEPC and the fire department and to establish a mutual assistance network with other facility managers in the community or region.

- Establish hot lines for telephone and e-mail communications between interested parties and your designated risk communication staff and, if feasible, a web site for posting useful information.

In all of these efforts, remember to use plain language and commonly understood terms; avoid the use of acronyms and technical and legal jargon. In addition, depending on your audience, keep in mind that the preparation of multilingual materials may be useful or even necessary.

Secondly, you may want to initiate or expand programs that more directly involve the community in your operations and safety programs. Traditional approaches include the following:

- • Arrange facility tours so that members of the public can view operations and discuss safety procedures with supervisors and employees.

- • Schedule drills and simulations of incidents to demonstrate how prevention and response programs work, with participation by community responders and other organizations (e.g., neighboring companies).

- • Conduct a "Safety Street" - a community forum generally sponsored by several industries in a locality, where your representatives present facility safety information, explain risks, and respond to public questions (see Section 11.4 for a reference to more information on this program).

- • Periodically reaffirm and demonstrate your commitment to safety in accordance with and beyond regulatory requirements and present data on your safety performance, using appropriate benchmarks or measures, in newsletters and by posting the information at your web site.

- • Publicly honor and reward managers and employees who have performed safety responsibilities in superior fashion and citizens who have made important contributions to the dialogue on safety.

If community interest is significant, you may also want to consider the following activities:

- • Invite public participation in monitoring implementation of your risk management program elements.

- • Invite public participation in auditing your performance in safety responsibilities, such as chemical handling and tracking procedures and analysis and follow-up on accidents and near misses.

- • Organize a committee comprised of representatives from the facility, other industry, emergency planning and response organizations, and community groups and chaired by a community leader to independently evaluate your safety and communication efforts (e.g., a Community Advisory Panel). You may also want to finance the committee to pay for an independent engineering consultant to assist with technical issues and learn what can be done to improve safety, and thereby share control with the community.

Your communication staff should review these examples, consider designing their own activities as well as joint efforts with other local organizations, and ultimately decide with the community on which set of practices are feasible and can best create a healthy risk communication process in your community. Once these decisions are made, you may want to integrate the chosen set of practices in an overall communication program for your facility, transform some into standard procedures, and monitor and evaluate them for continuous improvement.

## OTHER COMMUNICATION OPPORTUNITIES

By complying with the RMP rule and participating in the communications process with the community, you should have developed a comprehensive system for preventing, mitigating, and responding to chemical accidents at your facility. Why not share this knowledge with your staff, others you do business with (e.g., customers, distributors, contractors), and, perhaps through industry groups, others in your industry? If you transfer this knowledge to others, you can help improve their chemical safety management capabilities, enhance public safety beyond your community, and possibly gain economic benefits for your organization.

## 11.4    FOR MORE INFORMATION

Among the numerous publications on risk communication, the following may be particularly helpful:

*   *Improving Risk Communication*, National Academy Press, Washington, D.C., 1989

*   "Safety Street" and other materials on the Kanawha Valley Demonstration Program, Chemical Manufacturers Association, Arlington, VA

*   Community Awareness and Emergency Response Code of Management Practices and various Guidance, Chemical Manufacturers Association, Arlington, VA

*   *Communicating Risks to the Public*, R. Kasperson and P. Stallen, eds., Kluwer Publishing Co., 1991

*   "Challenges in Risk and Safety Communication with the Public," S. Maher, Risk Management Professionals, Mission Viejo, CA, April 1996

*   Primer on Health Risk Communication Principles and Practices, Agency for Toxic Substances and Disease Registry, on the World Wide Web at atsdr1.atsdr.cdc.gov:8080

*   *Risk Communication about Chemicals in Your Community: A Manual for Local Officials*, US Environmental Protection Agency, EPA EPCRA/Superfund/RCRA/CAA Hotline

- *Risk Communication about Chemicals in Your Community: Facilitator's Manual and Guide*, US Environmental Protection Agency, EPA EPCRA/Superfund/RCRA/CAA Hotline

- *Chemicals, the Press, and the Public: A Journalist's Guide to Reporting on Chemicals in the Community*, US Environmental Protection Agency, EPA EPCRA/Superfund/RCRA/CAA Hotline

**Appendix A**
**40 CFR part 68**

local agent, any noncompliance penalties owed by the source owner or operator shall be paid to the State or local agent.

APPENDIX A TO PART 67—TECHNICAL
SUPPORT DOCUMENT

NOTE: EPA will make copies of appendix A available from: Director, Stationary Source Compliance Division, EN–341, 401 M Street, SW., Washington, DC 20460.

[54 FR 25259, June 20, 1989]

APPENDIX B TO PART 67—INSTRUCTION
MANUAL

NOTE: EPA will make copies of appendix B available from: Director, Stationary Source Compliance Division, EN–341, 401 M Street, SW., Washington, DC 20460.

[54 FR 25259, June 20, 1989]

APPENDIX C TO PART 67—COMPUTER
PROGRAM

NOTE: EPA will make copies of appendix C available from: Director, Stationary Source Compliance Division, EN–341, 401 M Street, SW., Washington, DC 20460.

[54 FR 25259, June 20, 1989]

# PART 68—CHEMICAL ACCIDENT PREVENTION PROVISIONS

## Subpart A—General

AUTHORITY: 42 U.S.C. 7412(r), 7601(a)(1), 7661–7661f.

SOURCE: 59 FR 4493, Jan. 31, 1994, unless otherwise noted.

## Subpart A—General

### §68.1 Scope.

This part sets forth the list of regulated substances and thresholds, the petition process for adding or deleting substances to the list of regulated substances, the requirements for owners or operators of stationary sources concerning the prevention of accidental releases, and the State accidental release prevention programs approved under section 112(r). The list of substances, threshold quantities, and accident prevention regulations promulgated under this part do not limit in any way the general duty provisions under section 112(r)(1).

### §68.2 Stayed provisions.

(a) Notwithstanding any other provision of this part, the effectiveness of the following provisions is stayed from March 2, 1994 to December 22, 1997.

(1) In Sec. 68.3, the definition of "stationary source," to the extent that such definition includes naturally occurring hydrocarbon reservoirs or transportation subject to oversight or regulation under a state natural gas or hazardous liquid program for which the state has in effect a certification to DOT under 49 U.S.C. 60105;

(2) Section 68.115(b)(2) of this part, to the extent that such provision requires an owner or operator to treat as a regulated flammable substance:

(i) Gasoline, when in distribution or related storage for use as fuel for internal combustion engines;

(ii) Naturally occurring hydrocarbon mixtures prior to entry into a petroleum refining process unit or a natural gas processing plant. Naturally occurring hydrocarbon mixtures include any of the following: condensate, crude oil, field gas, and produced water, each as defined in paragraph (b) of this section;

(iii) Other mixtures that contain a regulated flammable substance and that do not have a National Fire Protection Association flammability hazard rating of 4, the definition of which is in the NFPA 704, Standard System for the Identification of the Fire Hazards of Materials, National Fire Protection Association, Quincy, MA, 1990, available from the National Fire Protection Association, 1 Batterymarch Park, Quincy, MA 02269–9101; and

(3) Section 68.130(a).

(b) From March 2, 1994 to December 22, 1997, the following definitions shall apply to the stayed provisions described in paragraph (a) of this section:

*Condensate* means hydrocarbon liquid separated from natural gas that condenses because of changes in temperature, pressure, or both, and remains liquid at standard conditions.

*Crude oil* means any naturally occurring, unrefined petroleum liquid.

*Field gas* means gas extracted from a production well before the gas enters a natural gas processing plant.

*Natural gas processing plant* means any processing site engaged in the extraction of natural gas liquids from field gas, fractionation of natural gas liquids to natural gas products, or both. A separator, dehydration unit, heater treater, sweetening unit, compressor, or similar equipment shall not be considered a "processing site" unless such equipment is physically located within a natural gas processing plant (gas plant) site.

*Petroleum refining process unit* means a process unit used in an establishment primarily engaged in petroleum refining as defined in the Standard Industrial Classification code for petroleum refining (2911) and used for the following: Producing transportation fuels (such as gasoline, diesel fuels, and jet fuels), heating fuels (such as kerosene, fuel gas distillate, and fuel oils), or lubricants; separating petroleum; or separating, cracking, reacting, or reforming intermediate petroleum streams. Examples of such units include, but are not limited to, petroleum based solvent units, alkylation units, catalytic hydrotreating, catalytic hydrorefining, catalytic hydrocracking, catalytic reforming, catalytic cracking, crude distillation, lube oil processing, hydrogen production, isomerization, polymerization, thermal processes, and blending, sweetening, and treating processes. Petroleum refining process units include sulfur plants.

*Produced water* means water extracted from the earth from an oil or natural gas production well, or that is separated from oil or natural gas after extraction.

(c) Notwithstanding any other provision of this part, the effectiveness of part 68 is stayed from June 21, 1999 to December 21, 1999 with respect to regulated flammable hydrocarbon substances when the substance is intended for use as a fuel and does not exceed 67,000 pounds in a process that is not manufacturing the fuel, does not contain greater than a threshold quantity of another regulated substance, and is not collocated or interconnected to another covered process.

[59 FR 4493, Jan. 31, 1994, as amended at 61 FR 31731, June 20, 1996; 64 FR 29170, May 28, 1999]

### § 68.3   Definitions.

For the purposes of this part:

*Accidental release* means an unanticipated emission of a regulated substance or other extremely hazardous substance into the ambient air from a stationary source.

*Act* means the Clean Air Act as amended (42 U.S.C. 7401 *et seq.*)

*Administrative controls* mean written procedural mechanisms used for hazard control.

*Administrator* means the administrator of the U.S. Environmental Protection Agency.

*AIChE/CCPS* means the American Institute of Chemical Engineers/Center for Chemical Process Safety.

*API* means the American Petroleum Institute.

*Article* means a manufactured item, as defined under 29 CFR 1910.1200(b), that is formed to a specific shape or design during manufacture, that has end use functions dependent in whole or in part upon the shape or design during end use, and that does not release or otherwise result in exposure to a regulated substance under normal conditions of processing and use.

*ASME* means the American Society of Mechanical Engineers.

*CAS* means the Chemical Abstracts Service.

*Catastrophic release* means a major uncontrolled emission, fire, or explosion, involving one or more regulated substances that presents imminent and substantial endangerment to public health and the environment.

*Classified information* means "classified information" as defined in the Classified Information Procedures Act, 18 U.S.C. App. 3, section 1(a) as "any information or material that has been determined by the United States Government pursuant to an executive order, statute, or regulation, to require protection against unauthorized disclosure for reasons of national security."

*Condensate* means hydrocarbon liquid separated from natural gas that condenses due to changes in temperature, pressure, or both, and remains liquid at standard conditions.

*Covered process* means a process that has a regulated substance present in more than a threshold quantity as determined under § 68.115.

*Crude oil* means any naturally occurring, unrefined petroleum liquid.

*Designated agency* means the state, local, or Federal agency designated by the state under the provisions of § 68.215(d).

*DOT* means the United States Department of Transportation.

*Environmental receptor* means natural areas such as national or state parks, forests, or monuments; officially designated wildlife sanctuaries, preserves, refuges, or areas; and Federal wilderness areas, that could be exposed at any time to toxic concentrations, radiant heat, or overpressure greater than or equal to the endpoints provided in § 68.22(a), as a result of an accidental release and that can be identified on local U. S. Geological Survey maps.

*Field gas* means gas extracted from a production well before the gas enters a natural gas processing plant.

*Hot work* means work involving electric or gas welding, cutting, brazing, or similar flame or spark-producing operations.

*Implementing agency* means the state or local agency that obtains delegation for an accidental release prevention program under subpart E, 40 CFR part 63. The implementing agency may, but is not required to, be the state or local air permitting agency. If no state or local agency is granted delegation, EPA will be the implementing agency for that state.

*Injury* means any effect on a human that results either from direct exposure to toxic concentrations; radiant heat; or overpressures from accidental releases or from the direct consequences of a vapor cloud explosion (such as flying glass, debris, and other projectiles) from an accidental release and that requires medical treatment or hospitalization.

*Major change* means introduction of a new process, process equipment, or regulated substance, an alteration of process chemistry that results in any change to safe operating limits, or other alteration that introduces a new hazard.

*Mechanical integrity* means the process of ensuring that process equipment is fabricated from the proper materials of construction and is properly installed, maintained, and replaced to prevent failures and accidental releases.

*Medical treatment* means treatment, other than first aid, administered by a physician or registered professional personnel under standing orders from a physician.

*Mitigation or mitigation system* means specific activities, technologies, or equipment designed or deployed to capture or control substances upon loss of containment to minimize exposure of the public or the environment. Passive mitigation means equipment, devices, or technologies that function without human, mechanical, or other energy input. Active mitigation means equipment, devices, or technologies that need human, mechanical, or other energy input to function.

*NAICS* means North American Industry Classification System.

*NFPA* means the National Fire Protection Association.

*Natural gas processing plant (gas plant)* means any processing site engaged in the extraction of natural gas liquids from field gas, fractionation of mixed natural gas liquids to natural gas products, or both, classified as North American Industrial Classification System (NAICS) code 211112 (previously Standard Industrial Classification (SIC) code 1321).

*Offsite* means areas beyond the property boundary of the stationary source, and areas within the property boundary to which the public has routine and unrestricted access during or outside business hours.

*OSHA* means the U.S. Occupational Safety and Health Administration. Owner or operator means any person who owns, leases, operates, controls, or supervises a stationary source.

*Petroleum refining process unit* means a process unit used in an establishment primarily engaged in petroleum refining as defined in NAICS code 32411 for petroleum refining (formerly SIC code 2911) and used for the following: Producing transportation fuels (such as gasoline, diesel fuels, and jet fuels), heating fuels (such as kerosene, fuel gas distillate, and fuel oils), or lubricants; Separating petroleum; or Separating, cracking, reacting, or reforming intermediate petroleum streams. Examples of such units include, but are not limited to, petroleum based solvent units, alkylation units, catalytic hydrotreating, catalytic hydrorefining, catalytic hydrocracking, catalytic reforming, catalytic cracking, crude distillation, lube oil processing, hydrogen production, isomerization, polymerization, thermal processes, and blending, sweetening, and treating processes. Petroleum refining process units include sulfur plants.

*Population* means the public.

*Process* means any activity involving a regulated substance including any use, storage, manufacturing, handling, or on-site movement of such substances, or combination of these activities. For the purposes of this definition, any group of vessels that are interconnected, or separate vessels that are located such that a regulated substance could be involved in a potential release, shall be considered a single process.

*Produced water* means water extracted from the earth from an oil or natural gas production well, or that is separated from oil or natural gas after extraction.

*Public* means any person except employees or contractors at the stationary source.

*Public receptor* means offsite residences, institutions (e.g., schools, hospitals), industrial, commercial, and office buildings, parks, or recreational areas inhabited or occupied by the public at any time without restriction by the stationary source where members of the public could be exposed to toxic concentrations, radiant heat, or over-pressure, as a result of an accidental release.

*Regulated substance* is any substance listed pursuant to section 112(r)(3) of the Clean Air Act as amended, in § 68.130.

*Replacement in kind* means a replacement that satisfies the design specifications.

*RMP* means the risk management plan required under subpart G of this part.

*Stationary source* means any buildings, structures, equipment, installations, or substance emitting stationary activities which belong to the same industrial group, which are located on one or more contiguous properties, which are under the control of the same person (or persons under common control), and from which an accidental release may occur. The term stationary source does not apply to transportation, including storage incident to transportation, of any regulated substance or any other extremely hazardous substance under the provisions of this part. A stationary source includes transportation containers used for storage not incident to transportation and transportation containers connected to equipment at a stationary source for loading or unloading. Transportation includes, but is not limited to, transportation subject to oversight or regulation under 49 CFR parts 192, 193, or 195, or a state natural gas or hazardous liquid program for which the state has in effect a certification to DOT under 49 U.S.C. section 60105. A stationary source does not include naturally occurring hydrocarbon reservoirs. Properties shall not be considered contiguous solely because of a railroad or pipeline right-of-way.

*Threshold quantity* means the quantity specified for regulated substances pursuant to section 112(r)(5) of the Clean Air Act as amended, listed in § 68.130 and determined to be present at a stationary source as specified in § 68.115 of this part.

*Typical meteorological conditions* means the temperature, wind speed, cloud cover, and atmospheric stability class, prevailing at the site based on data gathered at or near the site or from a local meteorological station.

*Vessel* means any reactor, tank, drum, barrel, cylinder, vat, kettle, boiler, pipe, hose, or other container.

*Worst-case release* means the release of the largest quantity of a regulated substance from a vessel or process line failure that results in the greatest distance to an endpoint defined in § 68.22(a).

[59 FR 4493, Jan. 31, 1994, as amended at 61 FR 31717, June 20, 1996; 63 FR 644, Jan. 6, 1998; 64 FR 979, Jan. 6, 1999]

§ 68.10  Applicability.

(a) An owner or operator of a stationary source that has more than a threshold quantity of a regulated substance in a process, as determined under § 68.115, shall comply with the requirements of this part no later than the latest of the following dates:

(1) June 21, 1999;

(2) Three years after the date on which a regulated substance is first listed under § 68.130; or

(3) The date on which a regulated substance is first present above a threshold quantity in a process.

(b) Program 1 eligibility requirements. A covered process is eligible for Program 1 requirements as provided in § 68.12(b) if it meets all of the following requirements:

(1) For the five years prior to the submission of an RMP, the process has not had an accidental release of a regulated substance where exposure to the substance, its reaction products, overpressure generated by an explosion involving the substance, or radiant heat generated by a fire involving the substance led to any of the following offsite:

(i) Death;

(ii) Injury; or

(iii) Response or restoration activities for an exposure of an environmental receptor;

(2) The distance to a toxic or flammable endpoint for a worst-case release assessment conducted under Subpart B and § 68.25 is less than the distance to any public receptor, as defined in § 68.30; and

(3) Emergency response procedures have been coordinated between the stationary source and local emergency planning and response organizations.

(c) Program 2 eligibility requirements. A covered process is subject to Program 2 requirements if it does not meet the eligibility requirements of either paragraph (b) or paragraph (d) of this section.

(d) Program 3 eligibility requirements. A covered process is subject to Program 3 if the process does not meet the requirements of paragraph (b) of this section, and if either of the following conditions is met:

(1) The process is in NAICS code 32211, 32411, 32511, 325181, 325188, 325192, 325199, 325211, 325311, or 32532; or

(2) The process is subject to the OSHA process safety management standard, 29 CFR 1910.119.

(e) If at any time a covered process no longer meets the eligibility criteria of its Program level, the owner or operator shall comply with the requirements of the new Program level that applies to the process and update the RMP as provided in §68.190.

(f) The provisions of this part shall not apply to an Outer Continental Shelf ("OCS") source, as defined in 40 CFR 55.2.

[61 FR 31717, June 20, 1996, as amended at 63 FR 645, Jan. 6, 1998; 64 FR 979, Jan. 6, 1999]

### §68.12 General requirements.

(a) General requirements. The owner or operator of a stationary source subject to this part shall submit a single RMP, as provided in §§68.150 to 68.185. The RMP shall include a registration that reflects all covered processes.

(b) Program 1 requirements. In addition to meeting the requirements of paragraph (a) of this section, the owner or operator of a stationary source with a process eligible for Program 1, as provided in §68.10(b), shall:

(1) Analyze the worst-case release scenario for the process(es), as provided in §68.25; document that the nearest public receptor is beyond the distance to a toxic or flammable endpoint defined in §68.22(a); and submit in the RMP the worst-case release scenario as provided in §68.165;

(2) Complete the five-year accident history for the process as provided in §68.42 of this part and submit it in the RMP as provided in §68.168;

(3) Ensure that response actions have been coordinated with local emergency planning and response agencies; and

(4) Certify in the RMP the following: "Based on the criteria in 40 CFR 68.10, the distance to the specified endpoint for the worst-case accidental release scenario for the following process(es) is less than the distance to the nearest public receptor: [list process(es)]. Within the past five years, the process(es) has (have) had no accidental release that caused offsite impacts provided in the risk management program rule (40 CFR 68.10(b)(1)). No additional measures are necessary to prevent offsite impacts from accidental releases. In the event of fire, explosion, or a release of a regulated substance from the process(es), entry within the distance to the specified endpoints may pose a danger to public emergency responders. Therefore, public emergency responders should not enter this area except as arranged with the emergency contact indicated in the RMP. The undersigned certifies that, to the best of my knowledge, information, and belief, formed after reasonable inquiry, the information submitted is true, accurate, and complete. [Signature, title, date signed]."

(c) Program 2 requirements. In addition to meeting the requirements of paragraph (a) of this section, the owner or operator of a stationary source with a process subject to Program 2, as provided in §68.10(c), shall:

(1) Develop and implement a management system as provided in §68.15;

(2) Conduct a hazard assessment as provided in §§68.20 through 68.42;

(3) Implement the Program 2 prevention steps provided in §§68.48 through 68.60 or implement the Program 3 prevention steps provided in §§68.65 through 68.87;

(4) Develop and implement an emergency response program as provided in §§68.90 to 68.95; and

(5) Submit as part of the RMP the data on prevention program elements for Program 2 processes as provided in §68.170.

(d) Program 3 requirements. In addition to meeting the requirements of paragraph (a) of this section, the owner or operator of a stationary source with

41

a process subject to Program 3, as provided in § 68.10(d) shall:

(1) Develop and implement a management system as provided in § 68.15;

(2) Conduct a hazard assessment as provided in §§ 68.20 through 68.42;

(3) Implement the prevention requirements of §§ 68.65 through 68.87;

(4) Develop and implement an emergency response program as provided in §§ 68.90 to 68.95 of this part; and

(5) Submit as part of the RMP the data on prevention program elements for Program 3 processes as provided in § 68.175.

[61 FR 31718, June 20, 1996]

## § 68.15  Management.

(a) The owner or operator of a stationary source with processes subject to Program 2 or Program 3 shall develop a management system to oversee the implementation of the risk management program elements.

(b) The owner or operator shall assign a qualified person or position that has the overall responsibility for the development, implementation, and integration of the risk management program elements.

(c) When responsibility for implementing individual requirements of this part is assigned to persons other than the person identified under paragraph (b) of this section, the names or positions of these people shall be documented and the lines of authority defined through an organization chart or similar document.

[61 FR 31718, June 20, 1996]

## Subpart B—Hazard Assessment

SOURCE: 61 FR 31718, June 20, 1996, unless otherwise noted.

## § 68.20  Applicability.

The owner or operator of a stationary source subject to this part shall prepare a worst-case release scenario analysis as provided in § 68.25 of this part and complete the five-year accident history as provided in § 68.42. The owner or operator of a Program 2 and 3 process must comply with all sections in this subpart for these processes.

## § 68.22  Offsite consequence analysis parameters.

(a) Endpoints. For analyses of offsite consequences, the following endpoints shall be used:

(1) Toxics. The toxic endpoints provided in appendix A of this part.

(2) Flammables. The endpoints for flammables vary according to the scenarios studied:

(i) Explosion. An overpressure of 1 psi.

(ii) Radiant heat/exposure time. A radiant heat of 5 kw/m[2] for 40 seconds.

(iii) Lower flammability limit. A lower flammability limit as provided in NFPA documents or other generally recognized sources.

(b) Wind speed/atmospheric stability class. For the worst-case release analysis, the owner or operator shall use a wind speed of 1.5 meters per second and F atmospheric stability class. If the owner or operator can demonstrate that local meteorological data applicable to the stationary source show a higher minimum wind speed or less stable atmosphere at all times during the previous three years, these minimums may be used. For analysis of alternative scenarios, the owner or operator may use the typical meteorological conditions for the stationary source.

(c) Ambient temperature/humidity. For worst-case release analysis of a regulated toxic substance, the owner or operator shall use the highest daily maximum temperature in the previous three years and average humidity for the site, based on temperature/humidity data gathered at the stationary source or at a local meteorological station; an owner or operator using the RMP Offsite Consequence Analysis Guidance may use 25 °C and 50 percent humidity as values for these variables. For analysis of alternative scenarios, the owner or operator may use typical temperature/humidity data gathered at the stationary source or at a local meteorological station.

(d) Height of release. The worst-case release of a regulated toxic substance shall be analyzed assuming a ground level (0 feet) release. For an alternative scenario analysis of a regulated toxic substance, release height may be determined by the release scenario.

(e) Surface roughness. The owner or operator shall use either urban or rural topography, as appropriate. Urban means that there are many obstacles in the immediate area; obstacles include buildings or trees. Rural means there are no buildings in the immediate area and the terrain is generally flat and unobstructed.

(f) Dense or neutrally buoyant gases. The owner or operator shall ensure that tables or models used for dispersion analysis of regulated toxic substances appropriately account for gas density.

(g) Temperature of released substance. For worst case, liquids other than gases liquified by refrigeration only shall be considered to be released at the highest daily maximum temperature, based on data for the previous three years appropriate for the stationary source, or at process temperature, whichever is higher. For alternative scenarios, substances may be considered to be released at a process or ambient temperature that is appropriate for the scenario.

## §68.25 Worst-case release scenario analysis.

(a) The owner or operator shall analyze and report in the RMP:

(1) For Program 1 processes, one worst-case release scenario for each Program 1 process;

(2) For Program 2 and 3 processes:

(i) One worst-case release scenario that is estimated to create the greatest distance in any direction to an endpoint provided in appendix A of this part resulting from an accidental release of regulated toxic substances from covered processes under worst-case conditions defined in §68.22;

(ii) One worst-case release scenario that is estimated to create the greatest distance in any direction to an endpoint defined in §68.22(a) resulting from an accidental release of regulated flammable substances from covered processes under worst-case conditions defined in §68.22; and

(iii) Additional worst-case release scenarios for a hazard class if a worst-case release from another covered process at the stationary source potentially affects public receptors different from those potentially affected by the worst-case release scenario developed under paragraphs (a)(2)(i) or (a)(2)(ii) of this section.

(b) *Determination of worst-case release quantity.* The worst-case release quantity shall be the greater of the following:

(1) For substances in a vessel, the greatest amount held in a single vessel, taking into account administrative controls that limit the maximum quantity; or

(2) For substances in pipes, the greatest amount in a pipe, taking into account administrative controls that limit the maximum quantity.

(c) *Worst-case release scenario—toxic gases.* (1) For regulated toxic substances that are normally gases at ambient temperature and handled as a gas or as a liquid under pressure, the owner or operator shall assume that the quantity in the vessel or pipe, as determined under paragraph (b) of this section, is released as a gas over 10 minutes. The release rate shall be assumed to be the total quantity divided by 10 unless passive mitigation systems are in place.

(2) For gases handled as refrigerated liquids at ambient pressure:

(i) If the released substance is not contained by passive mitigation systems or if the contained pool would have a depth of 1 cm or less, the owner or operator shall assume that the substance is released as a gas in 10 minutes;

(ii) If the released substance is contained by passive mitigation systems in a pool with a depth greater than 1 cm, the owner or operator may assume that the quantity in the vessel or pipe, as determined under paragraph (b) of this section, is spilled instantaneously to form a liquid pool. The volatilization rate (release rate) shall be calculated at the boiling point of the substance and at the conditions specified in paragraph (d) of this section.

(d) *Worst-case release scenario—toxic liquids.* (1) For regulated toxic substances that are normally liquids at ambient temperature, the owner or operator shall assume that the quantity in the vessel or pipe, as determined under paragraph (b) of this section, is spilled instantaneously to form a liquid pool.

43

(i) The surface area of the pool shall be determined by assuming that the liquid spreads to 1 centimeter deep unless passive mitigation systems are in place that serve to contain the spill and limit the surface area. Where passive mitigation is in place, the surface area of the contained liquid shall be used to calculate the volatilization rate.

(ii) If the release would occur onto a surface that is not paved or smooth, the owner or operator may take into account the actual surface characteristics.

(2) The volatilization rate shall account for the highest daily maximum temperature occurring in the past three years, the temperature of the substance in the vessel, and the concentration of the substance if the liquid spilled is a mixture or solution.

(3) The rate of release to air shall be determined from the volatilization rate of the liquid pool. The owner or operator may use the methodology in the RMP Offsite Consequence Analysis Guidance or any other publicly available techniques that account for the modeling conditions and are recognized by industry as applicable as part of current practices. Proprietary models that account for the modeling conditions may be used provided the owner or operator allows the implementing agency access to the model and describes model features and differences from publicly available models to local emergency planners upon request.

(e) *Worst-case release scenario—flammable gases.* The owner or operator shall assume that the quantity of the substance, as determined under paragraph (b) of this section and the provisions below, vaporizes resulting in a vapor cloud explosion. A yield factor of 10 percent of the available energy released in the explosion shall be used to determine the distance to the explosion endpoint if the model used is based on TNT equivalent methods.

(1) For regulated flammable substances that are normally gases at ambient temperature and handled as a gas or as a liquid under pressure, the owner or operator shall assume that the quantity in the vessel or pipe, as determined under paragraph (b) of this section, is released as a gas over 10 min-

utes. The total quantity shall be assumed to be involved in the vapor cloud explosion.

(2) For flammable gases handled as refrigerated liquids at ambient pressure:

(i) If the released substance is not contained by passive mitigation systems or if the contained pool would have a depth of one centimeter or less, the owner or operator shall assume that the total quantity of the substance is released as a gas in 10 minutes, and the total quantity will be involved in the vapor cloud explosion.

(ii) If the released substance is contained by passive mitigation systems in a pool with a depth greater than 1 centimeter, the owner or operator may assume that the quantity in the vessel or pipe, as determined under paragraph (b) of this section, is spilled instantaneously to form a liquid pool. The volatilization rate (release rate) shall be calculated at the boiling point of the substance and at the conditions specified in paragraph (d) of this section. The owner or operator shall assume that the quantity which becomes vapor in the first 10 minutes is involved in the vapor cloud explosion.

(f) *Worst-case release scenario—flammable liquids.* The owner or operator shall assume that the quantity of the substance, as determined under paragraph (b) of this section and the provisions below, vaporizes resulting in a vapor cloud explosion. A yield factor of 10 percent of the available energy released in the explosion shall be used to determine the distance to the explosion endpoint if the model used is based on TNT equivalent methods.

(1) For regulated flammable substances that are normally liquids at ambient temperature, the owner or operator shall assume that the entire quantity in the vessel or pipe, as determined under paragraph (b) of this section, is spilled instantaneously to form a liquid pool. For liquids at temperatures below their atmospheric boiling point, the volatilization rate shall be calculated at the conditions specified in paragraph (d) of this section.

(2) The owner or operator shall assume that the quantity which becomes vapor in the first 10 minutes is involved in the vapor cloud explosion.

44

(g) *Parameters to be applied.* The owner or operator shall use the parameters defined in §68.22 to determine distance to the endpoints. The owner or operator may use the methodology provided in the RMP Offsite Consequence Analysis Guidance or any commercially or publicly available air dispersion modeling techniques, provided the techniques account for the modeling conditions and are recognized by industry as applicable as part of current practices. Proprietary models that account for the modeling conditions may be used provided the owner or operator allows the implementing agency access to the model and describes model features and differences from publicly available models to local emergency planners upon request.

(h) *Consideration of passive mitigation.* Passive mitigation systems may be considered for the analysis of worst case provided that the mitigation system is capable of withstanding the release event triggering the scenario and would still function as intended.

(i) *Factors in selecting a worst-case scenario.* Notwithstanding the provisions of paragraph (b) of this section, the owner or operator shall select as the worst case for flammable regulated substances or the worst case for regulated toxic substances, a scenario based on the following factors if such a scenario would result in a greater distance to an endpoint defined in §68.22(a) beyond the stationary source boundary than the scenario provided under paragraph (b) of this section:

(1) Smaller quantities handled at higher process temperature or pressure; and

(2) Proximity to the boundary of the stationary source.

[61 FR 31718, June 20, 1996, as amended at 64 FR 28700, May 26, 1999]

§68.28 **Alternative release scenario analysis.**

(a) The number of scenarios. The owner or operator shall identify and analyze at least one alternative release scenario for each regulated toxic substance held in a covered process(es) and at least one alternative release scenario to represent all flammable substances held in covered processes.

(b) *Scenarios to consider.* (1) For each scenario required under paragraph (a) of this section, the owner or operator shall select a scenario:

(i) That is more likely to occur than the worst-case release scenario under §68.25; and

(ii) That will reach an endpoint offsite, unless no such scenario exists.

(2) Release scenarios considered should include, but are not limited to, the following, where applicable:

(i) Transfer hose releases due to splits or sudden hose uncoupling;

(ii) Process piping releases from failures at flanges, joints, welds, valves and valve seals, and drains or bleeds;

(iii) Process vessel or pump releases due to cracks, seal failure, or drain, bleed, or plug failure;

(iv) Vessel overfilling and spill, or overpressurization and venting through relief valves or rupture disks; and

(v) Shipping container mishandling and breakage or puncturing leading to a spill.

(c) Parameters to be applied. The owner or operator shall use the appropriate parameters defined in §68.22 to determine distance to the endpoints. The owner or operator may use either the methodology provided in the RMP Offsite Consequence Analysis Guidance or any commercially or publicly available air dispersion modeling techniques, provided the techniques account for the specified modeling conditions and are recognized by industry as applicable as part of current practices. Proprietary models that account for the modeling conditions may be used provided the owner or operator allows the implementing agency access to the model and describes model features and differences from publicly available models to local emergency planners upon request.

(d) Consideration of mitigation. Active and passive mitigation systems may be considered provided they are capable of withstanding the event that triggered the release and would still be functional.

(e) Factors in selecting scenarios. The owner or operator shall consider the following in selecting alternative release scenarios:

(1) The five-year accident history provided in §68.42; and

(2) Failure scenarios identified under §68.50 or §68.67.

## § 68.30 Defining offsite impacts—population.

(a) The owner or operator shall estimate in the RMP the population within a circle with its center at the point of the release and a radius determined by the distance to the endpoint defined in §68.22(a).

(b) *Population to be defined.* Population shall include residential population. The presence of institutions (schools, hospitals, prisons), parks and recreational areas, and major commercial, office, and industrial buildings shall be noted in the RMP.

(c) *Data sources acceptable.* The owner or operator may use the most recent Census data, or other updated information, to estimate the population potentially affected.

(d) *Level of accuracy.* Population shall be estimated to two significant digits.

## § 68.33 Defining offsite impacts—environment.

(a) The owner or operator shall list in the RMP environmental receptors within a circle with its center at the point of the release and a radius determined by the distance to the endpoint defined in §68.22(a) of this part.

(b) Data sources acceptable. The owner or operator may rely on information provided on local U.S. Geological Survey maps or on any data source containing U.S.G.S. data to identify environmental receptors.

## 68.36 Review and update.

(a) The owner or operator shall review and update the offsite consequence analyses at least once every five years.

(b) If changes in processes, quantities stored or handled, or any other aspect of the stationary source might reasonably be expected to increase or decrease the distance to the endpoint by a factor of two or more, the owner or operator shall complete a revised analysis within six months of the change and submit a revised risk management plan as provided in §68.190.

## § 68.39 Documentation.

The owner or operator shall maintain the following records on the offsite consequence analyses:

(a) For worst-case scenarios, a description of the vessel or pipeline and substance selected as worst case, assumptions and parameters used, and the rationale for selection; assumptions shall include use of any administrative controls and any passive mitigation that were assumed to limit the quantity that could be released. Documentation shall include the anticipated effect of the controls and mitigation on the release quantity and rate.

(b) For alternative release scenarios, a description of the scenarios identified, assumptions and parameters used, and the rationale for the selection of specific scenarios; assumptions shall include use of any administrative controls and any mitigation that were assumed to limit the quantity that could be released. Documentation shall include the effect of the controls and mitigation on the release quantity and rate.

(c) Documentation of estimated quantity released, release rate, and duration of release.

(d) Methodology used to determine distance to endpoints.

(e) Data used to estimate population and environmental receptors potentially affected.

## § 68.42 Five-year accident history.

(a) The owner or operator shall include in the five-year accident history all accidental releases from covered processes that resulted in deaths, injuries, or significant property damage on site, or known offsite deaths, injuries, evacuations, sheltering in place, property damage, or environmental damage.

(b) *Data required.* For each accidental release included, the owner or operator shall report the following information:

(1) Date, time, and approximate duration of the release;

(2) Chemical(s) released;

(3) Estimated quantity released in pounds and, for mixtures containing regulated toxic substances, percentage concentration by weight of the released regulated toxic substance in the liquid mixture;

(4) Five- or six-digit NAICS code that most closely corresponds to the process;

(5) The type of release event and its source;

(6) Weather conditions, if known;

(7) On-site impacts;

(8) Known offsite impacts;

(9) Initiating event and contributing factors if known;

(10) Whether offsite responders were notified if known; and

(11) Operational or process changes that resulted from investigation of the release.

(c) *Level of accuracy.* Numerical estimates may be provided to two significant digits.

[61 FR 31718, June 20, 1996, as amended at 64 FR 979, Jan. 6, 1999]

## Subpart C—Program 2 Prevention Program

Source: 61 FR 31721, June 20, 1996, unless otherwise noted.

### §68.48  Safety information.

(a) The owner or operator shall compile and maintain the following up-to-date safety information related to the regulated substances, processes, and equipment:

(1) Material Safety Data Sheets that meet the requirements of 29 CFR 1910.1200(g);

(2) Maximum intended inventory of equipment in which the regulated substances are stored or processed;

(3) Safe upper and lower temperatures, pressures, flows, and compositions;

(4) Equipment specifications; and

(5) Codes and standards used to design, build, and operate the process.

(b) The owner or operator shall ensure that the process is designed in compliance with recognized and generally accepted good engineering practices. Compliance with Federal or state regulations that address industry-specific safe design or with industry-specific design codes and standards may be used to demonstrate compliance with this paragraph.

(c) The owner or operator shall update the safety information if a major change occurs that makes the information inaccurate.

### §68.50  Hazard review.

(a) The owner or operator shall conduct a review of the hazards associated with the regulated substances, process, and procedures. The review shall identify the following:

(1) The hazards associated with the process and regulated substances;

(2) Opportunities for equipment malfunctions or human errors that could cause an accidental release;

(3) The safeguards used or needed to control the hazards or prevent equipment malfunction or human error; and

(4) Any steps used or needed to detect or monitor releases.

(b) The owner or operator may use checklists developed by persons or organizations knowledgeable about the process and equipment as a guide to conducting the review. For processes designed to meet industry standards or Federal or state design rules, the hazard review shall, by inspecting all equipment, determine whether the process is designed, fabricated, and operated in accordance with the applicable standards or rules.

(c) The owner or operator shall document the results of the review and ensure that problems identified are resolved in a timely manner.

(d) The review shall be updated at least once every five years. The owner or operator shall also conduct reviews whenever a major change in the process occurs; all issues identified in the review shall be resolved before startup of the changed process.

### §68.52  Operating procedures.

(a) The owner or operator shall prepare written operating procedures that provide clear instructions or steps for safely conducting activities associated with each covered process consistent with the safety information for that process. Operating procedures or instructions provided by equipment manufacturers or developed by persons or organizations knowledgeable about the process and equipment may be used as a basis for a stationary source's operating procedures.

(b) The procedures shall address the following:

(1) Initial startup;

(2) Normal operations;

(3) Temporary operations;

(4) Emergency shutdown and operations;

(5) Normal shutdown;

(6) Startup following a normal or emergency shutdown or a major change that requires a hazard review;

(7) Consequences of deviations and steps required to correct or avoid deviations; and

(8) Equipment inspections.

(c) The owner or operator shall ensure that the operating procedures are updated, if necessary, whenever a major change occurs and prior to startup of the changed process.

## § 68.54   Training.

(a) The owner or operator shall ensure that each employee presently operating a process, and each employee newly assigned to a covered process have been trained or tested competent in the operating procedures provided in § 68.52 that pertain to their duties. For those employees already operating a process on June 21, 1999, the owner or operator may certify in writing that the employee has the required knowledge, skills, and abilities to safely carry out the duties and responsibilities as provided in the operating procedures.

(b) Refresher training. Refresher training shall be provided at least every three years, and more often if necessary, to each employee operating a process to ensure that the employee understands and adheres to the current operating procedures of the process. The owner or operator, in consultation with the employees operating the process, shall determine the appropriate frequency of refresher training.

(c) The owner or operator may use training conducted under Federal or state regulations or under industry-specific standards or codes or training conducted by covered process equipment vendors to demonstrate compliance with this section to the extent that the training meets the requirements of this section.

(d) The owner or operator shall ensure that operators are trained in any updated or new procedures prior to startup of a process after a major change.

## § 68.56   Maintenance.

(a) The owner or operator shall prepare and implement procedures to maintain the on-going mechanical integrity of the process equipment. The owner or operator may use procedures or instructions provided by covered process equipment vendors or procedures in Federal or state regulations or industry codes as the basis for stationary source maintenance procedures.

(b) The owner or operator shall train or cause to be trained each employee involved in maintaining the on-going mechanical integrity of the process. To ensure that the employee can perform the job tasks in a safe manner, each such employee shall be trained in the hazards of the process, in how to avoid or correct unsafe conditions, and in the procedures applicable to the employee's job tasks.

(c) Any maintenance contractor shall ensure that each contract maintenance employee is trained to perform the maintenance procedures developed under paragraph (a) of this section.

(d) The owner or operator shall perform or cause to be performed inspections and tests on process equipment. Inspection and testing procedures shall follow recognized and generally accepted good engineering practices. The frequency of inspections and tests of process equipment shall be consistent with applicable manufacturers' recommendations, industry standards or codes, good engineering practices, and prior operating experience.

## § 68.58   Compliance audits.

(a) The owner or operator shall certify that they have evaluated compliance with the provisions of this subpart at least every three years to verify that the procedures and practices developed under the rule are adequate and are being followed.

(b) The compliance audit shall be conducted by at least one person knowledgeable in the process.

(c) The owner or operator shall develop a report of the audit findings.

(d) The owner or operator shall promptly determine and document an appropriate response to each of the findings of the compliance audit and document that deficiencies have been corrected.

(e) The owner or operator shall retain the two (2) most recent compliance audit reports. This requirement does not apply to any compliance audit report that is more than five years old.

### §68.60  Incident investigation.

(a) The owner or operator shall investigate each incident which resulted in, or could reasonably have resulted in a catastrophic release.

(b) An incident investigation shall be initiated as promptly as possible, but not later than 48 hours following the incident.

(c) A summary shall be prepared at the conclusion of the investigation which includes at a minimum:

(1) Date of incident;

(2) Date investigation began;

(3) A description of the incident;

(4) The factors that contributed to the incident; and,

(5) Any recommendations resulting from the investigation.

(d) The owner or operator shall promptly address and resolve the investigation findings and recommendations. Resolutions and corrective actions shall be documented.

(e) The findings shall be reviewed with all affected personnel whose job tasks are affected by the findings.

(f) Investigation summaries shall be retained for five years.

## Subpart D—Program 3 Prevention Program

SOURCE: 61 FR 31722, June 20, 1996, unless otherwise noted.

### §68.65  Process safety information.

(a) In accordance with the schedule set forth in §68.67, the owner or operator shall complete a compilation of written process safety information before conducting any process hazard analysis required by the rule. The compilation of written process safety information is to enable the owner or operator and the employees involved in operating the process to identify and understand the hazards posed by those processes involving regulated substances. This process safety information shall include information pertaining to the hazards of the regulated substances used or produced by the process, information pertaining to the technology of the process, and information pertaining to the equipment in the process.

(b) Information pertaining to the hazards of the regulated substances in the process. This information shall consist of at least the following:

(1) Toxicity information;

(2) Permissible exposure limits;

(3) Physical data;

(4) Reactivity data;

(5) Corrosivity data;

(6) Thermal and chemical stability data; and

(7) Hazardous effects of inadvertent mixing of different materials that could foreseeably occur.

NOTE TO PARAGRAPH (b): Material Safety Data Sheets meeting the requirements of 29 CFR 1910.1200(g) may be used to comply with this requirement to the extent they contain the information required by this subparagraph.

(c) Information pertaining to the technology of the process.

(1) Information concerning the technology of the process shall include at least the following:

(i) A block flow diagram or simplified process flow diagram;

(ii) Process chemistry;

(iii) Maximum intended inventory;

(iv) Safe upper and lower limits for such items as temperatures, pressures, flows or compositions; and,

(v) An evaluation of the consequences of deviations.

(2) Where the original technical information no longer exists, such information may be developed in conjunction with the process hazard analysis in sufficient detail to support the analysis.

(d) Information pertaining to the equipment in the process.

(1) Information pertaining to the equipment in the process shall include:

(i) Materials of construction;

(ii) Piping and instrument diagrams (P&ID's);

(iii) Electrical classification;

(iv) Relief system design and design basis;

(v) Ventilation system design;

(vi) Design codes and standards employed;

(vii) Material and energy balances for processes built after June 21, 1999; and

(viii) Safety systems (e.g. interlocks, detection or suppression systems).

(2) The owner or operator shall document that equipment complies with recognized and generally accepted good engineering practices.

(3) For existing equipment designed and constructed in accordance with codes, standards, or practices that are no longer in general use, the owner or operator shall determine and document that the equipment is designed, maintained, inspected, tested, and operating in a safe manner.

### § 68.67  Process hazard analysis.

(a) The owner or operator shall perform an initial process hazard analysis (hazard evaluation) on processes covered by this part. The process hazard analysis shall be appropriate to the complexity of the process and shall identify, evaluate, and control the hazards involved in the process. The owner or operator shall determine and document the priority order for conducting process hazard analyses based on a rationale which includes such considerations as extent of the process hazards, number of potentially affected employees, age of the process, and operating history of the process. The process hazard analysis shall be conducted as soon as possible, but not later than June 21, 1999. Process hazards analyses completed to comply with 29 CFR 1910.119(e) are acceptable as initial process hazards analyses. These process hazard analyses shall be updated and revalidated, based on their completion date.

(b) The owner or operator shall use one or more of the following methodologies that are appropriate to determine and evaluate the hazards of the process being analyzed.

(1) What-If;

(2) Checklist;

(3) What-If/Checklist;

(4) Hazard and Operability Study (HAZOP);

(5) Failure Mode and Effects Analysis (FMEA);

(6) Fault Tree Analysis; or

(7) An appropriate equivalent methodology.

(c) The process hazard analysis shall address:

(1) The hazards of the process;

(2) The identification of any previous incident which had a likely potential for catastrophic consequences.

(3) Engineering and administrative controls applicable to the hazards and their interrelationships such as appropriate application of detection methodologies to provide early warning of releases. (Acceptable detection methods might include process monitoring and control instrumentation with alarms, and detection hardware such as hydrocarbon sensors.);

(4) Consequences of failure of engineering and administrative controls;

(5) Stationary source siting;

(6) Human factors; and

(7) A qualitative evaluation of a range of the possible safety and health effects of failure of controls.

(d) The process hazard analysis shall be performed by a team with expertise in engineering and process operations, and the team shall include at least one employee who has experience and knowledge specific to the process being evaluated. Also, one member of the team must be knowledgeable in the specific process hazard analysis methodology being used.

(e) The owner or operator shall establish a system to promptly address the team's findings and recommendations; assure that the recommendations are resolved in a timely manner and that the resolution is documented; document what actions are to be taken; complete actions as soon as possible; develop a written schedule of when these actions are to be completed; communicate the actions to operating, maintenance and other employees whose work assignments are in the process and who may be affected by the recommendations or actions.

(f) At least every five (5) years after the completion of the initial process hazard analysis, the process hazard analysis shall be updated and revalidated by a team meeting the requirements in paragraph (d) of this section, to assure that the process hazard analysis is consistent with the current process. Updated and revalidated process hazard analyses completed to comply with 29 CFR 1910.119(e) are acceptable to meet the requirements of this paragraph.

(g) The owner or operator shall retain process hazards analyses and updates or revalidations for each process covered by this section, as well as the documented resolution of recommendations described in paragraph (e) of this section for the life of the process.

### §68.69 Operating procedures.

(a) The owner or operator shall develop and implement written operating procedures that provide clear instructions for safely conducting activities involved in each covered process consistent with the process safety information and shall address at least the following elements.

(1) Steps for each operating phase:

(i) Initial startup;

(ii) Normal operations;

(iii) Temporary operations;

(iv) Emergency shutdown including the conditions under which emergency shutdown is required, and the assignment of shutdown responsibility to qualified operators to ensure that emergency shutdown is executed in a safe and timely manner.

(v) Emergency operations;

(vi) Normal shutdown; and,

(vii) Startup following a turnaround, or after an emergency shutdown.

(2) Operating limits:

(i) Consequences of deviation; and

(ii) Steps required to correct or avoid deviation.

(3) Safety and health considerations:

(i) Properties of, and hazards presented by, the chemicals used in the process;

(ii) Precautions necessary to prevent exposure, including engineering controls, administrative controls, and personal protective equipment;

(iii) Control measures to be taken if physical contact or airborne exposure occurs;

(iv) Quality control for raw materials and control of hazardous chemical inventory levels; and,

(v) Any special or unique hazards.

(4) Safety systems and their functions.

(b) Operating procedures shall be readily accessible to employees who work in or maintain a process.

(c) The operating procedures shall be reviewed as often as necessary to assure that they reflect current operating practice, including changes that result from changes in process chemicals, technology, and equipment, and changes to stationary sources. The owner or operator shall certify annually that these operating procedures are current and accurate.

(d) The owner or operator shall develop and implement safe work practices to provide for the control of hazards during operations such as lockout/tagout; confined space entry; opening process equipment or piping; and control over entrance into a stationary source by maintenance, contractor, laboratory, or other support personnel. These safe work practices shall apply to employees and contractor employees.

### §68.71 Training.

(a) *Initial training.* (1) Each employee presently involved in operating a process, and each employee before being involved in operating a newly assigned process, shall be trained in an overview of the process and in the operating procedures as specified in §68.69. The training shall include emphasis on the specific safety and health hazards, emergency operations including shutdown, and safe work practices applicable to the employee's job tasks.

(2) In lieu of initial training for those employees already involved in operating a process on June 21, 1999 an owner or operator may certify in writing that the employee has the required knowledge, skills, and abilities to safely carry out the duties and responsibilities as specified in the operating procedures.

(b) *Refresher training.* Refresher training shall be provided at least every three years, and more often if necessary, to each employee involved in operating a process to assure that the employee understands and adheres to the current operating procedures of the process. The owner or operator, in consultation with the employees involved in operating the process, shall determine the appropriate frequency of refresher training.

(c) *Training documentation.* The owner or operator shall ascertain that each employee involved in operating a process has received and understood the training required by this paragraph. The owner or operator shall prepare a record which contains the identity of the employee, the date of training, and the means used to verify that the employee understood the training.

### § 68.73  Mechanical integrity.

(a) *Application.* Paragraphs (b) through (f) of this section apply to the following process equipment:

(1) Pressure vessels and storage tanks;

(2) Piping systems (including piping components such as valves);

(3) Relief and vent systems and devices;

(4) Emergency shutdown systems;

(5) Controls (including monitoring devices and sensors, alarms, and interlocks) and,

(6) Pumps.

(b) *Written procedures.* The owner or operator shall establish and implement written procedures to maintain the on-going integrity of process equipment.

(c) *Training for process maintenance activities.* The owner or operator shall train each employee involved in maintaining the on-going integrity of process equipment in an overview of that process and its hazards and in the procedures applicable to the employee's job tasks to assure that the employee can perform the job tasks in a safe manner.

(d) *Inspection and testing.* (1) Inspections and tests shall be performed on process equipment.

(2) Inspection and testing procedures shall follow recognized and generally accepted good engineering practices.

(3) The frequency of inspections and tests of process equipment shall be consistent with applicable manufacturers' recommendations and good engineering practices, and more frequently if determined to be necessary by prior operating experience.

(4) The owner or operator shall document each inspection and test that has been performed on process equipment.

The documentation shall identify the date of the inspection or test, the name of the person who performed the inspection or test, the serial number or other identifier of the equipment on which the inspection or test was performed, a description of the inspection or test performed, and the results of the inspection or test.

(e) *Equipment deficiencies.* The owner or operator shall correct deficiencies in equipment that are outside acceptable limits (defined by the process safety information in § 68.65) before further use or in a safe and timely manner when necessary means are taken to assure safe operation.

(f) *Quality assurance.* (1) In the construction of new plants and equipment, the owner or operator shall assure that equipment as it is fabricated is suitable for the process application for which they will be used.

(2) Appropriate checks and inspections shall be performed to assure that equipment is installed properly and consistent with design specifications and the manufacturer's instructions.

(3) The owner or operator shall assure that maintenance materials, spare parts and equipment are suitable for the process application for which they will be used.

### § 68.75  Management of change.

(a) The owner or operator shall establish and implement written procedures to manage changes (except for ''replacements in kind'') to process chemicals, technology, equipment, and procedures; and, changes to stationary sources that affect a covered process.

(b) The procedures shall assure that the following considerations are addressed prior to any change:

(1) The technical basis for the proposed change;

(2) Impact of change on safety and health;

(3) Modifications to operating procedures;

(4) Necessary time period for the change; and,

(5) Authorization requirements for the proposed change.

(c) Employees involved in operating a process and maintenance and contract employees whose job tasks will be affected by a change in the process shall be informed of, and trained in, the change prior to start-up of the process or affected part of the process.

(d) If a change covered by this paragraph results in a change in the process safety information required by § 68.65 of this part, such information shall be updated accordingly.

(e) If a change covered by this paragraph results in a change in the operating procedures or practices required by § 68.69, such procedures or practices shall be updated accordingly.

## § 68.77 Pre-startup review.

(a) The owner or operator shall perform a pre-startup safety review for new stationary sources and for modified stationary sources when the modification is significant enough to require a change in the process safety information.

(b) The pre-startup safety review shall confirm that prior to the introduction of regulated substances to a process:

(1) Construction and equipment is in accordance with design specifications;

(2) Safety, operating, maintenance, and emergency procedures are in place and are adequate;

(3) For new stationary sources, a process hazard analysis has been performed and recommendations have been resolved or implemented before startup; and modified stationary sources meet the requirements contained in management of change, § 68.75.

(4) Training of each employee involved in operating a process has been completed.

## § 68.79 Compliance audits.

(a) The owner or operator shall certify that they have evaluated compliance with the provisions of this subpart at least every three years to verify that procedures and practices developed under this subpart are adequate and are being followed.

(b) The compliance audit shall be conducted by at least one person knowledgeable in the process.

(c) A report of the findings of the audit shall be developed.

(d) The owner or operator shall promptly determine and document an appropriate response to each of the findings of the compliance audit, and document that deficiencies have been corrected.

(e) The owner or operator shall retain the two (2) most recent compliance audit reports.

[61 FR 31722, June 20, 1996, as amended at 64 FR 979, Jan. 6, 1999]

## § 68.81 Incident investigation.

(a) The owner or operator shall investigate each incident which resulted in, or could reasonably have resulted in a catastrophic release of a regulated substance.

(b) An incident investigation shall be initiated as promptly as possible, but not later than 48 hours following the incident.

(c) An incident investigation team shall be established and consist of at least one person knowledgeable in the process involved, including a contract employee if the incident involved work of the contractor, and other persons with appropriate knowledge and experience to thoroughly investigate and analyze the incident.

(d) A report shall be prepared at the conclusion of the investigation which includes at a minimum:

(1) Date of incident;

(2) Date investigation began;

(3) A description of the incident;

(4) The factors that contributed to the incident; and,

(5) Any recommendations resulting from the investigation.

(e) The owner or operator shall establish a system to promptly address and resolve the incident report findings and recommendations. Resolutions and corrective actions shall be documented.

(f) The report shall be reviewed with all affected personnel whose job tasks are relevant to the incident findings including contract employees where applicable.

(g) Incident investigation reports shall be retained for five years.

## § 68.83 Employee participation.

(a) The owner or operator shall develop a written plan of action regarding the implementation of the employee participation required by this section.

(b) The owner or operator shall consult with employees and their representatives on the conduct and development of process hazards analyses and on the development of the other elements of process safety management in this rule.

(c) The owner or operator shall provide to employees and their representatives access to process hazard analyses and to all other information required to be developed under this rule.

## § 68.85  Hot work permit.

(a) The owner or operator shall issue a hot work permit for hot work operations conducted on or near a covered process.

(b) The permit shall document that the fire prevention and protection requirements in 29 CFR 1910.252(a) have been implemented prior to beginning the hot work operations; it shall indicate the date(s) authorized for hot work; and identify the object on which hot work is to be performed. The permit shall be kept on file until completion of the hot work operations.

## § 68.87  Contractors.

(a) *Application.* This section applies to contractors performing maintenance or repair, turnaround, major renovation, or specialty work on or adjacent to a covered process. It does not apply to contractors providing incidental services which do not influence process safety, such as janitorial work, food and drink services, laundry, delivery or other supply services.

(b) *Owner or operator responsibilities.* (1) The owner or operator, when selecting a contractor, shall obtain and evaluate information regarding the contract owner or operator's safety performance and programs.

(2) The owner or operator shall inform contract owner or operator of the known potential fire, explosion, or toxic release hazards related to the contractor's work and the process.

(3) The owner or operator shall explain to the contract owner or operator the applicable provisions of subpart E of this part.

(4) The owner or operator shall develop and implement safe work practices consistent with § 68.69(d), to control the entrance, presence, and exit of the contract owner or operator and contract employees in covered process areas.

(5) The owner or operator shall periodically evaluate the performance of the contract owner or operator in fulfilling their obligations as specified in paragraph (c) of this section.

(c) *Contract owner or operator responsibilities.* (1) The contract owner or operator shall assure that each contract employee is trained in the work practices necessary to safely perform his/her job.

(2) The contract owner or operator shall assure that each contract employee is instructed in the known potential fire, explosion, or toxic release hazards related to his/her job and the process, and the applicable provisions of the emergency action plan.

(3) The contract owner or operator shall document that each contract employee has received and understood the training required by this section. The contract owner or operator shall prepare a record which contains the identity of the contract employee, the date of training, and the means used to verify that the employee understood the training.

(4) The contract owner or operator shall assure that each contract employee follows the safety rules of the stationary source including the safe work practices required by § 68.69(d).

(5) The contract owner or operator shall advise the owner or operator of any unique hazards presented by the contract owner or operator's work, or of any hazards found by the contract owner or operator's work.

## Subpart E—Emergency Response

SOURCE: 61 FR 31725, June 20, 1996, unless otherwise noted.

## § 68.90  Applicability.

(a) Except as provided in paragraph (b) of this section, the owner or operator of a stationary source with Program 2 and Program 3 processes shall comply with the requirements of § 68.95.

(b) The owner or operator of stationary source whose employees will not respond to accidental releases of regulated substances need not comply with §68.95 of this part provided that they meet the following:

(1) For stationary sources with any regulated toxic substance held in a process above the threshold quantity, the stationary source is included in the community emergency response plan developed under 42 U.S.C. 11003;

(2) For stationary sources with only regulated flammable substances held in a process above the threshold quantity, the owner or operator has coordinated response actions with the local fire department; and

(3) Appropriate mechanisms are in place to notify emergency responders when there is a need for a response.

### §68.95 Emergency response program.

(a) The owner or operator shall develop and implement an emergency response program for the purpose of protecting public health and the environment. Such program shall include the following elements:

(1) An emergency response plan, which shall be maintained at the stationary source and contain at least the following elements:

(i) Procedures for informing the public and local emergency response agencies about accidental releases;

(ii) Documentation of proper first-aid and emergency medical treatment necessary to treat accidental human exposures; and

(iii) Procedures and measures for emergency response after an accidental release of a regulated substance;

(2) Procedures for the use of emergency response equipment and for its inspection, testing, and maintenance;

(3) Training for all employees in relevant procedures; and

(4) Procedures to review and update, as appropriate, the emergency response plan to reflect changes at the stationary source and ensure that employees are informed of changes.

(b) A written plan that complies with other Federal contingency plan regulations or is consistent with the approach in the National Response Team's Integrated Contingency Plan Guidance ("One Plan") and that, among other matters, includes the elements provided in paragraph (a) of this section, shall satisfy the requirements of this section if the owner or operator

also complies with paragraph (c) of this section.

(c) The emergency response plan developed under paragraph (a)(1) of this section shall be coordinated with the community emergency response plan developed under 42 U.S.C. 11003. Upon request of the local emergency planning committee or emergency response officials, the owner or operator shall promptly provide to the local emergency response officials information necessary for developing and implementing the community emergency response plan.

## Subpart F—Regulated Substances for Accidental Release Prevention

SOURCE: 59 FR 4493, Jan. 31, 1994, unless otherwise noted. Redesignated at 61 FR 31717, June 20, 1996.

### §68.100 Purpose.

This subpart designates substances to be listed under section 112(r)(3), (4), and (5) of the Clean Air Act, as amended, identifies their threshold quantities, and establishes the requirements for petitioning to add or delete substances from the list.

### §68.115 Threshold determination.

(a) A threshold quantity of a regulated substance listed in §68.130 is present at a stationary source if the total quantity of the regulated substance contained in a process exceeds the threshold.

(b) For the purposes of determining whether more than a threshold quantity of a regulated substance is present at the stationary source, the following exemptions apply:

(1) *Concentrations of a regulated toxic substance in a mixture.* If a regulated substance is present in a mixture and the concentration of the substance is below one percent by weight of the mixture, the amount of the substance in the mixture need not be considered when determining whether more than a threshold quantity is present at the stationary source. Except for oleum, toluene 2,4-diisocyanate, toluene 2,6-diisocyanate, and toluene diisocyanate (unspecified isomer), if the concentration of the regulated substance in the mixture is one percent or greater by

55

weight, but the owner or operator can demonstrate that the partial pressure of the regulated substance in the mixture (solution) under handling or storage conditions in any portion of the process is less than 10 millimeters of mercury (mm Hg), the amount of the substance in the mixture in that portion of the process need not be considered when determining whether more than a threshold quantity is present at the stationary source. The owner or operator shall document this partial pressure measurement or estimate.

(2) *Concentrations of a regulated flammable substance in a mixture.* (i) *General provision.* If a regulated substance is present in a mixture and the concentration of the substance is below one percent by weight of the mixture, the mixture need not be considered when determining whether more than a threshold quantity of the regulated substance is present at the stationary source. Except as provided in paragraph (b)(2) (ii) and (iii) of this section, if the concentration of the substance is one percent or greater by weight of the mixture, then, for purposes of determining whether a threshold quantity is present at the stationary source, the entire weight of the mixture shall be treated as the regulated substance unless the owner or operator can demonstrate that the mixture itself does not have a National Fire Protection Association flammability hazard rating of 4. The demonstration shall be in accordance with the definition of flammability hazard rating 4 in the NFPA 704, Standard System for the Identification of the Hazards of Materials for Emergency Response, National Fire Protection Association, Quincy, MA, 1996. Available from the National Fire Protection Association, 1 Batterymarch Park, Quincy, MA 02269–9101. This incorporation by reference was approved by the Director of the Federal Register in accordance with 5 U.S.C. 552(a) and 1 CFR part 51. Copies may be inspected at the Environmental Protection Agency Air Docket (6102), Attn: Docket No. A–96–08, Waterside Mall, 401 M. St. SW., Washington DC; or at the Office of Federal Register at 800 North Capitol St., NW, Suite 700, Washington, DC. Boiling point and flash point shall be defined and determined in accordance with NFPA 30, Flammable and Combustible Liquids Code, National Fire Protection Association, Quincy, MA, 1996. Available from the National Fire Protection Association, 1 Batterymarch Park, Quincy, MA 02269–9101. This incorporation by reference was approved by the Director of the Federal Register in accordance with 5 U.S.C. 552(a) and 1 CFR part 51. Copies may be inspected at the Environmental Protection Agency Air Docket (6102), Attn: Docket No. A–96–08, Waterside Mall, 401 M. St. SW., Washington DC; or at the Office of Federal Register at 800 North Capitol St., NW., Suite 700, Washington, DC. The owner or operator shall document the National Fire Protection Association flammability hazard rating.

(ii) *Gasoline.* Regulated substances in gasoline, when in distribution or related storage for use as fuel for internal combustion engines, need not be considered when determining whether more than a threshold quantity is present at a stationary source.

(iii) *Naturally occurring hydrocarbon mixtures.* Prior to entry into a natural gas processing plant or a petroleum refining process unit, regulated substances in naturally occurring hydrocarbon mixtures need not be considered when determining whether more than a threshold quantity is present at a stationary source. Naturally occurring hydrocarbon mixtures include any combination of the following: condensate, crude oil, field gas, and produced water, each as defined in § 68.3 of this part.

(3) *Articles.* Regulated substances contained in articles need not be considered when determining whether more than a threshold quantity is present at the stationary source.

(4) *Uses.* Regulated substances, when in use for the following purposes, need not be included in determining whether more than a threshold quantity is present at the stationary source:

(i) Use as a structural component of the stationary source;

(ii) Use of products for routine janitorial maintenance;

(iii) Use by employees of foods, drugs, cosmetics, or other personal items containing the regulated substance; and

(iv) Use of regulated substances present in process water or non-contact cooling water as drawn from the environment or municipal sources, or use of regulated substances present in air used either as compressed air or as part of combustion.

(5) *Activities in laboratories.* If a regulated substance is manufactured, processed, or used in a laboratory at a stationary source under the supervision of a technically qualified individual as defined in §720.3(ee) of this chapter, the quantity of the substance need not be considered in determining whether a threshold quantity is present. This exemption does not apply to:

(i) Specialty chemical production;

(ii) Manufacture, processing, or use of substances in pilot plant scale operations; and

(iii) Activities conducted outside the laboratory.

[59 FR 4493, Jan. 31, 1994. Redesignated at 61 FR 31717, June 20, 1996, as amended at 63 FR 645, Jan. 6, 1998]

### §68.120 Petition process.

(a) Any person may petition the Administrator to modify, by addition or deletion, the list of regulated substances identified in §68.130. Based on the information presented by the petitioner, the Administrator may grant or deny a petition.

(b) A substance may be added to the list if, in the case of an accidental release, it is known to cause or may be reasonably anticipated to cause death, injury, or serious adverse effects to human health or the environment.

(c) A substance may be deleted from the list if adequate data on the health and environmental effects of the substance are available to determine that the substance, in the case of an accidental release, is not known to cause and may not be reasonably anticipated to cause death, injury, or serious adverse effects to human health or the environment.

(d) No substance for which a national primary ambient air quality standard has been established shall be added to the list. No substance regulated under title VI of the Clean Air Act, as amended, shall be added to the list.

(e) The burden of proof is on the petitioner to demonstrate that the criteria for addition and deletion are met. A petition will be denied if this demonstration is not made.

(f) The Administrator will not accept additional petitions on the same substance following publication of a final notice of the decision to grant or deny a petition, unless new data becomes available that could significantly affect the basis for the decision.

(g) Petitions to modify the list of regulated substances must contain the following:

(1) Name and address of the petitioner and a brief description of the organization(s) that the petitioner represents, if applicable;

(2) Name, address, and telephone number of a contact person for the petition;

(3) Common chemical name(s), common synonym(s), Chemical Abstracts Service number, and chemical formula and structure;

(4) Action requested (add or delete a substance);

(5) Rationale supporting the petitioner's position; that is, how the substance meets the criteria for addition and deletion. A short summary of the rationale must be submitted along with a more detailed narrative; and

(6) Supporting data; that is, the petition must include sufficient information to scientifically support the request to modify the list. Such information shall include:

(i) A list of all support documents;

(ii) Documentation of literature searches conducted, including, but not limited to, identification of the database(s) searched, the search strategy, dates covered, and printed results;

(iii) Effects data (animal, human, and environmental test data) indicating the potential for death, injury, or serious adverse human and environmental impacts from acute exposure following an accidental release; printed copies of the data sources, in English, should be provided; and

(iv) Exposure data or previous accident history data, indicating the potential for serious adverse human health or environmental effects from an accidental release. These data may

57

include, but are not limited to, physical and chemical properties of the substance, such as vapor pressure; modeling results, including data and assumptions used and model documentation; and historical accident data, citing data sources.

(h) Within 18 months of receipt of a petition, the Administrator shall publish in the FEDERAL REGISTER a notice either denying the petition or granting the petition and proposing a listing.

## § 68.125  Exemptions.

*Agricultural nutrients.* Ammonia used as an agricultural nutrient, when held by farmers, is exempt from all provisions of this part.

## § 68.130  List of substances.

(a) Regulated toxic and flammable substances under section 112(r) of the Clean Air Act are the substances listed in Tables 1, 2, 3, and 4. Threshold quantities for listed toxic and flammable substances are specified in the tables.

(b) The basis for placing toxic and flammable substances on the list of regulated substances are explained in the notes to the list.

TABLE 1 TO § 68.130.—LIST OF REGULATED TOXIC SUBSTANCES AND THRESHOLD QUANTITIES FOR ACCIDENTAL RELEASE PREVENTION
[Alphabetical Order—77 Substances]

| Chemical name | CAS No. | Threshold quantity (lbs) | Basis for listing |
|---|---|---|---|
| Acrolein [2-Propenal]. | 107–02–8 | 5,000 | b |
| Acrylonitrile [2-Propenenitrile]. | 107–13–1 | 20,000 | b |
| Acrylyl chloride [2-Propenoyl chloride]. | 814–68–6 | 5,000 | b |
| Allyl alcohol [2-Propen-l-ol]. | 107–18–01 | 15,000 | b |
| Allylamine [2-Propen-l-amine]. | 107–11–9 | 10,000 | b |
| Ammonia (anhydrous). | 7664–41–7 | 10,000 | a, b |
| Ammonia (conc 20% or greater). | 7664–41–7 | 20,000 | a, b |
| Arsenous trichloride. | 7784–34–1 | 15,000 | b |
| Arsine | 7784–42–1 | 1,000 | b |
| Boron trichloride [Borane, trichloro-]. | 10294–34–5 | 5,000 | b |
| Boron trifluoride [Borane, trifluoro-]. | 7637–07–2 | 5,000 | b |

TABLE 1 TO § 68.130.—LIST OF REGULATED TOXIC SUBSTANCES AND THRESHOLD QUANTITIES FOR ACCIDENTAL RELEASE PREVENTION—Continued
[Alphabetical Order—77 Substances]

| Chemical name | CAS No. | Threshold quantity (lbs) | Basis for listing |
|---|---|---|---|
| Boron trifluoride compound with methyl ether (1:1) [Boron, trifluoro [oxybis [metane]]-, T-4-. | 353–42–4 | 15,000 | b |
| Bromine | 7726–95–6 | 10,000 | a, b |
| Carbon disulfide | 75–15–0 | 20,000 | b |
| Chlorine | 7782–50–5 | 2,500 | a, b |
| Chlorine dioxide [Chlorine oxide (ClO2)]. | 10049–04–4 | 1,000 | c |
| Chloroform [Methane, trichloro-]. | 67–66–3 | 20,000 | b |
| Chloromethyl ether [Methane, oxybis[chloro-]. | 542–88–1 | 1,000 | b |
| Chloromethyl methyl ether [Methane, chloromethoxy-]. | 107–30–2 | 5,000 | b |
| Crotonaldehyde [2-Butenal]. | 4170–30–3 | 20,000 | b |
| Crotonaldehyde, (E)- [2-Butenal, (E)-]. | 123–73–9 | 20,000 | b |
| Cyanogen chloride. | 506–77–4 | 10,000 | c |
| Cyclohexylamine [Cyclohexanamine]. | 108–91–8 | 15,000 | b |
| Diborane | 19287–45–7 | 2,500 | b |
| Dimethyldichlorosilane [Silane, dichlorodimethyl-]. | 75–78–5 | 5,000 | b |
| 1,1-Dimethylhydrazine [Hydrazine, 1,1-dimethyl-]. | 57–14–7 | 15,000 | b |
| Epichlorohydrin [Oxirane, (chloromethyl)-]. | 106–89–8 | 20,000 | b |
| Ethylenediamine [1,2-Ethanediamine]. | 107–15–3 | 20,000 | b |
| Ethyleneimine [Aziridine]. | 151–56–4 | 10,000 | b |
| Ethylene oxide [Oxirane]. | 75–21–8 | 10,000 | a, b |
| Fluorine | 7782–41–4 | 1,000 | b |
| Formaldehyde (solution). | 50–00–0 | 15,000 | b |
| Furan | 110–00–9 | 5,000 | b |
| Hydrazine | 302–01–2 | 15,000 | b |
| Hydrochloric acid (conc 37% or greater). | 7647–01–0 | 15,000 | d |
| Hydrocyanic acid | 74–90–8 | 2,500 | a, b |

TABLE 1 TO §68.130.—LIST OF REGULATED TOXIC SUBSTANCES AND THRESHOLD QUANTITIES FOR ACCIDENTAL RELEASE PREVENTION—Continued

[Alphabetical Order—77 Substances]

| Chemical name | CAS No. | Threshold quantity (lbs) | Basis for listing |
|---|---|---|---|
| Hydrogen chloride (anhydrous) [Hydrochloric acid]. | 7647–01–0 | 5,000 | a |
| Hydrogen fluoride/ Hydrofluoric acid (conc 50% or greater) [Hydrofluoric acid]. | 7664–39–3 | 1,000 | a, b |
| Hydrogen selenide. | 7783–07–5 | 500 | b |
| Hydrogen sulfide | 7783–06–4 | 10,000 | a, b |
| Iron, pentacarbonyl- [Iron carbonyl (Fe(CO)5), (TB-5-11)-]. | 13463–40–6 | 2,500 | b |
| Isobutyronitrile [Propanenitrile, 2-methyl-]. | 78–82–0 | 20,000 | b |
| Isopropyl chloroformate [Carbonochloridic acid, 1-methylethyl ester]. | 108–23–6 | 15,000 | b |
| Methacrylonitrile [2-Propenenitrile, 2-methyl-]. | 126–98–7 | 10,000 | b |
| Methyl chloride [Methane, chloro-]. | 74–87–3 | 10,000 | a |
| Methyl chloroformate [Carbonochloridic acid, methylester]. | 79–22–1 | 5,000 | b |
| Methyl hydrazine [Hydrazine, methyl-]. | 60–34–4 | 15,000 | b |
| Methyl isocyanate [Methane, isocyanato-]. | 624–83–9 | 10,000 | a, b |
| Methyl mercaptan [Methanethiol]. | 74–93–1 | 10,000 | b |
| Methyl thiocyanate [Thiocyanic acid, methyl ester]. | 556–64–9 | 20,000 | b |
| Methyltrichlorosilane [Silane, trichloromethyl-]. | 75–79–6 | 5,000 | b |
| Nickel carbonyl | 13463–39–3 | 1,000 | b |
| Nitric acid (conc 80% or greater). | 7697–37–2 | 15,000 | b |
| Nitric oxide [Nitrogen oxide (NO)]. | 10102–43–9 | 10,000 | b |

TABLE 1 TO §68.130.—LIST OF REGULATED TOXIC SUBSTANCES AND THRESHOLD QUANTITIES FOR ACCIDENTAL RELEASE PREVENTION—Continued

[Alphabetical Order—77 Substances]

| Chemical name | CAS No. | Threshold quantity (lbs) | Basis for listing |
|---|---|---|---|
| Oleum (Fuming Sulfuric acid) [Sulfuric acid, mixture with sulfur trioxide] [1]. | 8014–95–7 | 10,000 | e |
| Peracetic acid [Ethaneperoxoic acid]. | 79–21–0 | 10,000 | b |
| Perchloromethylmercaptan [Methanesulfenyl chloride, trichloro-]. | 594–42–3 | 10,000 | b |
| Phosgene [Carbonic dichloride]. | 75–44–5 | 500 | a, b |
| Phosphine | 7803–51–2 | 5,000 | b |
| Phosphorus oxychloride [Phosphoryl chloride]. | 10025–87–3 | 5,000 | b |
| Phosphorus trichloride [Phosphorous trichloride]. | 7719–12–2 | 15,000 | b |
| Piperidine | 110–89–4 | 15,000 | b |
| Propionitrile [Propanenitrile]. | 107–12–0 | 10,000 | b |
| Propyl chloroformate [Carbonochloridic acid, propylester]. | 109–61–5 | 15,000 | b |
| Propyleneimine [Aziridine, 2-methyl-]. | 75–55–8 | 10,000 | b |
| Propylene oxide [Oxirane, methyl-]. | 75–56–9 | 10,000 | b |
| Sulfur dioxide (anhydrous). | 7446–09–5 | 5,000 | a, b |
| Sulfur tetrafluoride [Sulfur fluoride (SF4), (T-4)-]. | 7783–60–0 | 2,500 | b |
| Sulfur trioxide | 7446–11–9 | 10,000 | a, b |
| Tetramethyllead [Plumbane, tetramethyl-]. | 75–74–1 | 10,000 | b |
| Tetranitromethane [Methane, tetranitro-]. | 509–14–8 | 10,000 | b |
| Titanium tetrachloride [Titanium chloride (TiCl4) (T-4)-]. | 7550–45–0 | 2,500 | b |
| Toluene 2,4-diisocyanate [Benzene, 2,4-diisocyanato-1-methyl-] [1]. | 584–84–9 | 10,000 | a |

TABLE 1 TO §68.130.—LIST OF REGULATED TOXIC SUBSTANCES AND THRESHOLD QUANTITIES FOR ACCIDENTAL RELEASE PREVENTION—Continued

[Alphabetical Order—77 Substances]

| Chemical name | CAS No. | Threshold quantity (lbs) | Basis for listing |
|---|---|---|---|
| Toluene 2,6-diisocyanate [Benzene, 1,3-diisocyanato-2-methyl-][1]. | 91-08-7 | 10,000 | a |
| Toluene diisocyanate (unspecified isomer) [Benzene, 1,3-diisocyanatom-ethyl-][1]. | 26471-62-5 | 10,000 | a |
| Trimethylchlorosilane [Silane, chlorotrimethyl-]. | 75-77-4 | 10,000 | b |

TABLE 1 TO §68.130.—LIST OF REGULATED TOXIC SUBSTANCES AND THRESHOLD QUANTITIES FOR ACCIDENTAL RELEASE PREVENTION—Continued

[Alphabetical Order—77 Substances]

| Chemical name | CAS No. | Threshold quantity (lbs) | Basis for listing |
|---|---|---|---|
| Vinyl acetate monomer [Acetic acid ethenyl ester]. | 108-05-4 | 15,000 | b |

[1] The mixture exemption in §68.115(b)(1) does not apply to the substance.

NOTE: Basis for Listing:
  a   Mandated for listing by Congress.
  b   On EHS list, vapor pressure 10 mmHg or greater.
  c   Toxic gas.
  d   Toxicity of hydrogen chloride, potential to release hydrogen chloride, and history of accidents.
  e   Toxicity of sulfur trioxide and sulfuric acid, potential to release sulfur trioxide, and history of accidents.

TABLE 2 TO §68.130.—LIST OF REGULATED TOXIC SUBSTANCES AND THRESHOLD QUANTITIES FOR ACCIDENTAL RELEASE PREVENTION

[CAS Number Order—77 Substances]

| CAS No. | Chemical name | Threshold quantity (lbs) | Basis for listing |
|---|---|---|---|
| 50-00-0 | Formaldehyde (solution) | 15,000 | b |
| 57-14-7 | 1,1-Dimethylhydrazine [Hydrazine, 1,1-dimethyl-] | 15,000 | b |
| 60-34-4 | Methyl hydrazine [Hydrazine, methyl-] | 15,000 | b |
| 67-66-3 | Chloroform [Methane, trichloro-] | 20,000 | b |
| 74-87-3 | Methyl chloride [Methane, chloro-] | 10,000 | a |
| 74-90-8 | Hydrocyanic acid | 2,500 | a, b |
| 74-93-1 | Methyl mercaptan [Methanethiol] | 10,000 | b |
| 75-15-0 | Carbon disulfide | 20,000 | b |
| 75-21-8 | Ethylene oxide [Oxirane] | 10,000 | a, b |
| 75-44-5 | Phosgene [Carbonic dichloride] | 500 | a, b |
| 75-55-8 | Propyleneimine [Aziridine, 2-methyl-] | 10,000 | b |
| 75-56-9 | Propylene oxide [Oxirane, methyl-] | 10,000 | b |
| 75-74-1 | Tetramethyllead [Plumbane, tetramethyl-] | 10,000 | b |
| 75-77-4 | Trimethylchlorosilane [Silane, chlorotrimethyl-] | 10,000 | b |
| 75-78-5 | Dimethyldichlorosilane [Silane, dichlorodimethyl-] | 5,000 | b |
| 75-79-6 | Methyltrichlorosilane [Silane, trichloromethyl-] | 5,000 | b |
| 78-82-0 | Isobutyronitrile [Propanenitrile, 2-methyl-] | 20,000 | b |
| 79-21-0 | Peracetic acid [Ethaneperoxoic acid] | 10,000 | b |
| 79-22-1 | Methyl chloroformate [Carbonochloridic acid, methylester] | 5,000 | b |
| 91-08-7 | Toluene 2,6-diisocyanate [Benzene, 1,3-diisocyanato-2-methyl-][1] | 10,000 | a |
| 106-89-8 | Epichlorohydrin [Oxirane, (chloromethyl)-] | 20,000 | b |
| 107-02-8 | Acrolein [2-Propenal] | 5,000 | b |
| 107-11-9 | Allylamine [2-Propen-1-amine] | 10,000 | b |
| 107-12-0 | Propionitrile [Propanenitrile] | 10,000 | b |
| 107-13-1 | Acrylonitrile [2-Propenenitrile] | 20,000 | b |
| 107-15-3 | Ethylenediamine [1,2-Ethanediamine] | 20,000 | b |
| 107-18-6 | Allyl alcohol [2-Propen-1-ol] | 15,000 | b |
| 107-30-2 | Chloromethyl methyl ether [Methane, chloromethoxy-] | 5,000 | b |
| 108-05-4 | Vinyl acetate monomer [Acetic acid ethenyl ester] | 15,000 | b |
| 108-23-6 | Isopropyl chloroformate [Carbonochloridic acid, 1-methylethyl ester] | 15,000 | b |
| 108-91-8 | Cyclohexylamine [Cyclohexanamine] | 15,000 | b |
| 109-61-5 | Propyl chloroformate [Carbonochloridic acid, propylester] | 15,000 | b |
| 110-00-9 | Furan | 5,000 | b |
| 110-89-4 | Piperidine | 15,000 | b |
| 123-73-9 | Crotonaldehyde, (E)- [2-Butenal, (E)-] | 20,000 | b |
| 126-98-7 | Methacrylonitrile [2-Propenenitrile, 2-methyl-] | 10,000 | b |
| 151-56-4 | Ethyleneimine [Aziridine] | 10,000 | b |
| 302-01-2 | Hydrazine | 15,000 | b |
| 353-42-4 | Boron trifluoride compound with methyl ether (1:1) [Boron, trifluoro[oxybis[methane]]-, T-4-. | 15,000 | b |

# Environmental Protection Agency §68.130

TABLE 2 TO §68.130.—LIST OF REGULATED TOXIC SUBSTANCES AND THRESHOLD QUANTITIES FOR ACCIDENTAL RELEASE PREVENTION—Continued

[CAS Number Order—77 Substances]

| CAS No. | Chemical name | Threshold quantity (lbs) | Basis for listing |
|---|---|---|---|
| 506–77–4 | Cyanogen chloride | 10,000 | c |
| 509–14–8 | Tetranitromethane [Methane, tetranitro-] | 10,000 | b |
| 542–88–1 | Chloromethyl ether [Methane, oxybis[chloro-] | 1,000 | b |
| 556–64–9 | Methyl thiocyanate [Thiocyanic acid, methyl ester] | 20,000 | b |
| 584–84–9 | Toluene 2,4-diisocyanate [Benzene, 2,4-diisocyanato-1-methyl-][1] | 10,000 | a |
| 594–42–3 | Perchloromethylmercaptan [Methanesulfenyl chloride, trichloro-] | 10,000 | b |
| 624–83–9 | Methyl isocyanate [Methane, isocyanato-] | 10,000 | a, b |
| 814–68–6 | Acrylyl chloride [2-Propenoyl chloride] | 5,000 | b |
| 4170–30–3 | Crotonaldehyde [2-Butenal] | 20,000 | b |
| 7446–09–5 | Sulfur dioxide (anhydrous) | 5,000 | a, b |
| 7446–11–9 | Sulfur trioxide | 10,000 | a, b |
| 7550–45–0 | Titanium tetrachloride [Titanium chloride (TiCl4) (T-4)-] | 2,500 | b |
| 7637–07–2 | Boron trifluoride [Borane, trifluoro-] | 5,000 | b |
| 7647–01–0 | Hydrochloric acid (conc 37% or greater) | 15,000 | d |
| 7647–01–0 | Hydrogen chloride (anhydrous) [Hydrochloric acid] | 5,000 | a |
| 7664–39–3 | Hydrogen fluoride/Hydrofluoric acid (conc 50% or greater) [Hydrofluoric acid] | 1,000 | a, b |
| 7664–41–7 | Ammonia (anhydrous) | 10,000 | a, b |
| 7664–41–7 | Ammonia (conc 20% or greater) | 20,000 | a, b |
| 7697–37–2 | Nitric acid (conc 80% or greater) | 15,000 | b |
| 7719–12–2 | Phosphorus trichloride [Phosphorous trichloride] | 15,000 | b |
| 7726–95–6 | Bromine | 10,000 | a, b |
| 7782–41–4 | Fluorine | 1,000 | b |
| 7782–50–5 | Chlorine | 2,500 | a, b |
| 7783–06–4 | Hydrogen sulfide | 10,000 | a, b |
| 7783–07–5 | Hydrogen selenide | 500 | b |
| 7783–60–0 | Sulfur tetrafluoride [Sulfur fluoride (SF4), (T-4)-] | 2,500 | b |
| 7784–34–1 | Arsenous trichloride | 15,000 | b |
| 7784–42–1 | Arsine | 1,000 | b |
| 7803–51–2 | Phosphine | 5,000 | b |
| 8014–95–7 | Oleum (Fuming Sulfuric acid) [Sulfuric acid, mixture with sulfur trioxide][1] | 10,000 | e |
| 10025–87–3 | Phosphorus oxychloride [Phosphoryl chloride] | 5,000 | b |
| 10049–04–4 | Chlorine dioxide [Chlorine oxide (ClO₂)] | 1,000 | c |
| 10102–43–9 | Nitric oxide [Nitrogen oxide (NO)] | 10,000 | b |
| 10294–34–5 | Boron trichloride [Borane, trichloro-] | 5,000 | b |
| 13463–39–3 | Nickel carbonyl | 1,000 | b |
| 13463–40–6 | Iron, pentacarbonyl- [Iron carbonyl (Fe(CO)₅), (TB-5-11)-] | 2,500 | b |
| 19287–45–7 | Diborane | 2,500 | b |
| 26471–62–5 | Toluene diisocyanate (unspecified isomer) [Benzene, 1,3-diisocyanatomethyl-1][1] | 10,000 | a |

[1] The mixture exemption in §68.115(b)(1) does not apply to the substance.
NOTE: Basis for Listing:
a  Mandated for listing by Congress.
b  On EHS list, vapor pressure 10 mmHg or greater.
c  Toxic gas.
d  Toxicity of hydrogen chloride, potential to release hydrogen chloride, and history of accidents.
e  Toxicity of sulfur trioxide and sulfuric acid, potential to release sulfur trioxide, and history of accidents.

TABLE 3 TO §68.130.—LIST OF REGULATED FLAMMABLE SUBSTANCES AND THRESHOLD QUANTITIES FOR ACCIDENTAL RELEASE PREVENTION

[Alphabetical Order—63 Substances]

| Chemical name | CAS No. | Threshold quantity (lbs) | Basis for listing |
|---|---|---|---|
| Acetaldehyde | 75–07–0 | 10,000 | g |
| Acetylene [Ethyne] | 74–86–2 | 10,000 | f |
| Bromotrifluorethylene [Ethene, bromotrifluoro-] | 598–73–2 | 10,000 | f |
| 1,3-Butadiene | 106–99–0 | 10,000 | f |
| Butane | 106–97–8 | 10,000 | f |
| 1-Butene | 106–98–9 | 10,000 | f |
| 2-Butene | 107–01–7 | 10,000 | f |
| Butene | 25167–67–3 | 10,000 | f |
| 2-Butene-cis | 590–18–1 | 10,000 | f |
| 2-Butene-trans [2-Butene, (E)] | 624–64–6 | 10,000 | f |
| Carbon oxysulfide [Carbon oxide sulfide (COS)] | 463–58–1 | 10,000 | f |
| Chlorine monoxide [Chlorine oxide] | 7791–21–1 | 10,000 | f |
| 2-Chloropropylene [1-Propene, 2-chloro-] | 557–98–2 | 10,000 | g |

61

TABLE 3 TO § 68.130.—LIST OF REGULATED FLAMMABLE SUBSTANCES AND THRESHOLD QUANTITIES FOR ACCIDENTAL RELEASE PREVENTION—Continued

[Alphabetical Order—63 Substances]

| Chemical name | CAS No. | Threshold quantity (lbs) | Basis for listing |
|---|---|---|---|
| 1-Chloropropylene [1-Propene, 1-chloro-] | 590–21–6 | 10,000 | g |
| Cyanogen [Ethanedinitrile] | 460–19–5 | 10,000 | f |
| Cyclopropane | 75–19–4 | 10,000 | f |
| Dichlorosilane [Silane, dichloro-] | 4109–96–0 | 10,000 | f |
| Difluoroethane [Ethane, 1,1-difluoro-] | 75–37–6 | 10,000 | f |
| Dimethylamine [Methanamine, N-methyl-] | 124–40–3 | 10,000 | f |
| 2,2-Dimethylpropane [Propane, 2,2-dimethyl-] | 463–82–1 | 10,000 | f |
| Ethane | 74–84–0 | 10,000 | f |
| Ethyl acetylene [1-Butyne] | 107–00–6 | 10,000 | f |
| Ethylamine [Ethanamine] | 75–04–7 | 10,000 | f |
| Ethyl chloride [Ethane, chloro-] | 75–00–3 | 10,000 | f |
| Ethylene [Ethene] | 74–85–1 | 10,000 | f |
| Ethyl ether [Ethane, 1,1'-oxybis-] | 60–29–7 | 10,000 | g |
| Ethyl mercaptan [Ethanethiol] | 75–08–1 | 10,000 | g |
| Ethyl nitrite [Nitrous acid, ethyl ester] | 109–95–5 | 10,000 | f |
| Hydrogen | 1333–74–0 | 10,000 | f |
| Isobutane [Propane, 2-methyl] | 75–28–5 | 10,000 | f |
| Isopentane [Butane, 2-methyl-] | 78–78–4 | 10,000 | g |
| Isoprene [1,3-Butadiene, 2-methyl-] | 78–79–5 | 10,000 | g |
| Isopropylamine [2-Propanamine] | 75–31–0 | 10,000 | g |
| Isopropyl chloride [Propane, 2-chloro-] | 75–29–6 | 10,000 | g |
| Methane | 74–82–8 | 10,000 | f |
| Methylamine [Methanamine] | 74–89–5 | 10,000 | f |
| 3-Methyl-1-butene | 563–45–1 | 10,000 | f |
| 2-Methyl-1-butene | 563–46–2 | 10,000 | g |
| Methyl ether [Methane, oxybis-] | 115–10–6 | 10,000 | f |
| Methyl formate [Formic acid, methyl ester] | 107–31–3 | 10,000 | g |
| 2-Methylpropene [1-Propene, 2-methyl-] | 115–11–7 | 10,000 | f |
| 1,3-Pentadinene | 504–60–9 | 10,000 | f |
| Pentane | 109–66–0 | 10,000 | g |
| 1-Pentene | 109–67–1 | 10,000 | g |
| 2-Pentene, (E)- | 646–04–8 | 10,000 | g |
| 2-Pentene, (Z)- | 627–20–3 | 10,000 | g |
| Propadiene [1,2-Propadiene] | 463–49–0 | 10,000 | f |
| Propane | 74–98–6 | 10,000 | f |
| Propylene [1-Propene] | 115–07–1 | 10,000 | f |
| Propyne [1-Propyne] | 74–99–7 | 10,000 | f |
| Silane | 7803–62–5 | 10,000 | f |
| Tetrafluoroethylene [Ethene, tetrafluoro-] | 116–14–3 | 10,000 | f |
| Tetramethylsilane [Silane, tetramethyl-] | 75–76–3 | 10,000 | g |
| Trichlorosilane [Silane, trichloro-] | 10025–78–2 | 10,000 | g |
| Trifluorochloroethylene [Ethene, chlorotrifluoro-] | 79–38–9 | 10,000 | f |
| Trimethylamine [Methanamine, N,N-dimethyl-] | 75–50–3 | 10,000 | f |
| Vinyl acetylene [1-Buten-3-yne] | 689–97–4 | 10,000 | f |
| Vinyl chloride [Ethene, chloro-] | 75–01–4 | 10,000 | a, f |
| Vinyl ethyl ether [Ethene, ethoxy-] | 109–92–2 | 10,000 | g |
| Vinyl fluoride [Ethene, fluoro-] | 75–02–5 | 10,000 | f |
| Vinylidene chloride [Ethene, 1,1-dichloro-] | 75–35–4 | 10,000 | g |
| Vinylidene fluoride [Ethene, 1,1-difluoro-] | 75–38–7 | 10,000 | f |
| Vinyl methyl ether [Ethene, methoxy-] | 107–25–5 | 10,000 | f |

NOTE. Basis for Listing.
a   Mandated for listing by Congress.
f   Flammable gas.
g   Volatile flammable liquid.

TABLE 4 TO § 68.130.—LIST OF REGULATED FLAMMABLE SUBSTANCES AND THRESHOLD QUANTITIES FOR ACCIDENTAL RELEASE PREVENTION

[CAS Number Order—63 Substances]

| CAS No. | Chemical name | CAS No. | Threshold quantity (lbs) | Basis for listing |
|---|---|---|---|---|
| 60–29–7 | Ethyl ether [Ethane, 1,1'-oxybis-] | 60–29–7 | 10,000 | g |
| 74–82–8 | Methane | 74–82–8 | 10,000 | f |
| 74–84–0 | Ethane | 74–84–0 | 10,000 | f |
| 74–85–1 | Ethylene [Ethene] | 74–85–1 | 10,000 | f |

TABLE 4 TO §68.130.—LIST OF REGULATED FLAMMABLE SUBSTANCES AND THRESHOLD QUANTITIES
FOR ACCIDENTAL RELEASE PREVENTION—Continued

[CAS Number Order—63 Substances]

| CAS No. | Chemical name | CAS No. | Threshold quantity (lbs) | Basis for listing |
|---|---|---|---|---|
| 74–86–2 | Acetylene [Ethyne] | 74–86–2 | 10,000 | f |
| 74–89–5 | Methylamine [Methanamine] | 74–89–5 | 10,000 | f |
| 74–98–6 | Propane | 74–98–6 | 10,000 | f |
| 74–99–7 | Propyne [1-Propyne] | 74–99–7 | 10,000 | f |
| 75–00–3 | Ethyl chloride [Ethane, chloro-] | 75–00–3 | 10,000 | f |
| 75–01–4 | Vinyl chloride [Ethene, chloro-] | 75–01–4 | 10,000 | a, f |
| 75–02–5 | Vinyl fluoride [Ethene, fluoro-] | 75–02–5 | 10,000 | f |
| 75–04–7 | Ethylamine [Ethanamine] | 75–04–7 | 10,000 | f |
| 75–07–0 | Acetaldehyde | 75–07–0 | 10,000 | g |
| 75–08–1 | Ethyl mercaptan [Ethanethiol] | 75–08–1 | 10,000 | g |
| 75–19–4 | Cyclopropane | 75–19–4 | 10,000 | f |
| 75–28–5 | Isobutane [Propane, 2-methyl] | 75–28–5 | 10,000 | f |
| 75–29–6 | Isopropyl chloride [Propane, 2-chloro-] | 75–29–6 | 10,000 | g |
| 75–31–0 | Isopropylamine [2-Propanamine] | 75–31–0 | 10,000 | g |
| 75–35–4 | Vinylidene chloride [Ethene, 1,1-dichloro-] | 75–35–4 | 10,000 | g |
| 75–37–6 | Difluoroethane [Ethane, 1,1-difluoro-] | 75–37–6 | 10,000 | f |
| 75–38–7 | Vinylidene fluoride [Ethene, 1,1-difluoro-] | 75–38–7 | 10,000 | f |
| 75–50–3 | Trimethylamine [Methanamine, N, N-dimethyl-] | 75–50–3 | 10,000 | f |
| 75–76–3 | Tetramethylsilane [Silane, tetramethyl-] | 75–76–3 | 10,000 | g |
| 78–78–4 | Isopentane [Butane, 2-methyl-] | 78–78–4 | 10,000 | g |
| 78–79–5 | Isoprene [1,3,-Butadiene, 2-methyl-] | 78–79–5 | 10,000 | g |
| 79–38–9 | Trifluorochloroethylene [Ethene, chlorotrifluoro-] | 79–38–9 | 10,000 | f |
| 106–97–8 | Butane | 106–97–8 | 10,000 | f |
| 106–98–9 | 1-Butene | 106–98–9 | 10,000 | f |
| 196–99–0 | 1,3-Butadiene | 106–99–0 | 10,000 | f |
| 107–00–6 | Ethyl acetylene [1-Butyne] | 107–00–6 | 10,000 | f |
| 107–01–7 | 2-Butene | 107–01–7 | 10,000 | f |
| 107–25–5 | Vinyl methyl ether [Ethene, methoxy-] | 107–25–5 | 10,000 | f |
| 107–31–3 | Methyl formate [Formic acid, methyl ester] | 107–31–3 | 10,000 | g |
| 109–66–0 | Pentane | 109–66–0 | 10,000 | g |
| 109–67–1 | 1-Pentene | 109–67–1 | 10,000 | g |
| 109–92–2 | Vinyl ethyl ether [Ethene, ethoxy-] | 109–92–2 | 10,000 | g |
| 109–95–5 | Ethyl nitrite [Nitrous acid, ethyl ester] | 109–95–5 | 10,000 | f |
| 115–07–1 | Propylene [1-Propene] | 115–07–1 | 10,000 | f |
| 115–10–6 | Methyl ether [Methane, oxybis-] | 115–10–6 | 10,000 | f |
| 115–11–7 | 2-Methylpropene [1-Propene, 2-methyl-] | 115–11–7 | 10,000 | f |
| 116–14–3 | Tetrafluoroethylene [Ethene, tetrafluoro-] | 116–14–3 | 10,000 | f |
| 124–40–3 | Dimethylamine [Methanamine, N-methyl-] | 124–40–3 | 10,000 | f |
| 460–19–5 | Cyanogen [Ethanedinitrile] | 460–19–5 | 10,000 | f |
| 463–49–0 | Propadiene [1,2-Propadiene] | 463–49–0 | 10,000 | f |
| 463–58–1 | Carbon oxysulfide [Carbon oxide sulfide (COS)] | 463–58–1 | 10,000 | f |
| 463–82–1 | 2,2-Dimethylpropane [Propane, 2,2-dimethyl-] | 463–82–1 | 10,000 | f |
| 504–60–9 | 1,3-Pentadiene | 504–60–9 | 10,000 | f |
| 557–98–2 | 2-Chloropropylene [1-Propene, 2-chloro-] | 557–98–2 | 10,000 | g |
| 563–45–1 | 3-Methyl-1-butene | 563–45–1 | 10,000 | f |
| 563–46–2 | 2-Methyl-1-butene | 563–46–2 | 10,000 | g |
| 590–18–1 | 2-Butene-cis | 590–18–1 | 10,000 | f |
| 590–21–6 | 1-Chloropropylene [1-Propene, 1-chloro-] | 590–21–6 | 10,000 | g |
| 598–73–2 | Bromotrifluorethylene [Ethene, bromotrifluoro-] | 598–73–2 | 10,000 | f |
| 624–64–6 | 2-Butene-trans [2-Butene, (E)] | 624–64–6 | 10,000 | f |
| 627–20–3 | 2-Pentene, (Z)- | 627–20–3 | 10,000 | g |
| 646–04–8 | 2-Pentene, (E)- | 646–04–8 | 10,000 | g |
| 689–97–4 | Vinyl acetylene [1-Buten-3-yne] | 689–97–4 | 10,000 | f |
| 1333–74–0 | Hydrogen | 1333–74–0 | 10,000 | f |
| 4109–96–0 | Dichlorosilane [Silane, dichloro-] | 4109–96–0 | 10,000 | f |
| 7791–21–1 | Chlorine monoxide [Chlorine oxide] | 7791–21–1 | 10,000 | f |
| 7803–62–5 | Silane | 7803–62–5 | 10,000 | f |
| 10025–78–2 | Trichlorosilane [Silane, trichloro-] | 10025–78–2 | 10,000 | g |
| 25167–67–3 | Butene | 25167–67–3 | 10,000 | f |

**Note:** Basis for Listing:     a   Mandated for listing by Congress.     f   Flammable gas.     g   Volatile flammable liquid.

[59 FR 4493, Jan. 31, 1994. Redesignated at 61 FR 31717, June 20, 1996, as amended at 62 FR 45132, Aug. 25, 1997; 63 FR 645, Jan. 6, 1998]

## Subpart G—Risk Management Plan

SOURCE: 61 FR 31726, June 20, 1996, unless otherwise noted.

### § 68.150  Submission.

(a) The owner or operator shall submit a single RMP that includes the information required by §§ 68.155 through 68.185 for all covered processes. The RMP shall be submitted in a method and format to a central point as specified by EPA prior to June 21, 1999.

(b) The owner or operator shall submit the first RMP no later than the latest of the following dates:

(1) June 21, 1999;

(2) Three years after the date on which a regulated substance is first listed under § 68.130; or

(3) The date on which a regulated substance is first present above a threshold quantity in a process.

(c) Subsequent submissions of RMPs shall be in accordance with § 68.190.

(d) Notwithstanding the provisions of §§ 68.155 to 68.190, the RMP shall exclude classified information. Subject to appropriate procedures to protect such information from public disclosure, classified data or information excluded from the RMP may be made available in a classified annex to the RMP for review by Federal and state representatives who have received the appropriate security clearances.

(e) Procedures for asserting that information submitted in the RMP is entitled to protection as confidential business information are set forth in §§ 68.151 and 68.152.

[61 FR 31726, June 20, 1996, as amended at 64 FR 979, Jan. 6, 1999]

### § 68.151  Assertion of claims of confidential business information.

(a) Except as provided in paragraph (b) of this section, an owner or operator of a stationary source required to report or otherwise provide information under this part may make a claim of confidential business information for any such information that meets the criteria set forth in 40 CFR 2.301.

(b) Notwithstanding the provisions of 40 CFR part 2, an owner or operator of a stationary source subject to this part may not claim as confidential business information the following information:

(1) Registration data required by § 68.160(b)(1) through (b)(6) and (b)(8), (b)(10) through (b)(13) and NAICS code and Program level of the process set forth in § 68.160(b)(7);

(2) Offsite consequence analysis data required by § 68.165(b)(4), (b)(9), (b)(10), (b)(11), and (b)(12).

(3) Accident history data required by § 68.168;

(4) Prevention program data required by § 68.170(b), (d), (e)(1), (f) through (k);

(5) Prevention program data required by § 68.175(b), (d), (e)(1), (f) through (p); and

(6) Emergency response program data required by § 68.180.

(c) Notwithstanding the procedures specified in 40 CFR part 2, an owner or operator asserting a claim of CBI with respect to information contained in its RMP, shall submit to EPA at the time it submits the RMP the following:

(1) The information claimed confidential, provided in a format to be specified by EPA;

(2) A sanitized (redacted) copy of the RMP, with the notation "CBI" substituted for the information claimed confidential, except that a generic category or class name shall be substituted for any chemical name or identity claimed confidential; and

(3) The document or documents substantiating each claim of confidential business information, as described in § 68.152.

[64 FR 979, Jan. 6, 1999]

### § 68.152  Substantiating claims of confidential business information.

(a) An owner or operator claiming that information is confidential business information must substantiate that claim by providing documentation that demonstrates that the claim meets the substantive criteria set forth in 40 CFR 2.301.

(b) Information that is submitted as part of the substantiation may be claimed confidential by marking it as confidential business information. Information not so marked will be treated as public and may be disclosed without notice to the submitter. If information that is submitted as part of the substantiation is claimed confidential, the owner or operator must provide a sanitized and unsanitized version of the substantiation.

(c) The owner, operator, or senior official with management responsibility of the stationary source shall sign a certification that the signer has personally examined the information submitted and that based on inquiry of the persons who compiled the information, the information is true, accurate, and complete, and that those portions of the substantiation claimed as confidential business information would, if disclosed, reveal trade secrets or other confidential business information.

[64 FR 980, Jan. 6, 1999]

### §68.155 Executive summary.

The owner or operator shall provide in the RMP an executive summary that includes a brief description of the following elements:

(a) The accidental release prevention and emergency response policies at the stationary source;

(b) The stationary source and regulated substances handled;

(c) The worst-case release scenario(s) and the alternative release scenario(s), including administrative controls and mitigation measures to limit the distances for each reported scenario;

(d) The general accidental release prevention program and chemical-specific prevention steps;

(e) The five-year accident history;

(f) The emergency response program; and

(g) Planned changes to improve safety.

### §68.160 Registration.

(a) The owner or operator shall complete a single registration form and include it in the RMP. The form shall cover all regulated substances handled in covered processes.

(b) The registration shall include the following data:

(1) Stationary source name, street, city, county, state, zip code, latitude and longitude, method for obtaining latitude and longitude, and description of location that latitude and longitude represent;

(2) The stationary source Dun and Bradstreet number;

(3) Name and Dun and Bradstreet number of the corporate parent company;

(4) The name, telephone number, and mailing address of the owner or operator;

(5) The name and title of the person or position with overall responsibility for RMP elements and implementation;

(6) The name, title, telephone number, and 24-hour telephone number of the emergency contact;

(7) For each covered process, the name and CAS number of each regulated substance held above the threshold quantity in the process, the maximum quantity of each regulated substance or mixture in the process (in pounds) to two significant digits, the five- or six-digit NAICS code that most closely corresponds to the process, and the Program level of the process;

(8) The stationary source EPA identifier;

(9) The number of full-time employees at the stationary source;

(10) Whether the stationary source is subject to 29 CFR 1910.119;

(11) Whether the stationary source is subject to 40 CFR part 355;

(12) If the stationary source has a CAA Title V operating permit, the permit number; and

(13) The date of the last safety inspection of the stationary source by a Federal, state, or local government agency and the identity of the inspecting entity.

(14) Source or Parent Company E-Mail Address (Optional);

(15) Source Homepage address (Optional)

(16) Phone number at the source for public inquiries (Optional);

(17) Local Emergency Planning Committee (Optional);

(18) OSHA Voluntary Protection Program status (Optional);

[61 FR 31726, June 20, 1996, as amended at 64 FR 980, Jan. 6, 1999]

### §68.165 Offsite consequence analysis.

(a) The owner or operator shall submit in the RMP information:

(1) One worst-case release scenario for each Program 1 process; and

(2) For Program 2 and 3 processes, one worst-case release scenario to represent all regulated toxic substances

held above the threshold quantity and one worst-case release scenario to represent all regulated flammable substances held above the threshold quantity. If additional worst-case scenarios for toxics or flammables are required by § 68.25(a)(2)(iii), the owner or operator shall submit the same information on the additional scenario(s). The owner or operator of Program 2 and 3 processes shall also submit information on one alternative release scenario for each regulated toxic substance held above the threshold quantity and one alternative release scenario to represent all regulated flammable substances held above the threshold quantity.

(b) The owner or operator shall submit the following data:

(1) Chemical name;

(2) Percentage weight of the chemical in a liquid mixture (toxics only);

(3) Physical state (toxics only);

(4) Basis of results (give model name if used);

(5) Scenario (explosion, fire, toxic gas release, or liquid spill and evaporation);

(6) Quantity released in pounds;

(7) Release rate;

(8) Release duration;

(9) Wind speed and atmospheric stability class (toxics only);

(10) Topography (toxics only);

(11) Distance to endpoint;

(12) Public and environmental receptors within the distance;

(13) Passive mitigation considered; and

(14) Active mitigation considered (alternative releases only);

[61 FR 31726, June 20, 1996, as amended at 64 FR 980, Jan. 6, 1999]

## § 68.168 Five-year accident history.

The owner or operator shall submit in the RMP the information provided in § 68.42(b) on each accident covered by § 68.42(a).

## § 68.170 Prevention program/Program 2.

(a) For each Program 2 process, the owner or operator shall provide in the RMP the information indicated in paragraphs (b) through (k) of this section. If the same information applies to more than one covered process, the owner or operator may provide the information only once, but shall indicate to which processes the information applies.

(b) The five- or six-digit NAICS code that most closely corresponds to the process.

(c) The name(s) of the chemical(s) covered.

(d) The date of the most recent review or revision of the safety information and a list of Federal or state regulations or industry-specific design codes and standards used to demonstrate compliance with the safety information requirement.

(e) The date of completion of the most recent hazard review or update.

(1) The expected date of completion of any changes resulting from the hazard review;

(2) Major hazards identified;

(3) Process controls in use;

(4) Mitigation systems in use;

(5) Monitoring and detection systems in use; and

(6) Changes since the last hazard review.

(f) The date of the most recent review or revision of operating procedures.

(g) The date of the most recent review or revision of training programs;

(1) The type of training provided—classroom, classroom plus on the job, on the job; and

(2) The type of competency testing used.

(h) The date of the most recent review or revision of maintenance procedures and the date of the most recent equipment inspection or test and the equipment inspected or tested.

(i) The date of the most recent compliance audit and the expected date of completion of any changes resulting from the compliance audit.

(j) The date of the most recent incident investigation and the expected date of completion of any changes resulting from the investigation.

(k) The date of the most recent change that triggered a review or revision of safety information, the hazard review, operating or maintenance procedures, or training.

[61 FR 31726, June 20, 1996, as amended at 64 FR 980, Jan. 6, 1999]

§ 68.175 Prevention program/Program 3.

(a) For each Program 3 process, the owner or operator shall provide the information indicated in paragraphs (b) through (p) of this section. If the same information applies to more than one covered process, the owner or operator may provide the information only once, but shall indicate to which processes the information applies.

(b) The five- or six-digit NAICS code that most closely corresponds to the process.

(c) The name(s) of the substance(s) covered.

(d) The date on which the safety information was last reviewed or revised.

(e) The date of completion of the most recent PHA or update and the technique used.

(1) The expected date of completion of any changes resulting from the PHA;

(2) Major hazards identified;

(3) Process controls in use;

(4) Mitigation systems in use;

(5) Monitoring and detection systems in use; and

(6) Changes since the last PHA.

(f) The date of the most recent review or revision of operating procedures.

(g) The date of the most recent review or revision of training programs;

(1) The type of training provided—classroom, classroom plus on the job, on the job; and

(2) The type of competency testing used.

(h) The date of the most recent review or revision of maintenance procedures and the date of the most recent equipment inspection or test and the equipment inspected or tested.

(i) The date of the most recent change that triggered management of change procedures and the date of the most recent review or revision of management of change procedures.

(j) The date of the most recent pre-startup review.

(k) The date of the most recent compliance audit and the expected date of completion of any changes resulting from the compliance audit;

(l) The date of the most recent incident investigation and the expected date of completion of any changes resulting from the investigation;

(m) The date of the most recent review or revision of employee participation plans;

(n) The date of the most recent review or revision of hot work permit procedures;

(o) The date of the most recent review or revision of contractor safety procedures; and

(p) The date of the most recent evaluation of contractor safety performance.

[61 FR 31726, June 20, 1996, as amended at 64 FR 980, Jan. 6, 1999]

§ 68.180 Emergency response program.

(a) The owner or operator shall provide in the RMP the following information:

(1) Do you have a written emergency response plan?

(2) Does the plan include specific actions to be taken in response to an accidental releases of a regulated substance?

(3) Does the plan include procedures for informing the public and local agencies responsible for responding to accidental releases?

(4) Does the plan include information on emergency health care?

(5) The date of the most recent review or update of the emergency response plan;

(6) The date of the most recent emergency response training for employees.

(b) The owner or operator shall provide the name and telephone number of the local agency with which emergency response activities and the emergency response plan is coordinated.

(c) The owner or operator shall list other Federal or state emergency plan requirements to which the stationary source is subject.

[61 FR 31726, June 20, 1996, as amended at 64 FR 980, Jan. 6, 1999]

§ 68.185 Certification.

(a) For Program 1 processes, the owner or operator shall submit in the RMP the certification statement provided in § 68.12(b)(4).

(b) For all other covered processes, the owner or operator shall submit in the RMP a single certification that, to the best of the signer's knowledge, information, and belief formed after reasonable inquiry, the information submitted is true, accurate, and complete.

### § 68.190  Updates.

(a) The owner or operator shall review and update the RMP as specified in paragraph (b) of this section and submit it in a method and format to a central point specified by EPA prior to June 21, 1999.

(b) The owner or operator of a stationary source shall revise and update the RMP submitted under § 68.150 as follows:

(1) Within five years of its initial submission or most recent update required by paragraphs (b)(2) through (b)(7) of this section, whichever is later.

(2) No later than three years after a newly regulated substance is first listed by EPA;

(3) No later than the date on which a new regulated substance is first present in an already covered process above a threshold quantity;

(4) No later than the date on which a regulated substance is first present above a threshold quantity in a new process;

(5) Within six months of a change that requires a revised PHA or hazard review;

(6) Within six months of a change that requires a revised offsite consequence analysis as provided in § 68.36; and

(7) Within six months of a change that alters the Program level that applied to any covered process.

(c) If a stationary source is no longer subject to this part, the owner or operator shall submit a revised registration to EPA within six months indicating that the stationary source is no longer covered.

## Subpart H—Other Requirements

SOURCE: 61 FR 31728, June 20, 1996, unless otherwise noted.

### § 68.200  Recordkeeping.

The owner or operator shall maintain records supporting the implementation of this part for five years unless otherwise provided in subpart D of this part.

### § 68.210  Availability of information to the public.

(a) The RMP required under subpart G of this part shall be available to the public under 42 U.S.C. 7414(c).

(b) The disclosure of classified information by the Department of Defense or other Federal agencies or contractors of such agencies shall be controlled by applicable laws, regulations, or executive orders concerning the release of classified information.

### § 68.215  Permit content and air permitting authority or designated agency requirements.

(a) These requirements apply to any stationary source subject to this part 68 and parts 70 or 71 of this chapter. The 40 CFR part 70 or part 71 permit for the stationary source shall contain:

(1) A statement listing this part as an applicable requirement;

(2) Conditions that require the source owner or operator to submit:

(i) A compliance schedule for meeting the requirements of this part by the date provided in § 68.10(a) or;

(ii) As part of the compliance certification submitted under 40 CFR 70.6(c)(5), a certification statement that the source is in compliance with all requirements of this part, including the registration and submission of the RMP.

(b) The owner or operator shall submit any additional relevant information requested by the air permitting authority or designated agency.

(c) For 40 CFR part 70 or part 71 permits issued prior to the deadline for registering and submitting the RMP and which do not contain permit conditions described in paragraph (a) of this section, the owner or operator or air permitting authority shall initiate permit revision or reopening according to the procedures of 40 CFR 70.7 or 71.7 to incorporate the terms and conditions consistent with paragraph (a) of this section.

(d) The state may delegate the authority to implement and enforce the requirements of paragraph (e) of this section to a state or local agency or agencies other than the air permitting authority. An up-to-date copy of any delegation instrument shall be maintained by the air permitting authority.

The state may enter a written agreement with the Administrator under which EPA will implement and enforce the requirements of paragraph (e) of this section.

(e) The air permitting authority or the agency designated by delegation or agreement under paragraph (d) of this section shall, at a minimum:

(1) Verify that the source owner or operator has registered and submitted an RMP or a revised plan when required by this part;

(2) Verify that the source owner or operator has submitted a source certification or in its absence has submitted a compliance schedule consistent with paragraph (a)(2) of this section;

(3) For some or all of the sources subject to this section, use one or more mechanisms such as, but not limited to, a completeness check, source audits, record reviews, or facility inspections to ensure that permitted sources are in compliance with the requirements of this part; and

(4) Initiate enforcement action based on paragraphs (e)(1) and (e)(2) of this section as appropriate.

### §68.220 Audits.

(a) In addition to inspections for the purpose of regulatory development and enforcement of the Act, the implementing agency shall periodically audit RMPs submitted under subpart G of this part to review the adequacy of such RMPs and require revisions of RMPs when necessary to ensure compliance with subpart G of this part.

(b) The implementing agency shall select stationary sources for audits based on any of the following criteria:

(1) Accident history of the stationary source;

(2) Accident history of other stationary sources in the same industry;

(3) Quantity of regulated substances present at the stationary source;

(4) Location of the stationary source and its proximity to the public and environmental receptors;

(5) The presence of specific regulated substances;

(6) The hazards identified in the RMP; and

(7) A plan providing for neutral, random oversight.

(c) Exemption from audits. A stationary source with a Star or Merit ranking under OSHA's voluntary protection program shall be exempt from audits under paragraph (b)(2) and (b)(7) of this section.

(d) The implementing agency shall have access to the stationary source, supporting documentation, and any area where an accidental release could occur.

(e) Based on the audit, the implementing agency may issue the owner or operator of a stationary source a written preliminary determination of necessary revisions to the stationary source's RMP to ensure that the RMP meets the criteria of subpart G of this part. The preliminary determination shall include an explanation for the basis for the revisions, reflecting industry standards and guidelines (such as AIChE/CCPS guidelines and ASME and API standards) to the extent that such standards and guidelines are applicable, and shall include a timetable for their implementation.

(f) *Written response to a preliminary determination.* (1) The owner or operator shall respond in writing to a preliminary determination made in accordance with paragraph (e) of this section. The response shall state the owner or operator will implement the revisions contained in the preliminary determination in accordance with the timetable included in the preliminary determination or shall state that the owner or operator rejects the revisions in whole or in part. For each rejected revision, the owner or operator shall explain the basis for rejecting such revision. Such explanation may include substitute revisions.

(2) The written response under paragraph (f)(1) of this section shall be received by the implementing agency within 90 days of the issue of the preliminary determination or a shorter period of time as the implementing agency specifies in the preliminary determination as necessary to protect public health and the environment. Prior to the written response being due and upon written request from the owner or operator, the implementing agency may provide in writing additional time for the response to be received.

(g) After providing the owner or operator an opportunity to respond under paragraph (f) of this section, the implementing agency may issue the owner or operator a written final determination of necessary revisions to the stationary source's RMP. The final determination may adopt or modify the revisions contained in the preliminary determination under paragraph (e) of this section or may adopt or modify the substitute revisions provided in the response under paragraph (f) of this section. A final determination that adopts a revision rejected by the owner or operator shall include an explanation of the basis for the revision. A final determination that fails to adopt a substitute revision provided under paragraph (f) of this section shall include an explanation of the basis for finding such substitute revision unreasonable.

(h) Thirty days after completion of the actions detailed in the implementation schedule set in the final determination under paragraph (g) of this section, the owner or operator shall be in violation of subpart G of this part and this section unless the owner or operator revises the RMP prepared under subpart G of this part as required by the final determination, and submits the revised RMP as required under § 68.150.

(i) The public shall have access to the preliminary determinations, responses, and final determinations under this section in a manner consistent with § 68.210.

(j) Nothing in this section shall preclude, limit, or interfere in any way with the authority of EPA or the state to exercise its enforcement, investigatory, and information gathering authorities concerning this part under the Act.

APPENDIX A TO PART 68—TABLE OF TOXIC ENDPOINTS

[As defined in § 68.22 of this part]

| CAS No. | Chemical name | Toxic endpoint (mg/L) |
|---|---|---|
| 107–02–8 | Acrolein [2-Propenal] | 0.0011 |
| 107–13–1 | Acrylonitrile [2-Propenenitrile] | 0.076 |
| 814–68–6 | Acrylyl chloride [2-Propenoyl chloride] | 0.00090 |
| 107–18–6 | Allyl alcohol [2-Propen-1-ol] | 0.036 |
| 107–11–9 | Allylamine [2-Propen-1-amine] | 0.0032 |
| 7664–41–7 | Ammonia (anhydrous) | 0.14 |
| 7664–41–7 | Ammonia (conc 20% or greater) | 0.14 |
| 7784–34–1 | Arsenous trichloride | 0.010 |
| 7784–42–1 | Arsine | 0.0019 |
| 10294–34–5 | Boron trichloride [Borane, trichloro-] | 0.010 |
| 7637–07–2 | Boron trifluoride [Borane, trifluoro-] | 0.028 |
| 353–42–4 | Boron trifluoride compound with methyl ether (1:1) [Boron, trifluoro[oxybis[methane]]-, T-4 | 0.023 |
| 7726–95–6 | Bromine | 0.0065 |
| 75–15–0 | Carbon disulfide | 0.16 |
| 7782–50–5 | Chlorine | 0.0087 |
| 10049–04–4 | Chlorine dioxide [Chlorine oxide (ClO2)] | 0.0028 |
| 67–66–3 | Chloroform [Methane, trichloro-] | 0.49 |
| 542–88–1 | Chloromethyl ether [Methane, oxybis[chloro-] | 0.00025 |
| 107–30–2 | Chloromethyl methyl ether [Methane, chloromethoxy-] | 0.0018 |
| 4170–30–3 | Crotonaldehyde [2-Butenal] | 0.029 |
| 123–73–9 | Crotonaldehyde, (E)-, [2-Butenal, (E)-] | 0.029 |
| 506–77–4 | Cyanogen chloride | 0.030 |
| 108–91–8 | Cyclohexylamine [Cyclohexanamine] | 0.16 |
| 19287–45–7 | Diborane | 0.0011 |
| 75–78–5 | Dimethyldichlorosilane [Silane, dichlorodimethyl-] | 0.026 |
| 57–14–7 | 1,1-Dimethylhydrazine [Hydrazine, 1,1-dimethyl-] | 0.012 |
| 106–89–8 | Epichlorohydrin [Oxirane, (chloromethyl)-] | 0.076 |
| 151–56–4 | Ethylenediamine [1,2-Ethanediamine] | 0.49 |
| 75–21–8 | Ethyleneimine [Aziridine] | 0.018 |
| 7782–41–4 | Ethylene oxide [Oxirane] | 0.090 |
| 50–00–0 | Fluorine | 0.0039 |
| 110–00–9 | Formaldehyde (solution) | 0.012 |
| 302–01–2 | Furan | 0.0012 |
| 7647–01–0 | Hydrazine | 0.011 |
| 74–90–8 | Hydrochloric acid (conc 37% or greater) | 0.030 |
| 7647–01–0 | Hydrocyanic acid | 0.011 |
| 7664–39–3 | Hydrogen chloride (anhydrous) [Hydrochloric acid] | 0.030 |
| 7783–07–5 | Hydrogen fluoride/Hydrofluoric acid (conc 50% or greater) [Hydrofluoric acid] | 0.016 |
| 13463–40–6 | Hydrogen selenide | 0.00066 |
| 78–82–0 | Hydrogen sulfide | 0.042 |
| 108–23–6 | Iron, pentacarbonyl- [Iron carbonyl (Fe(CO)5), (TB-5-11)-] | 0.00044 |
| 126–98–7 | Isobutyronitrile [Propanenitrile, 2-methyl-] | 0.14 |
| | Isopropyl chloroformate [Carbonochloridic acid, 1-methylethyl ester] | 0.10 |
| | Methacrylonitrile [2-Propenenitrile, 2-methyl-] | 0.0027 |

APPENDIX A TO PART 68—TABLE OF TOXIC ENDPOINTS—Continued

[As defined in §68.22 of this part]

| CAS No. | Chemical name | Toxic end-point (mg/L) |
|---|---|---|
| 74-87-3 | Methyl chloride [Methane, chloro-] | 0.82 |
| 79-22-1 | Methyl chloroformate [Carbonochloridic acid, methylester] | 0.0019 |
| 60-34-4 | Methyl hydrazine [Hydrazine, methyl-] | 0.0094 |
| 624-83-9 | Methyl isocyanate [Methane, isocyanato-] | 0.0012 |
| 74-93-1 | Methyl mercaptan [Methanethiol] | 0.049 |
| 556-64-9 | Methyl thiocyanate [Thiocyanic acid, methyl ester] | 0.085 |
| 75-79-6 | Methyltrichlorosilane [Silane, trichloromethyl-] | 0.018 |
| 13463-39-3 | Nickel carbonyl | 0.00067 |
| 7697-37-2 | Nitric acid (conc 80% or greater) | 0.026 |
| 10102-43-9 | Nitric oxide [Nitrogen oxide (NO)] | 0.031 |
| 8014-95-7 | Oleum (Fuming Sulfuric acid) [Sulfuric acid, mixture with sulfur trioxide] | 0.010 |
| 79-21-0 | Peracetic acid [Ethaneperoxoic acid] | 0.0045 |
| 594-42-3 | Perchloromethylmercaptan [Methanesulfenyl chloride, trichloro-] | 0.0076 |
| 75-44-5 | Phosgene [Carbonic dichloride] | 0.00081 |
| 7803-51-2 | Phosphine | 0.0035 |
| 10025-87-3 | Phosphorus oxychloride [Phosphoryl chloride] | 0.0030 |
| 7719-12-2 | Phosphorus trichloride [Phosphorous trichloride] | 0.028 |
| 110-89-4 | Piperidine | 0.022 |
| 107-12-0 | Propionitrile [Propanenitrile] | 0.0037 |
| 109-61-5 | Propyl chloroformate [Carbonochloridic acid, propylester] | 0.010 |
| 75-55-8 | Propyleneimine [Aziridine, 2-methyl-] | 0.12 |
| 75-56-9 | Propylene oxide [Oxirane, methyl-] | 0.59 |
| 7446-09-5 | Sulfur dioxide (anhydrous) | 0.0078 |
| 7783-60-0 | Sulfur tetrafluoride [Sulfur fluoride (SF4), (T-4)-] | 0.0092 |
| 7446-11-9 | Sulfur trioxide | 0.010 |
| 75-74-1 | Tetramethyllead [Plumbane, tetramethyl-] | 0.0040 |
| 509-14-8 | Tetranitromethane [Methane, tetranitro-] | 0.0040 |
| 7750-45-0 | Titanium tetrachloride [Titanium chloride (TiCl4) (T-4)-] | 0.020 |
| 584-84-9 | Toluene 2,4-diisocyanate [Benzene, 2,4-diisocyanato-1-methyl-] | 0.0070 |
| 91-08-7 | Toluene 2,6-diisocyanate [Benzene, 1,3-diisocyanato-2-methyl-] | 0.0070 |
| 26471-62-5 | Toluene diisocyanate (unspecified isomer) [Benzene, 1,3-diisocyanatomethyl-] | 0.0070 |
| 75-77-4 | Trimethylchlorosilane [Silane, chlorotrimethyl-] | 0.050 |
| 108-05-4 | Vinyl acetate monomer [Acetic acid ethenyl ester] | 0.26 |

[61 FR 31729, June 20, 1996, as amended at 62 FR 45132, Aug. 25, 1997]

finding that notice and public procedure is impracticable, unnecessary or contrary to the public interest. This determination must be supported by a brief statement (5 U.S.C. 808(2)).

As stated previously, we have made such a good cause finding, including the reasons therefore, and established an effective date of March 13, 2000. The EPA will submit a report containing this rule and other required information to the U.S. Senate, the U.S. House of Representatives, and the Comptroller General of the United States prior to publication of the rule in the **Federal Register**. This action is not a "major rule" as defined by 5 U.S.C. 804(2).

## List of Subjects in 40 CFR Part 60

Environmental protection, Administrative practice and procedure, Air pollution control, Intergovernmental relations, Nitrogen oxides, Recordkeeping and reporting requirements.

Dated: March 2, 2000.

**Robert Perciasepe,**

*Assistant Administrator, Office of Air and Radiation.*

For the reasons set out in the preamble, title 40, chapter I, part 60, of the Code of Federal Regulations is amended as follows:

## PART 60—[AMENDED]

1. The authority citation for part 60 continues to read as follows:

**Authority:** 42 U.S.C. 7401–7601.

## Subpart Db—Standards of Performance for Industrial-Commercial-Institutional Steam Generating Units

2. Section 60.49b is amended by revising paragraph (s) and adding paragraph (w) to read as follows:

### § 60.49b  Reporting and recordkeeping requirements.

\*      \*      \*      \*      \*

(s) Facility specific nitrogen oxides standard for Cytec Industries Fortier Plant's C.AOG incinerator located in Westwego, Louisiana:

(1) *Definitions.*

*Oxidation zone* is defined as the portion of the C.AOG incinerator that extends from the inlet of the oxidizing zone combustion air to the outlet gas stack.

*Reducing zone* is defined as the portion of the C.AOG incinerator that extends from the burner section to the inlet of the oxidizing zone combustion air.

*Total inlet air* is defined as the total amount of air introduced into the C.AOG incinerator for combustion of natural gas and chemical by-product waste and is equal to the sum of the air flow into the reducing zone and the air flow into the oxidation zone.

(2) *Standard for nitrogen oxides.* (i) When fossil fuel alone is combusted, the nitrogen oxides emission limit for fossil fuel in § 60.44b(a) applies.

(ii) When natural gas and chemical by-product waste are simultaneously combusted, the nitrogen oxides emission limit is 289 ng/J (0.67 lb/million Btu) and a maximum of 81 percent of the total inlet air provided for combustion shall be provided to the reducing zone of the C.AOG incinerator.

(3) *Emission monitoring.* (i) The percent of total inlet air provided to the reducing zone shall be determined at least every 15 minutes by measuring the air flow of all the air entering the reducing zone and the air flow of all the air entering the oxidation zone, and compliance with the percentage of total inlet air that is provided to the reducing zone shall be determined on a 3-hour average basis.

(ii) The nitrogen oxides emission limit shall be determined by the compliance and performance test methods and procedures for nitrogen oxides in § 60.46b(i).

(iii) The monitoring of the nitrogen oxides emission limit shall be performed in accordance with § 60.48b.

(4) *Reporting and recordkeeping requirements.* (i) The owner or operator of the C.AOG incinerator shall submit a report on any excursions from the limits required by paragraph (a)(2) of this section to the Administrator with the quarterly report required by paragraph (i) of this section.

(ii) The owner or operator of the C.AOG incinerator shall keep records of the monitoring required by paragraph (a)(3) of this section for a period of 2 years following the date of such record.

(iii) The owner of operator of the C.AOG incinerator shall perform all the applicable reporting and recordkeeping requirements of this section.

\*      \*      \*      \*      \*

(w) The reporting period for the reports required under this subpart is each 6 month period. All reports shall be submitted to the Administrator and shall be postmarked by the 30th day following the end of the reporting period.

[FR Doc. 00–5797 Filed 3–10–00; 8:45 am]

**BILLING CODE 6560–50–P**

## ENVIRONMENTAL PROTECTION AGENCY

### 40 CFR Part 68

[FRL–6550–1]

RIN 2050–AE74

### Amendments to the List of Regulated Substances and Thresholds for Accidental Release Prevention; Flammable Substances Used as Fuel or Held for Sale as Fuel at Retail Facilities

**AGENCY:** Environmental Protection Agency (EPA).

**ACTION:** Final rule.

---

**SUMMARY:** EPA is modifying its chemical accident prevention regulations to conform to the fuels provision of the recently enacted Chemical Safety Information, Site Security and Fuels Regulatory Relief Act (Pub. L. 106–40). In accordance with the new law, today's rule revises the list of regulated flammable substances to exclude those substances when used as a fuel or held for sale as a fuel at a retail facility. EPA is also announcing there will be no further action on a previous proposal concerning flammable substances, since the new law resolves the issue addressed by the proposal.

**DATES:** Effective March 13, 2000.

**ADDRESSES:** Docket. Supporting material used in developing the final rule is contained in Docket No. A–99–36. The docket is available for public inspection and copying between 8:00 am and 5:30 pm, Monday through Friday (except government holidays) at EPA's Air Docket, Room 1500, Waterside Mall, 401 M Street, SW, Washington, DC 20460; phone number: 202–260–7548. A reasonable fee may be charged for copying.

**FOR FURTHER INFORMATION CONTACT:** Breeda Reilly, Chemical Emergency Preparedness and Prevention Office, Environmental Protection Agency, Ariel Rios Building, 1200 Pennsylvania Ave, NW (5104), Washington, DC 20460, (202) 260–0716.

**SUPPLEMENTARY INFORMATION:**

**Table of Contents**

## I. Introduction and Background

### A. Statutory Authority

This rule is being issued under section 112(r) of the Clean Air Act (CAA) as amended by the Chemical Safety Information, Site Security and Fuels Regulatory Relief Act (the Act), which President Clinton signed into law on August 5, 1999. Section 2 of the Act immediately removed EPA's authority to "list a flammable substance when used as a fuel or held for sale as a fuel at a retail facility * * * solely because of the explosive or flammable properties of the substance, unless a fire or explosion caused by the substance will result in acute adverse health effects from human exposure to the substance, including the unburned fuel or its combustion byproducts, other than those caused by the heat of the fire or impact of the explosion."

The Act defines "retail facility" as "a stationary source at which more than one-half of the income is obtained from direct sales to end users or at which more than one-half of the fuel sold, by volume, is sold through a cylinder exchange program."

### B. Background on Chemical Accident Prevention Regulations

CAA section 112(r) contains requirements for the prevention and mitigation of accidental chemical releases. The focus is on those chemicals that pose the greatest risk to public health and the environment in the event of an accidental release. Section 112(r)(3) mandates that EPA identify at least 100 such chemicals and promulgate a list of "regulated substances" with threshold quantities. Section 112(r)(7) directs EPA to issue regulations requiring stationary sources that contain more than a threshold quantity of a regulated substance to develop and implement a risk management program and submit a risk management plan (RMP).

EPA promulgated the initial list of regulated substances on January 31, 1994 (59 FR 4478) (the "List Rule"). The Agency identified two categories of regulated substances—toxic and flammable—and listed substances accordingly. EPA included 77 chemicals on the toxic substances list based on each chemical's acute toxicity and several other factors—the chemical's physical state, physical/chemical properties and accident history— relevant to the likelihood that an accidental release of the chemical would lead to significant offsite consequences. The Agency also placed 63 substances on the flammable substances list, including vinyl chloride, a substance mandated for listing by Congress. EPA selected chemicals for the flammable substances list based on their flammability rating and the other factors related to likelihood of significant offsite consequences.

Of the originally listed substances, 14 met the criteria for both toxic and flammable substances (arsine, cyanogen chloride, diborane, ethylene oxide, formaldehyde, furan, hydrocyanic acid, hydrogen selenide, hydrogen sulfide, methyl chloride, methyl mercaptan, phosphine, propyleneimine, and propylene oxide). EPA placed these 14 substances on only the toxic substances list, because their toxicity poses the greater threat to human health and the environment.

Following promulgation of the List Rule, EPA issued a rule establishing the accidental release prevention requirements on June 20, 1996 (61 FR 31668) ("the RMP Rule"). Together these rules are codified at 40 CFR part 68.

In accordance with section 112(r)(7), the RMP rule requires that any stationary source with more than a threshold quantity of a regulated substance in a process develop and implement a risk management program and submit an RMP describing the source's program as well as its five-year accident history and potential offsite consequences. The rule further provides that RMPs be submitted by June 21, 1999 for sources with more than a threshold quantity of a regulated substance in a process by that date, or within a specified time of the source first exceeding the applicable threshold.

EPA has amended the List and RMP Rules several times. On August 25, 1997 (62 FR 45132), EPA amended the List Rule to change the listed concentration of hydrochloric acid. On January 6, 1998 (63 FR 640), EPA again amended the List Rule to delist Division 1.1 explosives (classified by the Department

of Transportation (DOT)), to clarify certain provisions related to regulated flammable substances, and to clarify the transportation exemption. EPA amended the RMP Rule on January 6, 1999 (64 FR 964) to add several mandatory and optional RMP data elements, to establish procedures for protecting confidential business information, to adopt a new industry classification system and to make technical corrections and clarifications. EPA also amended the RMP Rule on May 26, 1999 (64 FR 28696) to modify the requirements for conducting worst case release scenario analyses for flammable substances and to clarify its interpretation of CAA sections 112(1) and 112(r)(11) as they relate to DOT requirements under the Federal Hazardous Transportation Law.

## II. Discussion of Modification

### A. Affected Substances

The new Act provides that EPA shall not list a flammable substance when used as a fuel,[1] or held for sale as a fuel at a retail facility solely because of its explosive or flammable properties, except under certain circumstances. The purpose of today's rule is to revise the List Rule as needed to conform to the Act.

As described above, the List Rule currently contains two lists—one of toxic substances and one of flammable substances. The toxic substances list contains those chemicals that meet the criteria listing as toxic substances, even if they also meet the criteria for listing as flammable substances. Accordingly, every chemical on the toxic substances list was listed for its toxicity at least and not solely because of its explosive or flammable properties. The substances on the toxics list are thus not affected by the new Act.

The substances on the flammables list, on the other hand, are listed "solely" because they meet a certain flammability rating, taking other risk factors into account. In deciding what flammable substances to list, EPA concentrated on those substances that have the potential to result in significant offsite consequences. Accidents involving flammable substances may lead to vapor cloud explosions, vapor cloud fires, boiling liquid expanding vapor explosions (BLEVEs), pool fires, and jet fires, depending on the type of substance involved and the

---

[1] EPA has received a number of questions as to whether the fuel use exclusion is available only to retail facilities. EPA believes that the statute and legislative history are clear that the fuel use exclusion is available to any facility that uses a flammable substance as a fuel.

circumstances of the accident. Historically, flammable substance accidents having significant offsite impacts involved either vapor cloud explosions at refineries and chemical plants, or BLEVEs at sources storing large quantities of flammable substances. Vapor cloud explosions produce blast waves that potentially can cause offsite damage and kill or injure people. High overpressure levels can cause death or injury as a direct result of an a explosion; such effects generally occur close to the site of an explosion. People can also be killed or injured because of indirect effects of the blast (e.g., collapse of buildings, flying glass or debris); these effects can occur farther from the site of the blast.

By contrast, the effects of vapor cloud fires, in which the vapor cloud burns but does not explode, are limited primarily to the area covered by the burning cloud. BLEVEs, which generally involve the rupture of a container, can cause container fragments to be thrown substantial distances; such fragments have the potential to cause damage and injury.

Thermal radiation is the primary hazard of pool and jet fires. The potential effects of thermal radiation generally do not extend for as great a distance as those of blast waves and are related to the duration of exposure; people at some distance from a fire would likely be able to escape.

Based on this analysis and available accident history data, the Agency concluded that vapor cloud explosions and BLEVEs pose the greatest potential hazard from flammable substances to the public and environment. For purposes of the List Rule, EPA consequently focused on those chemicals with the potential to result in vapor cloud explosions or BLEVEs in the event of an accidental release. The Agency determined that chemicals meeting the highest flammability rating of the National Fire Protection Agency (NFPA) had this potential and used that rating as the principal criterion for including chemicals on the flammable substances list.

The other factors EPA considered in listing flammable substances—physical state, physical/chemical properties and accident history—all relate to a chemical's potential to be accidentally released in a way that could lead to a vapor cloud explosion or BLEVE. In short, the Agency included chemicals on the flammable substances list "solely" because of their explosive potential, a basis now disallowed by the new Act for flammable substances when used as a fuel or held for sale as a fuel at a retail facility.

The new Act nevertheless allows EPA to list a flammable substance when used as a fuel, or held for sale as a fuel where a fire or explosion caused by the substance will result in acute adverse health effects from human exposure to the substance or its combustion byproducts. EPA believes, however, that no listed substances on the flammable substances list is a candidate for this exception. As noted above, flammable substances that meet the listing criteria for toxic substances are on the toxic substances list only. Therefore, none of the chemicals on the flammable substances list will qualify for the exception based on acute health effects from exposure to the substance itself.

Further, combustion byproducts are generally not relevant to listing flammable substances. For hydrocarbons, including the listed flammable substances commonly used as fuels, typical combustion products include water vapor, carbon dioxide, carbon monoxide, and relatively small amounts of other oxidized inorganic substances and do not meet the listing criteria for toxic substances. Several other listed flammable substances may result in combustion byproducts that meet the listing criteria for toxic substances, but these substances are not commonly used as fuels. Further, any toxic combustion byproducts will be a fraction of the total mass and not likely to exceed the applicable threshold for coverage by the RMP rule. Quantities below the threshold are unlikely to have significant offsite consequences.

For these reasons, EPA believes that none of the listed flammable substances meet the new statute's test for listing fuels. Consequently, all of the listed flammable substances are potentially affected by the Act.

*B. Use or Sale as a Fuel*

The Act prohibits the listing of flammable substances "when used as a fuel or held for sale as a fuel at a retail facility." In limiting EPA's authority to list flammable substances used as a fuel, or sold as a fuel at retail facilities, Congress sought greater consistency between the RMP program and the Process Safety Management (PSM) Standard implemented by the Occupational Health and Safety Administration (OSHA). OSHA's PSM Standard is the workplace counterpart of EPA's RMP program. PSM requirements protect workers from accidental releases of highly hazardous substances in the workplace, while the RMP rule protects the public and environment from the offsite consequences of those releases.

The PSM and RMP programs are similar in many ways, covering mostly the same chemicals. Establishments subject to the PSM Standard must comply with the prevention program requirements which are the same as the RMP rule's Program 3 requirements (subpart D of the Part 68 regulations). However, OSHA provides an exemption from the PSM Standard for hydrocarbon fuels used solely for workplace consumption as a fuel (e.g., propane used for comfort heating), if such fuels are not part of a process containing another highly hazardous chemical covered by the standard. It also exempts such substances when sold by retail facilities.

The two prongs of the limitation on EPA's authority to list flammable substances (i.e., use as a fuel or held for sale as a fuel by a retail facility) largely follow the OSHA exemptions relating to fuel. EPA will therefore look to OSHA precedent and coordinate with OSHA in interpreting and applying the limitations to the extent they parallel OSHA's exemptions. For example, the new Act does not define the term "fuel," but OSHA has given "fuel" its ordinary meaning in applying the PSM fuel-related exemptions. Webster's Ninth New Collegiate Dictionary (1990) defines fuel as "a material used to produce heat or power by burning," and EPA has no reason to believe that "fuel" as used by the new Act should be defined differently.

Using the ordinary meaning of fuel, EPA reviewed the chemicals on its flammable substances list to determine which are used as fuel. Several of the listed substances are typically used as fuel, including propane, liquified petroleum gas (propane and/or butane often with small amounts of propylene and butylene); hydrogen; and gaseous natural gas (methane). EPA is aware of the possibility of other flammable substances being used as a fuel in particular circumstances. The following is a list of regulated flammable substances that EPA believes have been used as a fuel.

TABLE 1.—LIST OF COMMON FUELS

| Chemical name | CAS No. |
| --- | --- |
| Acetylene [Ethyne] | 74–86–2 |
| Butane | 106–97–8 |
| 1-Butene | 106–98–9 |
| 2-Butene | 107–01–7 |
| Butene | 25167–67–3 |
| 2-Butene-cis | 590–18–1 |
| 2-Butene-trans [2-Butene, (E)] | 624–64–6 |
| Ethane | 74–84–0 |
| Ethylene [Ethene] | 74–85–1 |
| Hydrogen | 1333–74–0 |

TABLE 1.—LIST OF COMMON FUELS—
Continued

| Chemical name | CAS No. |
|---|---|
| Isobutane [Propane, 2-methyl-] | 75–28–5 |
| Isopentane [Butane, 2-methyl-] | 78–78–4 |
| Methane | 74–82–8 |
| Pentane | 109–66–0 |
| 1-Pentene | 109–67–1 |
| 2-Pentene, (E)- | 646–04–8 |
| 2-Pentene, (Z)- | 627–20–3 |
| Propane | 74–98–6 |
| Propylene | 115–07–1 |

At the same time, all of the substances listed above are sometimes used as feedstock chemicals instead of fuel. Further, every listed flammable substance has the potential to be used as fuel, since it may be burned to create heat or power. Consequently, the List Rule cannot be conformed to the new law by deleting particular chemicals from the flammable substances list. Instead, EPA has added a provision to part 68, Subpart F (listing regulated substances) that excludes flammable substances when used as a fuel, or held for sale as a fuel at a retail facility from the list of regulated substances. The Agency has also annotated both versions of the flammable substances list (one version lists the substances alphabetically, the other by Chemical Abstract Service (CAS) number) to indicate that any flammable substance, when used as a fuel, or held for sale as a fuel at a retail facility, is excluded from the list.

As previously mentioned, the Act defines a "retail facility" as a stationary source at which more than one-half of the income is obtained from direct sales to end users or at which more than one-half of the fuel sold, by volume, is sold through a cylinder exchange program. The income test portion of the definition follows the definition of "retail facility" used by the OSHA in enforcing its PSM Standard (OSHA Directive CPL2–2.45A CH–1-Process Safety Management of Highly Hazardous Chemicals—Compliance Guidelines and Enforcement Procedures): "an establishment that would otherwise be subject to the PSM standard at which more than half of the income is obtained from direct sales to end users."

The effect of the income test portion of the new Act's retail facility definition is to provide relief to the same facilities that qualify for OSHA's retail facility exemption, and conversely, to require facilities that do not qualify for OSHA's exemption, and thus are subject to the PSM program, to also be subject to the

RMP program, provided no other exemption applies. EPA will consequently coordinate its interpretation and application of the income test portion of the retail facility definition with OSHA.

The second portion of the retail facility definition—concerning cylinder exchange programs—goes beyond that developed by OSHA and so provides greater relief than the OSHA retail facility exemption. In general, cylinder exchange programs represent a link between major retailers (for example, hardware stores, home centers and convenience stores) and propane distributors. The retailer typically provides space outdoors and manages transactions with end users such as homeowners; the propane distributor typically provides racks, filled cylinders, promotional materials, and training to the retailer's employees. Propane distributors may have several markets, including cylinder exchange; temporary heat during construction; commercial cooking, heating, and water heating; fuel to power vehicles, forklifts, and tractors; agricultural drying and heating; and others.

For propane or other fuel distributors which meet the definition of retail facility through either direct sales to end users or a cylinder exchange program, the fuel they hold is no longer covered by the RMP rule. For propane or other fuel distributors that do not meet the definition, the fuel they hold is *not* exempted from the RMP rule by the new law or today's action. EPA has added to part 68 a definition of "retail facility" that mirrors the statutory definition.

### III. Previous Actions Related to Fuels

*A. Previous Proposed Rule and Administrative Stay*

After promulgating the RMP rule, EPA became aware that a significant number of small, commercial sources use regulated flammable substances, particularly propane, as fuel in quantities in excess of the applicable threshold quantity (10,000 lbs in a process). As a result, these small sources, including farms, restaurants, hotels, and other commercial operations, were covered by the RMP requirements. Many of these sources are in rural locations where accidental releases are less likely to have significant offsite consequences. In light of the purpose of section 112(r)—to focus comprehensive accident prevention requirements on the most potentially dangerous sources—EPA reexamined whether farms and other small fuel users should be covered by the RMP rule.

On May 28, 1999, EPA issued a proposed amendment to the List Rule to create an exemption from threshold quantity determinations for processes containing 67,000 pounds or less of a listed flammable hydrocarbon fuel (64 FR 29171). EPA estimated that the proposed amendment, if promulgated, would reduce the universe of regulated sources from 69,485 to 50,300. At the same time (64 FR 29167), EPA published a temporary stay of the effectiveness of the RMP rule for those sources that would be exempted under the proposal. This stay, which expired on December 21, 1999, was in addition to, and did not affect, a stay of the rule for propane processes entered by the U.S. Court of Appeals for the D.C. Circuit (See Litigation and Court Stay).

While EPA was seeking comment on the proposed rule, Congress also studied the fuel issue and considered ways to provide regulatory relief to fuel users and retailers. Congress was concerned that the RMP rule placed a significant regulatory burden on facilities that were not previously covered by the OSHA PSM Standard. Congress decided to amend section 112(r) of the CAA to remove EPA's authority to list any flammable substance when used as a fuel, or held for sale as a fuel at a retail facility, except under specified circumstances.

While the new law and EPA's proposed rule and temporary stay all offer regulatory relief with respect to fuels, the new law reaches farther than EPA's actions. The new law provides relief for all fuels, not just hydrocarbon fuels. It also removes fuels from the RMP program regardless of the amount a stationary source uses or holds for retail sale, whereas EPA's proposal and stay only affects sources having no more than 67,000 lbs of fuel in a process. The new law does limit relief for fuel sellers to fuel retailers, whereas EPA's stay does not distinguish between types of fuel sellers. However, EPA believes that virtually no fuel wholesaler qualifies for the Agency's stay because wholesalers typically hold fuel in quantities far greater than 67,000 lbs. Even if a few wholesalers would have benefitted from EPA's proposed rule, the Agency believes that Congress has addressed the issue of how to provide regulatory relief to fuel users and sellers, and that EPA should thus implement Congress' approach without making exceptions to it.

Therefore, EPA is today withdrawing the proposed rule as it takes final action to amend the List Rule to conform to the new law. As previously mentioned, EPA's temporary stay of effectiveness expired on December 21, 1999.

*B. Litigation and Court Stay*

Following promulgation of the RMP rule in 1996, several petitions for judicial review of the rule were filed, including one by the National Propane Gas Association (NPGA). At NPGA's request, the U.S. Court of Appeals for the District of Columbia Circuit entered a temporary stay of the RMP rule as it applies to propane (Chlorine Institute v. Environmental Protection Agency, No. 96–1279, and consolidated cases (Nos. 96–1284, 96–1288, and 96–1290), Order of April 27, 1999). The judicial stay meant that any stationary source, or process at a stationary source, subject to the RMP rule only by virtue of propane was not subject to the RMP rule requirements, including those calling for a hazard assessment, accident prevention program, emergency response planning, and submission of (or inclusion in) an RMP by June 21, 1999.

On Jan. 5, 2000, the Court lifted its temporary stay in response to a joint motion by EPA and NPGA to dismiss the case and lift the stay. As of that date, part 68, as revised by the Act, is in effect with respect to any facility having more than the 10,000 pounds of propane in a process unless the facility uses the propane as a fuel or sells the propane as a retail facility. Facilities that use propane in their manufacturing processes or hold propane for purposes other than on-site fuel use at a non-retail facility must immediately come into compliance with Section 112(r) of the CAA.

## IV. RMP's Submitted Prior to Today's Action

EPA has received about 1,966 RMP's that address one or more of the 19 listed flammable substances that EPA has identified as likely to be used as a fuel. EPA cannot unilaterally delete any of the RMP's submitted for flammable substances from the RMP database, however, because the determination of whether a facility is eligible for the exclusion is based on information which is not reported to EPA, namely, whether a facility uses the flammable substance as a fuel or holds it for retail sale. Instead, EPA plans to send a letter to each of the 1,966 facilities to notify them of the exclusion, to ask them to evaluate their eligibility for the exclusion, and to describe the process the facilities should use to request a withdrawal of or to update these RMP's.

For about 950 of the 1,966 RMP's that reported a potential flammable fuel, only one chemical is reported. For these cases, the facilities will be asked to evaluate whether they qualify for the exclusion based on use or retail sales. If they determine that they do not qualify, no further action is required. If they determine that they do qualify, they may request that EPA withdraw their submission and EPA will delete it from the RMP database. Facilities will have the option of using the form that EPA developed to facilitate the withdrawal or simply stating their request in a letter. Alternatively, facilities can leave the RMP as a voluntary submission in the database and need not take further action.

The balance of the RMP's reported more than one substance. About 200 RMP's reported a toxic chemical substance in addition to the potential flammable fuel. For these cases, the facilities will be asked to evaluate whether their flammable substance qualifies for the exclusion based on use or retail sales. If they determine that they do not qualify, no further action is required. If they determine that they do qualify, they may resubmit their RMP, reporting only on the toxic substances. Alternatively, facilities can leave the original RMP including the flammable fuel submission in the database and need not take further action.

About 745 RMP's reported multiple flammable substances. For these cases, the facilities will be asked to evaluate whether each reported flammable substance qualifies for the exclusion based on use or retail sales. If they determine that none of their reported flammable substances qualify, no further action is required. If they determine that all of the reported substances qualify, they may request that EPA withdraw their submission and EPA will delete it from the RMP database. Facilities will have the option of using the formal withdrawal process or simply sending a letter. Alternatively, facilities can leave the RMP as a voluntary submission in the database and need not take further action. If they determine that only some of the flammable substances reported qualify, they will need to check their flammable worst case scenario and off-site consequence analysis (OCA). If their original worst case analysis is based on a flammable substance that is excluded, the facility should revise their RMP to provide appropriate OCA. Within its enforcement discretion, EPA plans to treat this similarly to the existing requirement to revise RMP's within 6 months of a process change, giving facilities 6 months to revise their RMP's. If their original worst case analysis is based on a flammable substance that is not excluded, the facility won't need to update their RMP, except as part of the regular reporting cycle.

## V. Rationale for Issuance of Rule Without Prior Notice

Section 553 of the Administrative Procedure Act, 5 U.S.C. 553(b)(B), provides that, when an agency for good cause finds that notice and public procedure are impracticable, unnecessary or contrary to the public interest, the agency may issue a rule without providing notice and an opportunity for public comment.

EPA is taking this action without prior notice and opportunity to comment. As previously mentioned, section 2 of the new Act, which took effect on August 5, 1999, immediately removed EPA's authority to list flammable substances when used as a fuel, or held for sale as a fuel at a retail facility. Consequently, EPA's regulation containing the list of regulated substances subject to the RMP rule needs to be modified to reflect the new law.

EPA has determined that there is good cause for making today's rule final without prior proposal and opportunity for comment because the Agency is codifying legislation which focuses clearly on a particular set of regulations and requires little interpretation by the Agency. In addition, EPA believes it is in the public interest to issue the revised list as soon as possible, to avoid confusion about the coverage of the RMP rule. As of August 5, 1999, there is no statutory basis for extending the RMP rule to listed flammable substances when used as a fuel, or held for sale as a fuel at a retail facility, except under certain circumstances. The Agency's rule should therefore be revised to reflect the change in authority as soon as possible. A comment period is unnecessary because today's action is nondiscretionary. A comment period would also be contrary to the public interest because the resulting delay would contribute to confusion about coverage of the RMP rule. Thus, notice and public procedure are unnecessary and contrary to the public interest. EPA finds that this constitutes good cause under 5 U.S.C. 553(b)(B).

The Agency is also issuing this rule with an immediate effective date. Since its effect is to relieve a restriction (*i.e.,* the requirement to comply with the RMP rule), EPA may make it effective upon promulgation. Further, EPA believes it is in the public interest to make it immediately effective, for the same reasons given above for dispensing with prior notice and comment.

## VI. Summary of Revisions to Rule

This section summarizes the changes to the rule.

Section 68.3, Definitions, has been revised to add a definition of retail facility, as defined in the new law.

Section 68.126 has been added to create an exclusion for regulated flammable substances used as fuel or held for sale as fuel at retail facilities. The exclusion is derived from the new law.

In Section 68.130, footnotes have been added to Tables 3 and 4. These two tables list the regulated flammable substances and their threshold quantities. Table 3 lists the regulated flammable substances in alphabetical order while Table 4 lists them in CAS number order. The footnotes remind the reader of the exclusion for regulated flammable substances. The reference to each footnote appears as an asterisk following the term "flammable substance" in the titles of Tables 3 and 4.

## VII. Administrative Requirements

### A. Docket

The docket is an organized and complete file of all the information considered by the EPA in the development of this rulemaking. The docket is a dynamic file, because it allows members of the public and industries involved to readily identify and locate documents so that they can effectively participate in the rulemaking process. Along with the proposed and promulgated rules and their preambles, the contents of the docket serve as the record in the case of judicial review. (See section 307(d)(7)(A) of the CAA.) The official record for this rulemaking has been established under Docket A–99–36, and is available for inspection from 8:00 a.m. to 5:30 p.m., Monday through Friday, excluding legal holidays. The official rulemaking record is located at the address in **ADDRESSES** at the beginning of this document.

### B. Executive Order 12866

Under Executive Order 12866 (58 FR 51735, October 4, 1993), the Agency must determine whether the regulatory action is "significant" and therefore subject to OMB review and the requirements of the Executive Order. The Order defines "significant regulatory action" as one that is likely to result in a rule that may:

(1) Have an annual effect on the economy of $100 million or more or adversely affect in a material way the economy, a sector of the economy, productivity, competition, jobs, the environment, public health or safety, or State, local, or tribal governments or communities;

(2) Create a serious inconsistency or otherwise interfere with an action taken or planned by another agency;

(3) Materially alter the budgetary impact of entitlements, grants, user fees, or loan programs or the rights and obligations of recipients thereof; or

(4) Raise novel legal or policy issues arising out of legal mandates, the President's priorities, or the principles set forth in the Executive Order.

It has been determined that this rule is not a "significant regulatory action" under the terms of Executive Order 12866 and is therefore not subject to OMB review.

### C. Executive Order 13045

Executive Order 13045: "Protection of Children from Environmental Health Risks and Safety Risks," (62 FR 19885, April 23, 1997), applies to any rule that: (1) Is determined to be "economically significant" as defined under E.O. 12866, and (2) concerns an environmental health or safety risk that EPA has reason to believe may have a disproportionate effect on children. If the regulatory action meets both criteria, the Agency must evaluate the environmental health or safety effects of the planned rule on children, and explain why the planned regulation is preferable to other potentially effective and reasonably feasible alternatives considered by the Agency.

EPA interprets E.O. 13045 as applying only to those regulatory actions that are based on health or safety risks, such that the analysis required under Section 5–501 of the Order has the potential to influence the regulation.

This action is not subject to this Executive Order because it is not economically significant as defined in E.O. 12866, and because it does not establish an environmental standard intended to mitigate health or safety risks.

### D. Executive Order 13084

Under Executive Order 13084, EPA may not issue a regulation that is not required by statute, that significantly or uniquely affects the communities of Indian tribal governments, and that imposes substantial direct compliance costs on those communities, unless the Federal government provides the funds necessary to pay the direct compliance costs incurred by the tribal governments, or EPA consults with those governments.

If EPA complies by consulting, Executive Order 13084 requires EPA to provide to the Office of Management and Budget, in a separately identified section of the preamble to the rule, a description of the extent of EPA's prior consultation with representatives of affected tribal governments, a summary of the nature of their concerns, and a statement supporting the need to issue the regulation. In addition, Executive Order 13084 requires EPA to develop an effective process permitting elected officials and other representatives of Indian tribal governments "to provide meaningful and timely input in the development of regulatory policies on matters that significantly or uniquely affect their communities."

Today's rule does not significantly or uniquely affect the communities of Indian tribal governments. This action reduces burden on flammable fuel users, which may include some sources owned or operated by Indian tribal governments. Accordingly, the requirements of section 3(b) of Executive Order 13084 do not apply to this rule.

### E. Executive Order 13132

Executive Order 13132, entitled "Federalism" (64 FR 43255, August 10, 1999), requires EPA to develop an accountable process to ensure "meaningful and timely input by State and local officials in the development of regulatory policies that have federalism implications." "Policies that have federalism implications" is defined in the Executive Order to include regulations that have "substantial direct effects on the States, on the relationship between the national government and the States, or on the distribution of power and responsibilities among the various levels of government."

Under Section 6 of Executive Order 13132, EPA may not issue a regulation that has federalism implications, that imposes substantial direct compliance costs, and that is not required by statute, unless the Federal government provides the funds necessary to pay the direct compliance costs incurred by State and local governments, or EPA consults with State and local officials early in the process of developing the proposed regulation. EPA also may not issue a regulation that has federalism implications and that preempts State law, unless the Agency consults with State and local officials early in the process of developing the proposed regulation.

This final rule does not have federalism implications. It will not have substantial direct effects on the States, on the relationship between the national government and the States, or on the distribution of power and responsibilities among the various levels of government, as specified in Executive Order 13132. Today's rule reduces the burden for those state, local, or tribal governments that may own or operate sources that use flammable fuels. Thus, the requirements of section 6 of the Executive Order do not apply to this rule.

*F. Regulatory Flexibility Act (RFA), as Amended by the Small Business Regulatory Enforcement Fairness Act of 1996 (SBREFA), 5 U.S.C. 601 et seq.*

Under the Regulatory Flexibility Act (RFA) of 1980 (5 U.S.C. 601, *et seq.*), as amended by the Small Business Regulatory Enforcement Fairness Act of 1996 (SBREFA), the Agency is required to give special consideration to the effect of Federal regulations on small entities and to consider regulatory options that might mitigate any such impacts. Small entities include small businesses, small not-for-profit enterprises, and small governmental jurisdictions.

Today's final rule is not subject to RFA, which generally requires an agency to prepare a regulatory flexibility analysis for any rule that will have a significant economic impact on a substantial number of small entities. The RFA applies only to rules subject to notice-and-comment rulemaking requirements under the Administrative Procedure Act (APA) or any other statute. The rule is subject to the APA, but as described in Section IV of this preamble, the Agency has invoked the "good cause" exemption under APA Section 553(b), which does not require notice and comment. Although this final rule is not subject to the RFA, EPA nonetheless has assessed the potential of this rule to adversely impact small entities subject to the rule. EPA does not believe the rule will adversely impact small entities. This action excludes flammable substances when used as a fuel, or held for sale as a fuel at a retail facility from the list of substances regulated by 40 CFR part 68, which will reduce burden on many small entities that otherwise would be covered by these requirements.

*G. Paperwork Reduction Act*

This action does not impose any new information collection burden. The Office of Management and Budget (OMB) has previously approved the information collection requirements contained in the existing regulations 40 CFR part 68 under the provisions of the *Paperwork Reduction Act*, 44 U.S.C. 3501 *et seq.* and has assigned OMB control number 2050–0144 (EPA ICR No.1656.06). EPA estimates a burden hour reduction of 70,400 hours.

Burden means the total time, effort, or financial resources expended by persons to generate, maintain, retain, or disclose or provide information to or for a Federal agency. This includes the time needed to review instructions; develop, acquire, install, and utilize technology and systems for the purposes of collecting, validating, and verifying information, processing and maintaining information, and disclosing and providing information; adjust the existing ways to comply with any previously applicable instructions and requirements; train personnel to be able to respond to a collection of information; search data sources; complete and review the collection of information; and transmit or otherwise disclose the information. An Agency may not conduct or sponsor, and a person is not required to respond to a collection of information unless it displays a currently valid OMB control number. The OMB control numbers for EPA's regulations are listed in 40 CFR part 9 and 48 CFR Chapter 15.

*H. Unfunded Mandates Reform Act*

Title II of the Unfunded Mandates Reform Act of 1995 (UMRA), Public Law 104–4, establishes requirements for Federal agencies to assess the effects of their regulatory actions on State, local, and tribal governments and the private sector. Under section 202 of the UMRA, EPA generally must prepare a written statement, including a cost-benefit analysis, for proposed and final rules with "Federal mandates" that may result in expenditures to State, local, and tribal governments, in the aggregate, or to the private sector, of $100 million or more in any one year. Before promulgating an EPA rule for which a written statement is needed, section 205 of the UMRA generally requires EPA to identify and consider a reasonable number of regulatory alternatives and adopt the least costly, most cost-effective or least burdensome alternative that achieves the objectives of the rule. The provisions of section 205 do not apply when they are inconsistent with applicable law. Moreover, section 205 allows EPA to adopt an alternative other than the least costly, most cost-effective or least burdensome alternative if the Administrator publishes with the final rule an explanation why that alternative was not adopted.

Before EPA establishes any regulatory requirements that may significantly or uniquely affect small governments, including tribal governments, it must have developed under section 203 of the UMRA a small government agency plan. The plan must provide for notifying potentially affected small governments, enabling officials of affected small governments to have meaningful and timely input in the development of EPA regulatory proposals with significant Federal intergovernmental mandates, and informing, educating, and advising small governments on compliance with the regulatory requirements.

Because the Agency has made a "good cause" finding that this action is not subject to notice-and-comment requirements under the Administrative Procedures Act or any or any other statute (see Section IV of this preamble), it is not subject to sections 202 and 205 of the Unfunded Mandates Reform Act of 1995 (UMRA) (Public Law 104–4).

Pursuant to Section 203 of UMRA, EPA has determined that this rule contains no regulatory requirements that might significantly or uniquely affect small governments. This rule does not contain any additional requirements, rather it reduces the burden on small governement sources that use flammable substances as fuel.

*I. National Technology Transfer and Advancement Act*

Section 12(d) of the National Technology Transfer and Advancement Act of 1995 ("NTTAA"), Public Law 104–113, section 12(d) (15 U.S.C. 272 note) directs EPA to use voluntary consensus standards in its regulatory activities unless to do so would be inconsistent with applicable law or otherwise impractical. Voluntary consensus standards are technical standards (*e.g.*, materials specifications, test methods, sampling procedures, and business practices) that are developed or adopted by voluntary consensus standards bodies. The NTTAA directs EPA to provide Congress, through OMB, explanations when the Agency decides not to use available and applicable voluntary consensus standards.

This action does not involve technical standards. Therefore, EPA did not consider the use of any voluntary consensus standards.

*J. Congressional Review Act*

The Congressional Review Act, 5 U.S.C. 801 *et seq.*, as added by the Small Business Regulatory Enforcement Fairness Act of 1996, generally provides that before a rule may take effect, the agency promulgating the rule must submit a rule report, which includes a copy of the rule, to each House of the Congress and to the Comptroller General of the United States. EPA will submit a report containing this rule and other required information to the U.S. Senate, the U.S. House of Representatives, and the Comptroller General of the United States prior to publication of the rule in the **Federal Register**. A "major rule" cannot take effect until 60 days after it is published in the **Federal Register**.

This action is not a "major rule" as defined by 5 U.S.C. 804(2). It takes effect today.

## List of Subjects in 40 CFR Part 68

Environmental protection, Chemicals, Chemical accident prevention.

Dated: March 3, 2000.

**Carol M. Browner,**

*Administrator.*

For the reasons stated in the preamble, EPA amends 40 CFR part 68 as follows:

## PART 68—[AMENDED]

1. The authority section for part 68 is revised to read as follows:

**Authority:** 42 U.S.C 7412(r), 7601 (a) (1).

## Subpart A—[Amended]

2. Section 68.3 is amended to add the following definition in alphabetical order:

## § 68.3   Definitions.

\*   \*   \*   \*   \*

*Retail facility* means a stationary source at which more than one-half of the income is obtained from direct sales to end users or at which more than one-half of the fuel sold, by volume, is sold through a cylinder exchange program.

\*   \*   \*   \*   \*

## Subpart F—[Amended]

3. Section 68.126 is added to subpart F to read as follows:

## § 68.126   Exclusion.

*Flammable Substances Used as Fuel or Held for Sale as Fuel at Retail Facilities.* A flammable substance listed in Tables 3 and 4 of § 68.130 is nevertheless excluded from all provisions of this part when the substance is used as a fuel or held for sale as a fuel at a retail facility.

4. Section 68.130 is amended by:
A. Revising the heading of Table 3;
B. Revising the notes to Table 3 and adding a new footnote 1;
C. Revising the heading to Table 4; and
D. Revising the notes to Table 4 and adding a new footnote 1.

The revisions and additions read as follows:

## § 68.130   List of substances.

\*   \*   \*   \*   \*

TABLE 3 TO § 68.130.—LIST OF REGU-LATED FLAMMABLE SUBSTANCES [1] AND THRESHOLD QUANTITIES FOR ACCIDENTAL RELEASE PREVENTION

[Alphabetical Order–63 Substances]

---

\*        \*        \*        \*        \*

---

[1] A flammable substance when used as a fuel or held for sale as a fuel at a retail facility is excluded from all provisions of this part (see § 68.126).

> **Note:** Basis for Listing:
> [a] Mandated for listing by Congress.
> [f] Flammable gas.
> [g] Volatile flammable liquid.

TABLE 4 TO § 68.130.—LIST OF REGU-LATED FLAMMABLE SUBSTANCES [1] AND THRESHOLD QUANTITIES FOR ACCIDENTAL RELEASE PREVENTION

[CAS Number Order–63 Substances]

---

\*        \*        \*        \*        \*

---

[1] A flammable substance when used as a fuel or held for sale as a fuel at a retail facility is excluded from all provisions of this part (see § 68.126).

> **Note:** Basis for Listing:
> [a] Mandated for listing by Congress.
> [f] Flammable gas.
> [g] Volatile flammable liquid.

[FR Doc. 00–5935 Filed 3–10–00; 8:45 am]

**BILLING CODE 6560–50–P**

---

## FEDERAL COMMUNICATIONS COMMISSION

### 47 CFR Part 73

[DA No. 00–494, MM Docket No. 99–256; RM–9527]

### Radio Broadcasting Services; Refugio and Taft, TX

**AGENCY:** Federal Communications Commission.

**ACTION:** Final rule.

**SUMMARY:** This document substitutes Channel 293C2 for Channel 291C3 at Refugio, Texas, reallots Channel 293C2 from Refugio, Texas, to Taft, Texas, and modifies the license for Station

KTKY(FM) to specify operation on Channel 293C2 at Taft in response to a petition filed by Pacific Broadcasting of Missouri, L.L.C. *See* 64 FR 39963, July 23, 1999. The coordinates for Channel 293C2 at Taft are 27–52–00 and 97–13–08. We shall also allot Channel 291A to Refugio, Texas, at coordinates 28–21–58 and 97–19–11. Mexican concurrence has been received for the allotments at Refugio and Taft, Texas. With this action, this proceeding is terminated.

**EFFECTIVE DATE:** April 17, 2000.

**FOR FURTHER INFORMATION CONTACT:** Kathleen Scheuerle, Mass Media Bureau, (202) 418–2180.

**SUPPLEMENTARY INFORMATION:** This is a summary of the Commission's Report and Order, MM Docket No. 99–256, adopted February 23, 2000, and released March 3, 2000. The full text of this Commission decision is available for inspection and copying during normal business hours in the Commission's Reference Center, 445 12th Street, SW, Washington, DC. The complete text of this decision may also be purchased from the Commission's copy contractors, International Transcription Services, Inc., 1231 20th Street, NW., Washington, DC 20036, (202) 857–3800, facsimile (202) 857–3805.

## List of Subjects in 47 CFR Part 73

Radio broadcasting.

Part 73 of title 47 of the Code of Federal Regulations is amended as follows:

## PART 73—[AMENDED]

1. The authority citation for Part 73 continues to read as follows:

**Authority:** 47 U.S.C. 154, 303, 334 and 336.

## § 73.202   [Amended]

2. Section 73.202(b), the Table of FM Allotments under Texas, is amended by removing Channel 291C3 and adding Channel 291A at Refugio and adding Taft, Channel 293C2.

Federal Communications Commission.

**John A. Karousos,**

*Chief, Allocations Branch, Policy and Rules Division, Mass Media Bureau.*

[FR Doc. 00–6052 Filed 3–10–00; 8:45 am]

**BILLING CODE 6712–01–U**

# Appendix B
## Selected NAICS Codes

# SELECTED 1997 NAICS CODES

**11 Agriculture**
11111 Soybean Farming
11113 Dry Pea and Bean Farming
11114 Wheat Farming
11115 Corn Farming
111191 Oilseed and Grain Farming
111199 All Other Grain Farming
111211 Potato Farming
111219 Other Vegetable and Melon Farming
11131 Orange Groves
11132 Other Citrus
111331 Apple Orchards
111332 Grape Vineyards
111339 Other Non Citrus Fruit Farming
111422 Floriculture Production
11191 Tobacco Farming
11192 Cotton Farming
11199 All Other Crop Farming
11211 Beef Cattle Ranching and Farming
11213 Dual Purpose Cattle Ranching and Farming
11221 Hog and Pig Farming
11231 Chicken Egg Production
11232 Broilers and Other Chicken Production
11233 Turkey Production
11234 Poultry Hatcheries
11239 Other Poultry Production
112511 Finfish Farming and Fish Hatcheries
11291 Apiculture
11299 All Other Animal Production
115111 Cotton Ginning
115112 Soil Preparation
115114 Post Harvest Crop Activities
11521 Support for Animal Production

**21 Mining**
211 Oil and Gas Extraction
211111 Crude Petroleum and Natural Gas Extraction
211112 Natural Gas Liquid Extraction
21211 Coal Mining
21221 Iron Ore Mining
21222 Gold and Silver Ore Mining
21223 Copper, Nickel, Lead, and Zinc Mining
21229 Other Metal Ore Mining
21231 Stone Mining and Quarrying
212322 Industrial Sand Mining
212324 Kaolin and Bal Clay Mining
21239 Other Non-Metallic Mineral Mining
21311 Support Activities for Mining

**22 Utilities**
22111 Electric Power Generation
221111 Hydroelectric Power Generation
221112 Fossil Fuel Electric Power Generation
221113 Nuclear Electric Power Generation

221119 Other Electric Power Generation
2213 Water, Sewage and Other Systems
22131 Water Supply and Irrigation Systems
22132 Sewage Treatment Facilities
22133 Steam and Air Conditioning Supply

**23 Constuction**

2333 Nonresidential Building Construction

**31-33 Manufacturing**
**311 Food Manufacturing**
3111 Animal Food Manufacturing
311111 Dog and Cat Food Manufacturing
311119 Other Animal Food Manufacturing
31121 Flour Milling and Malt Manufacturing
311211 Flour Milling
31122 Starch and Vegetable Fats and Oils Manufacturing
311221 Wet Corn Milling
311222 Soybean Processing
311223 Other Oilseed Processing
311225 Fats and Oils Refining and Blending
31123 Breakfast Cereal Manufacturing
311313 Beet Sugar Manufacturing
31132 Chocolate and Confectionery Manufacturing from Cacao Beans
31133 Confectionery Manufacturing from Purchased Chocolate
311411 Frozen Fruit, Juice and Vegetable Manufacturing
311412 Frozen Specialty Food Manufacturing
311421 Fruit and Vegetable Canning
311422 Specialty Canning
311423 Dried and Dehydrated Food Manufacturing
311511 Fluid Milk Manufacturing
311512 Creamery Butter Manufacturing
311513 Cheese Manufacturing
311514 Dry, Condensed, and Evaporated Dairy Product Manufacturing
31152 Ice Cream and Frozen Dessert Manufacturing
311611 Animal (except Poultry) Slaughtering
311612 Meat Processed from Carcasses
311613 Rendering and Meat By-product Processing
311615 Poultry Processing
311711 Seafood Canning
311712 Fresh and Frozen Seafood Processing
311811 Retail Bakeries
311812 Commercial Bakeries
311813 Frozen Cakes, Pies, and Other Pastries Manufacturing
311821 Cookie and Cracker Manufacturing
311822 Flour Mixes and Dough Manufacturing from Purchased Flour
311823 Dry Pasta Manufacturing
31191 Snack Food Manufacturing

311911   Roasted Nuts and Peanut Butter Manufacturing
311919   Other Snack Food Manufacturing
31192   Coffee and Tea Manufacturing
31193   Flavoring Syrup and Concentrate Manufacturing
311941      Mayonnaise, Dressing and Other Prepared
              Sauce Manufacturing
311991   Perishable Prepared Food Manufacturing
311999   All Other Miscellaneous Food Manufacturing

**312   Beverage and Tobacco Product Manufacturing**
312111   Soft Drink Manufacturing
312113   Ice Manufacturing
31212   Breweries
31213   Wineries
31214   Distilleries
31222   Tobacco Product Manufacturing

**313   Textile Mills**
313111   Yarn Spinning Mills
31323   Nonwoven Fabric Mills
31324   Knit Fabric Mills
313241   Weft Knit Fabric Mills
31331   Textile and Fabric Finishing Mills
313311   Broadwoven Fabric Finishing Mills

**314   Textile Product Mills**
31411   Carpet and Rug Mills
31499   All Other Textile Product  Mills
314992   Tire Cord and Tire Fabric Mills
314999   All Other Miscellaneous Textile Product Mills

**315   Apparel Manufacturing**
315111   Sheer Hosiery Mills
31522   Men's and Boys' Cut and Sew Apparel

**321   Wood Product Manufacturing**
321219      Reconstituted Wood Product Manufacturing

**322   Paper Manufacturing**
32211   Pulp Mills
32212   Paper Mills
322121   Paper (except Newsprint) Mills
322122   Newsprint Mills
32213   Paperboard Mills

**323   Printing and Related Support Activities**
323111   Commercial Gravure Printing
323117   Book Printing
323119      Other Commercial Printing

**324   Petroleum and Coal Products Manufacturing**
32411   Petroleum Refineries

324121      Asphalt Paving Mixture and Block
              Manufacturing

324191      Petroleum Lubricating Oil and Grease
              Manufacturing
324199      All Other Petroleum and Coal Products
              Manufacturing

**325   Chemical Manufacturing**
3251   Basic Chemical Manufacturing
32511   Petrochemical Manufacturing
32512   Industrial Gas Manufacturing
32513   Synthetic Dye and Pigment Manufacturing
325131   Inorganic Dye and Pigment Manufacturing
325132      Synthetic Organic Dye and Pigment
              Manufacturing
32518   Other Basic Inorganic Chemical Manufacturing
325181   Alkalies and Chlorine Manufacturing
325182   Carbon Black Manufacturing
325188      All Other Basic Inorganic Chemical
              Manufacturing
32519   Other Basic Organic Chemical Manufacturing
325191   Gum and Wood Chemical Manufacturing
325192   Cyclic Crude and Intermediate Manufacturing
325193   Ethyl Alcohol Manufacturing
325199      All Other Basic Organic Chemical
              Manufacturing
3252      Resin, Synthetic Rubber, and Artificial and
              Synthetic Fibers and Filaments Manufacturing
32521   Resin and Synthetic Rubber Manufacturing
325211   Plastics Material and Resin Manufacturing
325212   Synthetic Rubber Manufacturing
32522      Artificial and Synthetic Fibers and Filaments
              Manufacturing
325221      Cellulosic Organic Fiber Manufacturing
325222      Noncellulosic Organic Fiber Manufacturing
3253      Pesticide, Fertilizer and Other Agricultural
              Chemical Manufacturing
32531   Fertilizer Manufacturing
325311   Nitrogenous Fertilizer Manufacturing
325312   Phosphatic Fertilizer Manufacturing
325314   Fertilizer (Mixing Only) Manufacturing
32532      Pesticide and Other Agricultural Chemical
              Manufacturing
3254   Pharmaceutical and Medicine Manufacturing
32541   Pharmaceutical and Medicine Manufacturing
325411   Medicinal and Botanical Manufacturing
325412   Pharmaceutical Preparation Manufacturing
325413   In-Vitro Diagnostic Substance Manufacturing
325414      Biological Product (except Diagnostic)
              Manufacturing
3255   Paint, Coating, and Adhesive Manufacturing
32551   Paint and Coating Manufacturing
32552   Adhesive Manufacturing
3256      Soap, Cleaning Compound and Toilet Preparation
              Manufacturing
32561   Soap and Cleaning Compound Manufacturing
325611   Soap and Other Detergent  Manufacturing

325612    Polish and Other Sanitation Good Manufacturing
325613    Surface Active Agent Manufacturing
32562    Toilet Preparation Manufacturing
3259 Other Chemical Product Manufacturing
32591    Printing Ink Manufacturing
32592    Explosives Manufacturing
32599        All Other Chemical Product and Preparation
        Manufacturing
325991    Custom Compounding of Purchased Resin
325992        Photographic Film, Paper, Plate and Chemical
        Manufacturing
325998        All Other Miscellaneous Chemical Product
        and Preparation Manufacturing

## 326 Plastics and Rubber Products Manufacturing

32611        Unsupported Plastics Film, Sheet and Bag
        Manufacturing
326113        Unsupported Plastics Film and Sheet (except
        Packaging) Manufacturing
326121        Unsupported Plastics Profile Shape
        Manufacturing
32613        Laminated Plastics Plate, Sheet and Shape
        Manufacturing
32614    Polystyrene Foam Product Manufacturing
32615        Urethane and Other Foam Product (except
        Polystyrene) Manufacturing
32616    Plastics Bottle Manufacturing
32619    Other Plastics Product Manufacturing
326192    Resilient Floor Covering Manufacturing
326199    All Other Plastics Product Manufacturing
3262 Rubber Product Manufacturing
326211    Tire Manufacturing (except Retreading)
32629    Other Rubber Product Manufacturing
326299    All Other Rubber Product Manufacturing

## 327 Nonmetallic Mineral Product Manufacturing

32711        Pottery, Ceramics, and Plumbing Fixture
        Manufacturing
327111        Vitreous China Plumbing Fixtures and China
        and Earthenware Bathroom Accessories
        Manufacturing
327125        Nonclay Refractory Manufacturing
32721 Glass and Glass Product Manufacturing
327211        Flat Glass Manufacturing
327212        Other Pressed and Blown Glass and Glassware
        Manufacturing
327213        Glass Container Manufacturing
327215        Glass Product Manufacturing Made of
        Purchased Glass
32731    Cement Manufacturing
32732    Ready-Mix Concrete Manufacturing
32739    Other Concrete Product Manufacturing
32742    Gypsum Product Manufacturing
32791    Abrasive Product Manufacturing
327992        Ground or Treated Mineral and Earth

Manufacturing
327993        Mineral Wool Manufacturing
327999        All Other Miscellaneous Nonmetallic Mineral
        Product Manufacturing

## 331 Primary Metal Manufacturing

33111    Iron and Steel Mills and Ferroalloy Manufacturing
331111    Iron and Steel Mills
331312        Primary Aluminum Production
331314        Secondary Smelting and Alloying of Aluminum
331315        Aluminum Sheet, Plate and Foil Manufacturing
331316        Aluminum Extruded Product Manufacturing
331319        Other Aluminum Rolling and Drawing
33141    Nonferrous Metal (except Aluminum) Smelting
        and Refining
331411    Primary Smelting and Refining of Copper
331419        Primary Smelting and Refining of Nonferrous
        Metal (except Copper and Aluminum)
331421        Copper Rolling, Drawing and Extruding
331423        Secondary Smelting, Refining, and Alloying of
        Copper
33149        Nonferrous Metal (except Copper and
        Aluminum) Rolling, Drawing, Extruding and
        Alloying
331491        Nonferrous Metal (except Copper and
        Aluminum) Rolling, Drawing and Extruding
331492        Secondary Smelting, Refining, and Alloying of
        Nonferrous Metal (except Copper and
        Aluminum)
33151    Ferrous Metal Foundries
331511    Iron Foundries
331513    Steel Foundries, (except Investment)
33152    Nonferrous Metal Foundries
331521    Aluminum Die-Casting Foundries
331522        Nonferrous (except Aluminum) Die-Casting
        Foundries
331524    Aluminum Foundries (except Die-Casting)
331525    Copper Foundries (except Die-Casting)
331528        Other Nonferrous Foundries (except Die-
        Casting)

## 332 Fabricated Metal Product Manufacturing

33211    Forging and Stamping
332111    Iron and Steel Forging
332112    Nonferrous Forging
332116    Metal Stamping
332117    Powder Metallurgy Part Manufacturing
33221    Cutlery and Hand Tool Manufacturing
332211        Cutlery and Flatware (except Precious)
        Manufacturing
332321    Metal Window and Door Manufacturing
332322    Sheet Metal Work Manufacturing
33243        Metal Can, Box, and Other Metal Container
        (Light Gauge) Manufacturing
33251    Hardware Manufacturing

332612   Spring (Light Gauge) Manufacturing

33281   Coating, Engraving, Heat Treating, and Allied Activities

332811   Metal Heat Treating

332812   Metal Coating, Engraving (except Jewelry and Silverware), and Allied Services to Manufacturers

332813   Electroplating, Plating, Polishing, Anodizing and Coloring

332912   Fluid Power Valve and Hose Fitting Manufacturing

332919   Other Metal Valve and Pipe Fitting Manufacturing

33299   All Other Fabricated Metal Product Manufacturing

332991   Ball and Roller Bearing Manufacturing

332992   Small Arms Ammunition Manufacturing

332999   All Other Miscellaneous Fabricated Metal Product Manufacturing

## 333 Machinery Manufacturing

33311   Agricultural Implement Manufacturing

333111   Farm Machinery and Equipment Manufacturing

333112   Lawn and Garden Tractor and Home Lawn and Garden Equipment Manufacturing

33312   Construction Machinery Manufacturing

333295   Semiconductor Machinery Manufacturing

333298   All Other Industrial Machinery Manufacturing

333311   Automatic Vending Machine Manufacturing

333314   Optical Instrument and Lens Manufacturing

333315   Photographic and Photocopying Equipment Manufacturing

333319   Other Commercial and Service Industry Machinery Manufacturing

333415   Air-Conditioning and Warm Air Heating Equipment and Commercial and Industrial Refrigeration Equipment Manufacturing

33351   Metalworking Machinery Manufacturing

333511   Industrial Mold Manufacturing

333512   Machine Tool (Metal Cutting Types) Manufacturing

333515   Cutting Tool and Machine Tool Accessory Manufacturing

333611   Turbine and Turbine Generator Set Unit Manufacturing

333613   Mechanical Power Transmission Equipment Manufacturing

333618   Other Engine Equipment Manufacturing

333911   Pump and Pumping Equipment Manufacturing

333924   Industrial Truck, Tractor, Trailer and Stacker Machinery Manufacturing

333995   Fluid Power Cylinder and Actuator Manufacturing

333996   Fluid Power Pump and Motor Manufacturing

333999   All Other Miscellaneous General Purpose Machinery Manufacturing

## 334 Computer and Electronic Product Manufacturing

33411   Computer and Peripheral Equipment Manufacturing

334111   Electronic Computer Manufacturing

334112   Computer Storage Device Manufacturing

334113   Computer Terminal Manufacturing

334119   Other Computer Peripheral Equipment Manufacturing

33422   Radio and Television Broadcasting and Wireless Communications Equipment Manufacturing

33441   Semiconductor and Other Electronic Component Manufacturing

334411   Electron Tube Manufacturing

334412   Bare Printed Circuit Board Manufacturing

334413   Semiconductor and Related Device Manufacturing

334414   Electronic Capacitor Manufacturing

334415   Electronic Resistor Manufacturing

334416   Electronic Coil, Transformer, and Other Inductor Manufacturing

334417   Electronic Connector Manufacturing

334418   Printed Circuit Assembly (Electronic Assembly) Manufacturing

334419   Other Electronic Component Manufacturing

334519   Other Measuring and Controlling Device Manufacturing

334613   Magnetic and Optical Recording Media Manufacturing

## 335 Electrical Equipment, Appliance and Component Manufacturing

33511   Electric Lamp Bulb and Part Manufacturing

335122   Commercial, Industrial and Institutional Electric Lighting Fixture Manufacturing

335129   Other Lighting Equipment Manufacturing

33522   Major Appliance Manufacturing

335222   Household Refrigerator and Home Freezer Manufacturing

33531   Electrical Equipment Manufacturing

335311   Power, Distribution and Specialty Transformer Manufacturing

335312   Motor and Generator Manufacturing

33591   Battery Manufacturing

335911   Storage Battery Manufacturing

335912   Primary Battery Manufacturing

335921   Fiber Optic Cable Manufacturing

33599   All Other Electrical Equipment and Component Manufacturing

335991   Carbon and Graphite Product Manufacturing

335999   All Other Miscellaneous Electrical Equipment and Component Manufacturing

## 336 Transportation Equipment Manufacturing

33611   Automobile and Light Duty Motor Vehicle

Manufacturing
336111 Automobile Manufacturing
336112 Light Truck and Utility Vehicle Manufacturing
33612 Heavy Duty Truck Manufacturing
33621 Motor Vehicle Body and Trailer Manufacturing
336211 Motor Vehicle Body Manufacturing
336212 Truck Trailer Manufacturing
336213 Motor Home Manufacturing
336214 Travel Trailer and Camper Manufacturing
33631 Motor Vehicle Gasoline Engine and Engine Parts Manufacturing
336311 Carburetor, Piston, Piston Ring and Valve Manufacturing
336312 Gasoline Engine and Engine Parts Manufacturing
33632 Motor Vehicle Electrical and Electronic Equipment Manufacturing
336321 Vehicular Lighting Equipment Manufacturing
336322 Other Motor Vehicle Electrical and Electronic Equipment Manufacturing
33633 Motor Vehicle Steering and Suspension Components (except Spring) Manufacturing
33634 Motor Vehicle Brake System Manufacturing
33635 Motor Vehicle Transmission and Power Train Parts Manufacturing
33636 Motor Vehicle Seating and Interior Trim Manufacturing
33637 Motor Vehicle Metal Stamping
33639 Other Motor Vehicle Parts Manufacturing
336391 Motor Vehicle Air-Conditioning Manufacturing
336399 All Other Motor Vehicle Parts Manufacturing
33641 Aerospace Product and Parts Manufacturing
336411 Aircraft Manufacturing
336412 Aircraft Engine and Engine Parts Manufacturing
336413 Other Aircraft Part and Auxiliary Equipment Manufacturing
336414 Guided Missile and Space Vehicle Manufacturing
336415 Guided Missile and Space Vehicle Propulsion Unit and Propulsion Unit Parts Manufacturing
336419 Other Guided Missile and Space Vehicle Parts and Auxiliary Equipment Manufacturing
33651 Railroad Rolling Stock Manufacturing
33661 Ship and Boat Building
336611 Ship Building and Repairing
336612 Boat Building
33699 Other Transportation Equipment Manufacturing
336991 Motorcycle, Bicycle and Parts Manufacturing
336992 Military Armored Vehicle, Tank and Tank Component Manufacturing
336999 All Other Transportation Equipment Manufacturing

**337 Furniture and Related Product Manufacturing**
33712 Household and Institutional Furniture Manufacturing

337211 Wood Office Furniture Manufacturing

**339 Miscellaneous Manufacturing**
33911 Medical Equipment and Supplies Manufacturing
339112 Surgical and Medical Instrument Manufacturing
339113 Surgical Appliance and Supplies Manufacturing
339114 Dental Equipment and Supplies Manufacturing
3399 Other Miscellaneous Manufacturing
33991 Jewelry and Silverware Manufacturing
339911 Jewelry (except Costume) Manufacturing
339912 Silverware and Plated Ware Manufacturing
339913 Jewelers' Material and Lapidary Work Manufacturing
339914 Costume Jewelry and Novelty Manufacturing
339991 Gasket, Packing, and Sealing Device Manufacturing
339994 Broom, Brush and Mop Manufacturing
339999 All Other Miscellaneous Manufacturing

**42 Wholesale Trade**
421 Wholesale Trade, Durable Goods
42149 Other Professional Equipment and Supplies
42171 Hardware Wholesalers
42181 Construction and Mining Machinery
42184 Industrial Supplies

**422 Wholesale Trade, Nondurable Goods**
42211 Printing and Writing Paper Wholesalers
4224 Grocery and Related Product Wholesalers
42241 General Line Grocery Wholesalers
42242 Packaged Frozen Food Wholesalers
42243 Dairy Product (except Dried or Canned) Wholesalers
42244 Poultry and Poultry Product Wholesalers
42246 Fish and Seafood Wholesalers
42247 Meat and Meat Product Wholesalers
42248 Fresh Fruit and Vegetable Wholesalers
42249 Other Grocery and Related Products Wholesalers
4225 Farm Product Raw Material Wholesalers
42251 Grain and Field Bean Wholesalers
42252 Livestock Wholesalers
42259 Other Farm Product Raw Material Wholesalers
4226 Chemical and Allied Products Wholesalers
42261 Plastics Materials and Basic Forms and Shapes Wholesalers
42269 Other Chemical and Allied Products Wholesalers
42271 Petroleum Bulk Stations and Terminals
42272 Petroleum and Petroleum Products Wholesalers (except Bulk Stations and Terminals)
42281 Beer and Ale Wholesalers
42282 Wine and Distilled Alcoholic Beverage Wholesalers
4229 Miscellaneous Nondurable Goods Wholesalers
42291 Farm Supplies Wholesalers
42299 Other Miscellaneous Nondurable Goods

**44-45   Retail Trade**
4411 Automobline Dealers
442291Window Treatment Stores
4441  Building Material and Supplies Dealers
44422  Nursery and Garden Centers
44511 Grocery Stores
44523 Fruit and Vegetable Markets
44711 Gasoline Stations
45291 Warehouse Clubs and Superstores
45399 All Other Miscellaneous Store Retailers

**48-49   Transportation and Warehousing**
**488  Support Activities for Transportation**
48211 Rail Transportation
48311 Water Transportation
4842  Specialized Freight Trucking
48511 Urban Transit Systems
486 Pipeline Transportation
48811  Airport Operations
488119  Other Airport Operations
48819  Other Support Activities for Air Transportation
48821  Support Activities for Rail Transportation
48832  Marine Cargo Handling
48839  Other Support Activities for Water Transportation

**493  Warehousing and Storage**
49311  General  Warehousing and Storage
49312  Refrigerated Warehousing and Storage
49313  Farm Product Warehousing and Storage
49319  Other Warehousing and Storage

**54  Professional, Scientific, and Technical Services**
54138 Testing Labs
54171    Research and Development in the Physical,
            Engineering, and Life Sciences

**56 Administrative and Support , Waste Management
and Remediation Services**
561431 Private Mail Centers
56179 Other Services to Buildings
56221  Waste Treatment and Disposal
562211      Hazardous Waste Treatment and Disposal
562212      Solid Waste Landfill
562213      Solid Waste Combustors and Incinerators
562219       Other Nonhazardous Waste Treatment and
            Disposal
5629      Remediation  and Other Waste Management
            Services
56291   Remediation Services
56292   Materials Recovery Facilities
56299   All Other Waste Management Services
562998       All Other Miscellaneous Waste Management
            Services

**61  Educational Services**
6111 Elementary and Secondary Schools
61131 Colleges, Universities, Professional Schools

**62 Health Care and Social Assistance**
62151   Medical and Diagnostic Laboratories
621511   Medical Laboratories
62211   General Medical and Surgical Hospitals
6222  Psychiatric and Substance Abuse Hospitals
62221   Psychiatric and Substance Abuse Hospitals
6223      Specialty (except Psychiatric and Substance
            Abuse) Hospitals
62231      Specialty (except Psychiatric and Substance
            Abuse) Hospitals

# APPENDIX C: TECHNICAL ASSISTANCE

This appendix outlines the resources that are available to warehousing facilities in complying with 40 CFR part 68, organized by the group providing the assistance.

## U.S. ENVIRONMENTAL PROTECTION AGENCY

EPA's Chemical Emergency Preparedness and Prevention Office (CEPPO) administers the RMP program at the national level.

| | |
|---|---|
| Street Address: | 1200 Pennsylvania Ave, NW (Mailcode 5104) Washington, DC 20460 |
| Phone Number: | (202) 260-8600 |
| WWW Address: | www.epa.gov/ceppo/ |

The **CEPPO homepage** on the Internet provides access to downloadable versions of numerous risk management program documents, including factsheets, questions and answer documents, the list of regulated substances and thresholds, the text of the regulatory requirements, general and industry-specific guidance, risk communication, and RMP data elements and instructions. It also has links to sources of Material Safety Data Sheets and other hazardous chemical information, as well as up-to-date lists of LEPCs and SERCs.

If you do not have Internet access, these documents can be ordered from EPA's **EPCRA/Superfund/RCRA/CAA Hotline**. In addition, the Hotline responds to factual questions on a variety of federal EPA regulations, including those developed under Clean Air Act section 112(r).

| | |
|---|---|
| Phone Number: | Toll-Free: (800) 424-9346 Local: (703) 412-9810 TDD: (800) 553-7672 TDD Local: (703) 412-3323 Monday - Friday, 9:00 am - 6:00 pm EST |
| E-Mail: | epahotline@bah.com |

The Clean Air Act Amendments of 1990 requires that all States develop a program to assist small businesses in meeting the requirements of the Act. EPA has established its own Small Business Assistance Program (SBAP) to provide technical assistance to these State small business programs. The SBAP Internet site at **www.epa.gov/ttn/sbap** was developed to allow State and EPA programs to share information about their small business assistance materials and activities.

## OCCUPATIONAL SAFETY AND HEALTH ADMINISTRATION (OSHA)

OSHA administers the Process Safety Management Standard (29 CFR 1910.119), which mandates actions similar to that of EPA's prevention program. In about half of the states, OSHA programs are run by state agencies. For specific points of contact for OSHA regional offices, OSHA state consultative programs, and OSHA state plan states, see the OSHA web site.

### OSHA Process Safety Management Homepage
www.osha-slc.gov/SLTC/processsafetymanagement/index.html

The PSM homepage on the Internet provides access to downloadable versions of numerous process safety management documents.

The **OSHA Publications Office** provides single copies of various documents related to risk management, process safety, accident prevention, and emergency planning and response.

Street Address:       U.S. Department of Labor
Room N3101
200 Constitution Avenue, NW
Washington, DC  20210

Phone Number:      (202) 523-9667

WWW Address:      www.osha.gov

The **OSHA Computerized Information System** at www.osha-slc.gov/ provides downloadable versions of OSHA Standards and related documents, including OSHA Regulations, Federal Register notices, Interpretations and Compliance Letters, OSHA Regulations (preambles to final rules), and OSHA Directives and Fact Sheets. OSHA also publishes a quarterly CD-ROM with all of this information, available for Windows and Macintosh computers from:

> U.S. Government Printing Office
> Stock # 729-013-00000-5
> Phone:  (202) 512-1800
> Fax:  (202) 512-2250
> Price:  $38/year (four quarterly releases), $15 (single copy)

**OSHA's Small Business Outreach Training Program** has prepared an Instructional Guide on various topics in occupational safety and health, designed to provide ideas and organizational assistance to an instructor who wishes to present these topics, which specifically focus on the needs of small business. The text of the Guide is available at www.osha-slc.gov/SLTC/SmallBusiness.

Employers who take advantage of **OSHA's Consultation Services** can find out about potential hazards at their worksites, improve their occupational safety and health management systems, and even qualify for a one-year exemption from routine OSHA inspections. Targeted for smaller businesses, this safety and health consultation program is completely separate from the OSHA inspection effort. For more information, check in at www.osha.gov/oshprogs/consult.html.

## AMERICAN INSTITUTE OF CHEMICAL ENGINEERS (AIChE)

| | |
|---|---|
| Street Address: | 345 E. 47th St. |
| | New York, NY 10017-2395 |
| Phone Number: | (212) 705-7338 |
| WWW Address: | www.aiche.org/ |

AIChE and its Center for Chemical Process Safety publish a variety of documents including a Continuing Education catalog for its educational and training programs and an annual Publications Catalog from which documents can be purchased. They are available from:

**AIChExpress Service Center**
(800) AIC-HEME (242-4363)
Monday - Friday, 9:00 am - 5:00 pm EST
E-Mail: xpress@aiche.org

Some of the specific documents available include:

Guidelines for Safe Storage and Handling of Highly Toxic Hazardous Materials (119 pages)
Publication G-3
Order No. 0-8169-0400-6
Price: $65

Guidelines for Safe Warehousing of Chemicals (forthcoming)
Publication G-33
Order No. 0-8169-0659-9

## CHEMICAL MANUFACTURERS ASSOCIATION (CMA)

Street Address:        1300 Wilson Blvd.
                       Arlington, VA 22209

Phone Number:          (703) 741-5000

WWW Address:           www.cmahq.com/

CMA documents, including those listed below, are available from:

**CMA Publications Fulfillment Office**
341 Victory Drive
Herndon, VA 22070
Phone: (703) 709-0166

CMA Responsible Care: Handling and Storage: Warehouse Assessment
Protocol

Community Awareness and Emergency Response (CAER) Code Resource
Guide
Order No. 024012
Price: $50 members, $75 non-members

Safe Warehousing of Chemicals
Order No. 022003
Price: $10 members, $15 non-members

CMA also maintains **Chemtrec (Chemical Transportation Emergency Center)** to
provide a centralized information and assistance center for individuals responding to
chemical emergencies and carriers of hazardous materials:

Emergency: (800) 424-9300
Non-Emergency: (800) 262-8200

## INTERNATIONAL WAREHOUSE LOGISTICS ASSOCIATION

Street Address:        1300 W. Higgins Rd., Ste. 111
                       Park Ridge, IL 60068

Phone Number:          (847) 292-1891

WWW address            www.warehouselogistics.org/

## NATIONAL ASSOCIATION OF CHEMICAL DISTRIBUTORS

Street Address:        Chemical Educational Foundation
                       1525 Wilson Blvd., Ste. 750
                       Arlington, VA 22209

Phone Number:          (703) 527-6223

Fax Number:            (703) 527-7747

WWW Address:           www.nacd.com/

The **Chemical Educational Foundation** publishes an Educational Aids and Training Catalog and a Product Stewardship Resource Guide. CEF has published the following document related to risk management:

Making It Easy: Community Outreach Ideas and Examples

NACD Product Stewardship Bulletin #1: Chemical Use, Handling, Storage, and Transportation

NACD Product Stewardship Bulletin #5: Employee Hazardous Materials Safety Training

NACD Product Stewardship Bulletin #8: Risk Management Program

## NATIONAL FIRE PROTECTION ASSOCIATION

Street Address:        1 Batterymarch Park
                       P.O. Box 9101
                       Quincy, MA 02269-9101

Phone Number:          (617) 770-3000

Fax Number:            (617) 770-0700

E-mail:                library@NFPA.org.

WWW Address:           www.nfpa.org/

NFPA publishes standards related to fire safety, prevention, training, planning, and response that have been adopted as the official fire code in many states, as well as guidance on how to implement its standards. The collection of NFPA Standards is available at many libraries, but NFPA also publishes a bimonthly catalog of its standards and fire safety products, which are available from:

**NFPA Fulfillment Center**
11 Tracy Drive
Avon, MA  02322-9908
Phone:  (800) 344-3555
Monday - Friday, 8:30 am - 8:00 pm EST
Fax:  (800) 593-NFPA (6372)
E-mail:  custserv@NFPA.org.

A listing of these products and subscription service information (as well as a Spanish version of the catalog) is provided at the NFPA homepage.

## SMALL BUSINESS ADMINISTRATION

Street Address:          409 Third Street, SW
                         Washington, DC  20416

Phone Number:            (800) 827-5722

WWW Address:             www.sba.gov/

SBA was created to help America's entrepreneurs form successful small enterprises. SBA's program offices in every state offer financing, training and advocacy for small firms.  In addition, the SBA works with thousands of lending, educational, and training institutions nationwide.

**APPENDIX D**
**OSHA GUIDANCE ON PSM**

# APPENDIX D
# OSHA GUIDANCE ON PSM

The following text is taken directly from OSHA's non-mandatory appendix C to the PSM standard (29 CFR 1910.119). The only change has been to rearrange the sections to track the order of part 68.

## PROCESS SAFETY INFORMATION

Complete and accurate written information concerning process chemicals, process technology, and process equipment is essential to an effective process safety management program and to a process hazards analysis. The compiled information will be a necessary resource to a variety of users including the team that will perform the process hazards analysis; those developing the training programs and the operating procedures; contractors whose employees will be working with the process; those conducting the pre-startup reviews; local emergency preparedness planners; and insurance and enforcement officials.

The information to be compiled about the chemicals, including process intermediates, needs to be comprehensive enough for an accurate assessment of the fire and explosion characteristics, reactivity hazards, the safety and health hazards to workers, and the corrosion and erosion effects on the process equipment and monitoring tools. Current material safety data sheet (MSDS) information can be used to help meet this requirement, which must be supplemented with process chemistry information including runaway reaction and over pressure hazards if applicable.

Process technology information will be a part of the process safety information package and it is expected that it will include diagrams as well as employer established criteria for maximum inventory levels for process chemicals; limits beyond which would be considered upset conditions; and a qualitative estimate of the consequences or results of deviation that could occur if operating beyond the established process limits. Employers are encouraged to use diagrams which will help users understand the process.

A block flow diagram is used to show the major process equipment and interconnecting process flow lines and show flow rates, stream composition, temperatures, and pressures when necessary for clarity. The block flow diagram is a simplified diagram.

Process flow diagrams are more complex and will show all main flow streams including valves to enhance the understanding of the process, as well as pressures and temperatures on all feed and product lines within all major vessels, in and out of headers and heat exchangers, and points of pressure and temperature control. Also, materials of construction information, pump capacities and pressure heads, compressor horsepower and vessel design pressures and temperatures are shown when necessary for clarity. In addition, major components of control loops are usually shown along with key utilities on process flow diagrams.

Piping and instrument diagrams (P&IDS) may be the more appropriate type of diagrams to show some of the above details and to display the information for the piping designer and engineering staff. The P&IDS are to be used to describe the relationships between equipment and instrumentation as well as other relevant information that will enhance clarity. Computer software programs which do P&IDS or other diagrams useful to the information package, may be used to help meet this requirement.

The information pertaining to process equipment design must be documented. In other words, what were the codes and standards relied on to establish good engineering practice. These codes and standards are published by such organizations as the American Society of Mechanical Engineers, American Petroleum Institute, American National Standards Institute, National Fire Protection Association, American Society for Testing and Materials, National Board of Boiler and Pressure Vessel Inspectors, National Association of Corrosion Engineers, American Society of Exchange Manufacturers Association, and model building code groups. In addition, various engineering societies issue technical reports which impact process design. For example, the American Institute of Chemical Engineers has published technical reports on topics such as two phase flow for venting devices. This type of technically recognized report would constitute good engineering practice.

For existing equipment designed and constructed many years ago in accordance with the codes and standards available at that time and no longer in general use today, the employer must document which codes and standards were used and that the design and construction along with the testing, inspection and operation are still suitable for the intended use. Where the process technology requires a design which departs from the applicable codes and standards, the employer must document that the design and construction is suitable for the intended purpose.

## PROCESS HAZARD ANALYSIS

A process hazard analysis (PHA), sometimes called a process hazard evaluation, is one of the most important elements of the process safety management program. A PHA is an organized and systematic effort to identify and analyze the significance of potential hazards associated with the processing or handling of highly hazardous chemicals. A PHA provides information which will assist employers and employees in making decisions for improving safety and reducing the consequences of unwanted or unplanned releases of hazardous chemicals.

A PHA is directed toward analyzing potential causes and consequences of fires, explosions, releases of toxic or flammable chemicals and major spills of hazardous chemicals. The PHA focuses on equipment, instrumentation, utilities, human actions (routine and non-routine), and external factors that might impact the process. These considerations assist in determining the hazards and potential failure points or failure modes in a process.

The selection of a PHA methodology or technique will be influenced by many factors including the amount of existing knowledge about the process. Is it a process that has been operated for a long period of time with little or no innovation and extensive experience has been generated with its use? Or, is it a new process or one which has been changed frequently by the inclusion of innovative features? Also, the size and complexity of the process will influence the decision as to the appropriate PHA methodology to use. All PHA methodologies are subject to certain limitations. For example, the checklist methodology works well when the process is very stable and no changes are made, but it is not as effective when the process has undergone extensive change. The checklist may miss the most recent changes and consequently the changes would not be evaluated. Another limitation to be considered concerns the assumptions made by the team or analyst. The PHA is dependent on good judgment and the assumptions made during the study need to be documented and understood by the team and reviewer and kept for a future PHA.

The team conducting the PHA need to understand the methodology that is going to be used. A PHA team can vary in size from two people to a number of people with varied operational and technical backgrounds. Some team members may only be a part of the team for a limited time. The team leader needs to be fully knowledgeable in the proper implementation of the PHA methodology that is to be used and should be impartial in the evaluation. The other full or part time team members need to provide the team with expertise in areas such as process technology, process design, operating procedures and practices, including how the work is actually performed, alarms, emergency procedures, instrumentation, maintenance procedures, both routine and non-routine tasks, including how the tasks are authorized, procurement of parts and supplies, safety and health, and any other relevant subject as the need dictates. At least one team member must be familiar with the process.

The ideal team will have an intimate knowledge of the standards, codes, specifications and regulations applicable to the process being studied. The selected team members need to be compatible and the team leader needs to be able to manage the team and the PHA study. The team needs to be able to work together while benefiting from the expertise of others on the team or outside the team, to resolve issues, and to forge a consensus on the findings of the study and the recommendations.

The application of a PHA to a process may involve the use of different methodologies for various parts of the process. For example, a process involving a series of unit operations of varying sizes, complexities, and ages may use different methodologies and team members for each operation. Then the conclusions can be integrated into one final study and evaluation.

A more specific example is the use of a checklist PHA for a standard boiler or heat exchanger and the use of a Hazard and Operability PHA for the overall process. Also, for batch type processes like custom batch operations, a generic PHA of a representative batch may be used where there are only small changes of monomer or other ingredient ratios and the chemistry is documented for the full range and ratio of batch ingredients. Another process that might consider using a generic type of PHA is a gas plant. Often these plants are simply moved from site to site and therefore, a generic PHA may be used for these movable plants. Also, when an employer has several similar size gas plants and no sour gas is being processed at the site, then a generic PHA is feasible as long as the variations of the individual sites are accounted for in the PHA.

Finally, when an employer has a large continuous process which has several control rooms for different portions of the process such as for a distillation tower and a blending operation, the employer may wish to do each segment separately and then integrate the final results.

Additionally, small businesses which are covered by this rule, will often have processes that have less storage volume, less capacity, and less complicated than processes at a large facility. Therefore, OSHA would anticipate that the less complex methodologies would be used to meet the process hazard analysis criteria in the standard. These process hazard analyses can be done in less time and with a few people being involved. A less complex process generally means that less data, P&IDS, and process information is needed to perform a process hazard analysis.

Many small businesses have processes that are not unique, such as cold storage lockers or water treatment facilities. Where employer associations have a number of members with such facilities, a generic PHA, evolved from a checklist or what-if questions, could be developed and used by each employer effectively to reflect his/her particular process; this would simplify compliance for them.

When the employer has a number of processes which require a PHA, the employer must set up a priority system of which PHAs to conduct first. A preliminary or gross hazard analysis may be useful in prioritizing the processes that the employer has determined are subject to coverage by the process safety management standard. Consideration should first be given to those processes with the potential of adversely affecting the largest number of employees. This prioritizing should consider the potential severity of a chemical release, the number of potentially affected employees, the operating history of the process such as the frequency of chemical releases, the age of the process and any other relevant factors. These factors would suggest a ranking order and would suggest either using a weighing factor system or a systematic ranking method. The use of a preliminary hazard analysis would assist an employer in determining which process should be of the highest priority and thereby the employer would obtain the greatest improvement in safety at the facility.

## OPERATING PROCEDURES

Operating procedures describe tasks to be performed, data to be recorded, operating conditions to be maintained, samples to be collected, and safety and health precautions to be taken. The procedures need to be technically accurate, understandable to employees, and revised periodically to ensure that they reflect current operations. The process safety information package is to be used as a resource to better assure that the operating procedures and practices are consistent with the known hazards of the chemicals in the process and that the operating parameters are accurate. Operating procedures should be reviewed by engineering staff and operating personnel to ensure that they are accurate and provide practical instructions on how to actually carry out job duties safely.

Operating procedures will include specific instructions or details on what steps are to be taken or followed in carrying out the stated procedures. These operating instructions for each procedure should include the applicable safety precautions and should contain appropriate information on safety implications. For example, the operating procedures addressing operating parameters will contain operating instructions about pressure limits, temperature ranges, flow rates, what to do when an upset condition occurs, what alarms and instruments are pertinent if an upset condition occurs, and other subjects. Another example of using operating instructions to properly implement operating procedures is in starting up or shutting down the process. In these cases, different parameters will be required from those of normal operation. These operating instructions need to clearly indicate the distinctions between startup and normal operations such as the appropriate allowances for heating up a unit to reach the normal operating parameters. Also the operating instructions need to describe the proper method for increasing the temperature of the unit until the normal operating temperature parameters are achieved.

Computerized process control systems add complexity to operating instructions. These operating instructions need to describe the logic of the software as well as the relationship between the equipment and the control system; otherwise, it may not be apparent to the operator.

Operating procedures and instructions are important for training operating personnel. The operating procedures are often viewed as the standard operating practices (SOPs) for operations. Control room personnel and operating staff, in general, need to have a full understanding of operating procedures. If workers are not fluent in English then procedures and instructions need to be prepared in a second language understood by the workers. In addition, operating procedures need to be changed when there is a change in the process as a result of the management of change procedures. The consequences of operating procedure changes need to be fully evaluated and the information conveyed to the personnel.

For example, mechanical changes to the process made by the maintenance department (like changing a valve from steel to brass or other subtle changes) need to be evaluated to determine if operating procedures and practices also need to be changed. All management of change actions must be coordinated and integrated with current operating procedures and operating personnel must be oriented to the changes in procedures before the change is made. When the process is shutdown to make a change, then the operating procedures must be updated before startup of the process.

Training in how to handle upset conditions must be accomplished as well as what operating personnel are to do in emergencies such as when a pump seal fails or a pipeline ruptures. Communication between operating personnel and workers performing work within the process area, such as non-routine tasks, also must be maintained. The hazards of the tasks are to be conveyed to operating personnel in accordance with established procedures and to those performing the actual tasks. When the work is completed, operating personnel should be informed to provide closure on the job.

## TRAINING

All employees, including maintenance and contractor employees, involved with highly hazardous chemicals need to fully understand the safety and health hazards of the chemicals and processes they work with for the protection of themselves, their fellow employees and the citizens of nearby communities. Training conducted in compliance with 1910.1200, the Hazard Communication standard, will help employees to be more knowledgeable about the chemicals they work with as well as familiarize them with reading and understanding MSDS. However, additional training in subjects such as operating procedures and safety work practices, emergency evacuation and response, safety procedures, routine and non-routine work authorization activities, and other areas pertinent to process safety and health will need to be covered by an employer's training program.

In establishing their training programs, employers must clearly define the employees to be trained and what subjects are to be covered in their training. Employers in setting up their training program will need to clearly establish the goals and objectives they wish to achieve with the training that they provide to their employees. The learning goals or objectives should be written in clear measurable terms before the training begins. These goals and objectives need to be tailored to each of the specific training modules or segments. Employers should describe the important actions and conditions under which the employee will demonstrate competence or knowledge as well as what is acceptable performance.

Hands-on-training where employees are able to use their senses beyond listening, will enhance learning. For example, operating personnel, who will work in a control room or at control panels, would benefit by being trained at a simulated control panel or panels. Upset conditions of various types could be displayed on the simulator, and then the employee could go through the proper operating procedures to bring the simulator panel back to the normal operating parameters. A training environment could be created to help the trainee feel the full reality of the situation but, of course, under controlled conditions. This realistic type of training can be very effective in teaching employees correct procedures while allowing them to also see the consequences of what might happens if they do not follow established operating procedures. Other training techniques using videos or on-the-job training can also be very effective for teaching other job tasks, duties, or other important information. An effective training program will allow the employee to fully participate in the training process and to practice their skill or knowledge.

Employers need to periodically evaluate their training programs to see if the necessary skills, knowledge, and routines are being properly understood and implemented by their trained employees. The means or methods for evaluating the training should be developed along with the training program goals and objectives. Training program evaluation will help employers to determine the amount of training their employees understood, and whether the desired results were obtained. If, after the evaluation, it appears that the trained employees are not at the level of knowledge and skill that was expected, the employer will need to revise the training program, provide retraining, or provide more frequent refresher training sessions until the deficiency is resolved. Those who conducted the training and those who received the training should also be consulted as to how best to improve the training process. If there is a language barrier, the language known to the trainees should be used to reinforce the training messages and information.

Careful consideration must be given to assure that employees including maintenance and contract employees receive current and updated training. For example, if changes are made to a process, impacted employees must be trained in the changes and understand the effects of the changes on their job tasks (e.g., any new operating procedures pertinent to their tasks). Additionally, as already discussed the evaluation of the employee's absorption of training will certainly influence the need for training.

## MECHANICAL INTEGRITY

Employers will need to review their maintenance programs and schedules to see if there are areas where "breakdown" maintenance is used rather than an on-going mechanical integrity program. Equipment used to process, store, or handle highly hazardous chemicals needs to be designed, constructed, installed and maintained to minimize the risk of releases of such chemicals. This requires that a mechanical integrity program be in place to assure the continued integrity of process equipment.

Elements of a mechanical integrity program include the identification and categorization of equipment and instrumentation, inspections and tests, testing and inspection frequencies, development of maintenance procedures, training of maintenance personnel, the establishment of criteria for acceptable test results, documentation of test and inspection results, and documentation of manufacturer recommendations as to meantime to failure for equipment and instrumentation.

The first line of defense an employer has available is to operate and maintain the process as designed, and to keep the chemicals contained. This line of defense is backed up by the next line of defense which is the controlled release of chemicals through venting to scrubbers or flares, or to surge or overflow tanks which are designed to receive such chemicals, etc. These lines of defense are the primary lines of defense or means to prevent unwanted releases. The secondary lines of defense would include fixed fire protection systems like sprinklers, water spray, or deluge systems, monitor guns, etc., dikes, designed drainage systems, and other systems which would control or mitigate hazardous chemicals once an unwanted release occurs. These primary and secondary lines of defense are what the mechanical integrity program needs to protect and strengthen these primary and secondary lines of defenses where appropriate.

The first step of an effective mechanical integrity program is to compile and categorize a list of process equipment and instrumentation for inclusion in the program. This list would include pressure vessels, storage tanks, process piping, relief and vent systems, fire protection system components, emergency shutdown systems and alarms and interlocks and pumps. For the categorization of instrumentation and the listed equipment the employer would prioritize which pieces of equipment require closer scrutiny than others.

Meantime to failure of various instrumentation and equipment parts would be known from the manufacturer's data or the employer's experience with the parts, which would then influence the inspection and testing frequency and associated procedures. Also, applicable codes and standards such as the National Board Inspection Code, or those from the American Society for Testing and Material, American Petroleum Institute, National Fire Protection Association, American National Standards Institute, American Society of Mechanical Engineers, and other groups, provide information to help establish an effective testing and inspection frequency, as well as appropriate methodologies.

The applicable codes and standards provide criteria for external inspections for such items as foundation and supports, anchor bolts, concrete or steel supports, guy wires, nozzles and sprinklers, pipe hangers, grounding connections, protective coatings and insulation, and external metal surfaces of piping and vessels, etc. These codes and standards also provide information on methodologies for internal inspection, and a frequency formula based on the corrosion rate of the materials of construction. Also, erosion both internal and external needs to be considered along with corrosion effects for piping and valves. Where the corrosion rate is not known, a maximum inspection frequency is recommended, and methods of developing the corrosion rate are available in the codes. Internal inspections need to cover items such as vessel shell, bottom and head; metallic linings; nonmetallic linings; thickness measurements for vessels and piping; inspection for erosion, corrosion, cracking and bulges; internal equipment like trays, baffles, sensors and screens for erosion, corrosion or cracking and other deficiencies. Some of these inspections may be performed by state or local government inspectors under state and local statutes. However, each employer needs to develop procedures to ensure that tests and inspections are conducted properly and that consistency is maintained even where different employees may be involved. Appropriate training is to be provided to maintenance personnel to ensure that they understand the preventive maintenance program procedures, safe practices, and the proper use and application of special equipment or unique tools that may be required. This training is part of the overall training program called for in the standard.

A quality assurance system is needed to help ensure that the proper materials of construction are used, that fabrication and inspection procedures are proper, and that installation procedures recognize field installation concerns. The quality assurance program is an essential part of the mechanical integrity program and will help to maintain the primary and secondary lines of defense that have been designed into the process to prevent unwanted chemical releases or those which control or mitigate a release. "As built" drawings, together with certifications of coded vessels and other equipment, and materials of construction need to be verified and retained in the quality assurance documentation.

Equipment installation jobs need to be properly inspected in the field for use of proper materials and procedures and to assure that qualified craftsmen are used to do the job. The use of appropriate gaskets, packing, bolts, valves, lubricants and welding rods need to be verified in the field. Also, procedures for installation of safety devices need to be verified, such as the torque on the bolts on ruptured disc installations, uniform torque on flange bolts, proper installation of pump seals, etc. If the quality of parts is a problem, it may be appropriate to conduct audits of the equipment supplier's facilities to better assure proper purchases of required equipment which is suitable for its intended service. Any changes in equipment that may become necessary will need to go through the management of change procedures.

**MANAGEMENT OF CHANGE**

To properly manage changes to process chemicals, technology, equipment and facilities, one must define what is meant by change. In this process safety management standard, change includes all modifications to equipment, procedures, raw materials and processing conditions other than "replacement in kind." These changes need to be properly managed by identifying and reviewing them prior to implementation of the change. For example, the operating procedures contain the operating parameters (pressure limits, temperature ranges, flow rates, etc.) and the importance of operating within these limits. While the operator must have the flexibility to maintain safe operation within the established parameters, any operation outside of these parameters requires review and approval by a written management of change procedure. Management of change covers changes in process technology and changes to equipment and instrumentation. Changes in process technology can result from changes in production rates, raw materials, experimentation, equipment unavailability, new equipment, new product development, change in catalyst and changes in operating conditions to improve yield or quality. Equipment changes include among others change in materials of construction, equipment specifications, piping pre-arrangements, experimental equipment, computer program revisions and changes in alarms and interlocks. Employers need to establish means and methods to detect both technical changes and mechanical changes.

Temporary changes have caused a number of catastrophes over the years, and employers need to establish ways to detect temporary changes as well as those that are permanent. It is important that a time limit for temporary changes be established and monitored since, without control, these changes may tend to become permanent. Temporary changes are subject to the management of change provisions. In addition, the management of change procedures are used to insure that the equipment and procedures are returned to their original or designed conditions at the end of the temporary change. Proper documentation and review of these changes is invaluable in assuring that the safety and health considerations are being incorporated into the operating procedures and the process. Employers may wish to develop a form or clearance sheet to facilitate the processing of changes through the management of change procedures. A typical change form may include a description and the purpose of the change, the technical basis for the change, safety and health considerations, documentation of changes for the operating procedures, maintenance procedures, inspection and testing, P&IDS, electrical classification, training and communications, pre-startup inspection, duration if a temporary change, approvals and authorization. Where the impact of the change is minor and well understood, a check list reviewed by an authorized person with proper communication to others who are affected may be sufficient.

However, for a more complex or significant design change, a hazard evaluation procedure with approvals by operations, maintenance, and safety departments may be appropriate. Changes in documents such as P&IDS, raw materials, operating procedures, mechanical integrity programs, electrical classifications, etc., need to be noted so that these revisions can be made permanent when the drawings and procedure manuals are updated. Copies of process changes need to be kept in an accessible location to ensure that design changes are available to operating personnel as well as to PHA team members when a PHA is being done or one is being updated.

**PRE-STARTUP REVIEW**

For new processes, the employer will find a PHA helpful in improving the design and construction of the process from a reliability and quality point of view. The safe operation of the new process will be enhanced by making use of the PHA recommendations before final installations are completed. P&IDs

are to be completed along with having the operating procedures in place and the operating staff trained to run the process before startup. The initial startup procedures and normal operating procedures need to be fully evaluated as part of the pre-startup review to assure a safe transfer into the normal operating mode for meeting the process parameters.

For existing processes that have been shutdown for turnaround, or modification, etc., the employer must assure that any changes other than "replacement in kind" made to the process during shutdown go through the management of change procedures. P&IDS will need to be updated as necessary, as well as operating procedures and instructions. If the changes made to the process during shutdown are significant and impact the training program, then operating personnel as well as employees engaged in routine and non-routine work in the process area may need some refresher or additional training in light of the changes. Any incident investigation recommendations, compliance audits or PHA recommendations need to be reviewed as well to see what impacts they may have on the process before beginning the startup.

## COMPLIANCE AUDITS

Employers need to select a trained individual or assemble a trained team of people to audit the process safety management system and program. A small process or plant may need only one knowledgeable person to conduct an audit. The audit is to include an evaluation of the design and effectiveness of the process safety management system and a field inspection of the safety and health conditions and practices to verify that the employer's systems are effectively implemented. The audit should be conducted or led by a person knowledgeable in audit techniques and who is impartial towards the facility or area being audited. The essential elements of an audit program include planning, staffing, conducting the audit, evaluation and corrective action, follow-up and documentation.

Planning in advance is essential to the success of the auditing process. Each employer needs to establish the format, staffing, scheduling and verification methods prior to conducting the audit. The format should be designed to provide the lead auditor with a procedure or checklist which details the requirements of each section of the standard. The names of the audit team members should be listed as part of the format as well. The checklist, if properly designed, could serve as the verification sheet which provides the auditor with the necessary information to expedite the review and assure that no requirements of the standard are omitted. This verification sheet format could also identify those elements that will require evaluation or a response to correct deficiencies. This sheet could also be used for developing the follow-up and documentation requirements.

The selection of effective audit team members is critical to the success of the program. Team members should be chosen for their experience, knowledge, and training and should be familiar with the processes and with auditing techniques, practices and procedures. The size of the team will vary depending on the size and complexity of the process under consideration. For a large, complex, highly instrumented plant, it may be desirable to have team members with expertise in process engineering and design, process chemistry, instrumentation and computer controls, electrical hazards and classifications, safety and health disciplines, maintenance, emergency preparedness, warehousing or shipping, and process safety auditing. The team may use part-time members to provide for the depth of expertise required as well as for what is actually done or followed, compared to what is written.

An effective audit includes a review of the relevant documentation and process safety information, inspection of the physical facilities, and interviews with all levels of plant personnel. Using the audit procedure and checklist developed in the preplanning stage, the audit team can systematically analyze compliance with the provisions of the standard and any other corporate policies that are relevant. For example, the audit team will review all aspects of the training program as part of the overall audit. The team will review the written training program for adequacy of content, frequency of training, effectiveness of training in terms of its goals and objectives as well as to how it fits into meeting the standard's requirements, documentation, etc. Through interviews, the team can determine the employee's knowledge and awareness of the safety procedures, duties, rules, emergency response assignments, etc. During the inspection, the team can observe actual practices such as safety and health policies, procedures, and work authorization practices. This approach enables the team to identify deficiencies and determine where corrective actions or improvements are necessary.

An audit is a technique used to gather sufficient facts and information, including statistical information, to verify compliance with standards. Auditors should select as part of their preplanning a sample size sufficient to give a degree of confidence that the audit reflects the level of compliance with the standard. The audit team, through this systematic analysis, should document areas which require corrective action as well as those areas where the process safety management system is effective and working in an effective manner. This provides a record of the audit procedures and findings, and serves as a baseline of operation data for future audits. It will assist future auditors in determining changes or trends from previous audits.

Corrective action is one of the most important parts of the audit. It includes not only addressing the identified deficiencies, but also planning, follow up, and documentation. The corrective action process normally begins with a management review of the audit findings. The purpose of this review is to determine what actions are appropriate, and to establish priorities, timetables, resource allocations and requirements and responsibilities. In some cases, corrective action may involve a simple change in procedure or minor maintenance effort to remedy the concern. Management of change procedures need to be used, as appropriate, even for what may seem to be a minor change. Many of the deficiencies can be acted on promptly, while some may require engineering studies or in-depth review of actual procedures and practices. There may be instances where no action is necessary and this is a valid response to an audit finding. All actions taken, including an explanation where no action is taken on a finding, needs to be documented as to what was done and why.

It is important to assure that each deficiency identified is addressed, the corrective action to be taken noted, and the audit person or team responsible be properly documented by the employer.

To control the corrective action process, the employer should consider the use of a tracking system. This tracking system might include periodic status reports shared with affected levels of management, specific reports such as completion of an engineering study, and a final implementation report to provide closure for audit findings that have been through management of change, if appropriate, and then shared with affected employees and management. This type of tracking system provides the employer with the status of the corrective action. It also provides the documentation required to verify that appropriate corrective actions were taken on deficiencies identified in the audit.

## INCIDENT INVESTIGATION

Incident investigation is the process of identifying the underlying causes of incidents and implementing steps to prevent similar events from occurring. The intent of an incident investigation is for employers to learn from past experiences and thus avoid repeating past mistakes. Some of the events are sometimes referred to as "near misses," meaning that a serious consequence did not occur, but could have.

Employers need to develop in-house capability to investigate incidents that occur in their facilities. A team needs to be assembled by the employer and trained in the techniques of investigation including how to conduct interviews of witnesses, needed documentation and report writing. A multi-disciplinary team is better able to gather the facts of the event and to analyze them and develop plausible scenarios as to what happened, and why. Team members should be selected on the basis of their training, knowledge and ability to contribute to a team effort to fully investigate the incident.

Employees in the process area where the incident occurred should be consulted, interviewed or made a member of the team. Their knowledge of the events form a significant set of facts about the incident which occurred. The report, its findings and recommendations are to be shared with those who can benefit from the information. The cooperation of employees is essential to an effective incident investigation. The focus of the investigation should be to obtain facts, and not to place blame. The team and the investigation process should clearly deal with all involved individuals in a fair, open and consistent manner.

## EMPLOYEE PARTICIPATION

Section 304 of the Clean Air Act Amendments states that employers are to consult with their employees and their representatives regarding the employers efforts in the development and implementation of the process safety management program elements and hazard assessments. Section 304 also requires employers to train and educate their employees and to inform affected employees of the findings from incident investigations required by the process safety management program. Many employers, under their safety and health programs, have already established means and methods to keep employees and their representatives informed about relevant safety and health issues and employers may be able to adapt these practices and procedures to meet their obligations under this standard. Employers who have not implemented an occupational safety and health program may wish to form a safety and health committee of employees and management representatives to help the employer meet the obligations specified by this standard. These committees can become a significant ally in helping the employer to implement and maintain an effective process safety management program for all employees.

## HOT WORK PERMIT

Non-routine work which is conducted in process areas needs to be controlled by the employer in a consistent manner. The hazards identified involving the work that is to be accomplished must be communicated to those doing the work, but also to those operating personnel whose work could affect the safety of the process. A work authorization notice or permit must have a procedure that describes the steps the maintenance supervisor, contractor representative or other person needs to follow to obtain the necessary clearance to get the job started. The work authorization procedures need to reference and coordinate, as applicable, lockout/tagout procedures, line breaking procedures, confined space entry procedures and hot work authorizations. This procedure also needs to provide clear steps to follow once the job is completed to provide closure for those that need to know the job is now completed and equipment can be returned to normal.

## CONTRACTORS

Employers who use contractors to perform work in and around processes that involve highly hazardous chemicals, will need to establish a screening process so that they hire and use contractors who accomplish the desired job tasks without compromising the safety and health of employees at a facility. For contractors, whose safety performance on the job is not known to the hiring employer, the employer will need to obtain information on injury and illness rates and experience and should obtain contractor references. Additionally, the employer must assure that the contractor has the appropriate job skills, knowledge and certifications (such as for pressure vessel welders). Contractor work methods and experiences should be evaluated. For example, does the contractor conducting demolition work swing loads over operating processes or does the contractor avoid such hazards?

Contract employees must perform their work safely. Considering that contractors often perform very specialized and potentially hazardous tasks such as confined space entry activities and non-routine repair activities it is quite important that their activities be controlled while they are working on or near a covered process. A permit system or work authorization system for these activities would also be helpful to all affected employers. The use of a work authorization system keeps an employer informed of contract employee activities, and as a benefit the employer will have better coordination and more management control over the work being performed in the process area. A well run and well maintained process where employee safety is fully recognized will benefit all of those who work in the facility whether they be contract employees or employees of the owner.

www.ingramcontent.com/pod-product-compliance
Lightning Source LLC
Chambersburg PA
CBHW080634180526
45168CB00008B/3168